Lecture Notes in Economics and Mathematical Systems 658

Founding Editors:

M. Beckmann
H.P. Künzi

Managing Editors:

Prof. Dr. G. Fandel
Fachbereich Wirtschaftswissenschaften
Fernuniversität Hagen
Feithstr. 140/AVZ II, 58084 Hagen, Germany

Prof. Dr. W. Trockel
Institut für Mathematische Wirtschaftsforschung (IMW)
Universität Bielefeld
Universitätsstr. 25, 33615 Bielefeld, Germany

Editorial Board:

H. Dawid, D. Dimitrow, A. Gerber, C-J. Haake, C. Hofmann, T. Pfeiffer,
R. Slowiński, W.H.M. Zijm

For further volumes:
http://www.springer.com/series/300

Yuri Ermoliev • Marek Makowski • Kurt Marti
Editors

Managing Safety of Heterogeneous Systems

Decisions under Uncertainties and Risks

Editors
Prof. Dr. Yuri Ermoliev
Dr. Marek Makowski
International Institute for
Applied Systems Analysis (IIASA)
Schloßplatz 1
2361 Laxenburg
Austria
ermoliev@iiasa.ac.at
marek@iiasa.ac.at

Prof. Dr. Kurt Marti
Federal Armed Forces University Munich
Aerospace Engineering and Technology
Werner-Heisenberg-Weg 39
85577 Neubiberg/München
Germany
kurt.marti@unibw-muenchen.de

ISSN 0075-8442
ISBN 978-3-642-22883-4   e-ISBN 978-3-642-22884-1
DOI 10.1007/978-3-642-22884-1
Springer Heidelberg Dordrecht London New York

Library of Congress Control Number: 2011944256

© Springer-Verlag Berlin Heidelberg 2012
This work is subject to copyright. All rights are reserved, whether the whole or part of the material is concerned, specifically the rights of translation, reprinting, reuse of illustrations, recitation, broadcasting, reproduction on microfilm or in any other way, and storage in data banks. Duplication of this publication or parts thereof is permitted only under the provisions of the German Copyright Law of September 9, 1965, in its current version, and permission for use must always be obtained from Springer. Violations are liable to prosecution under the German Copyright Law.
The use of general descriptive names, registered names, trademarks, etc. in this publication does not imply, even in the absence of a specific statement, that such names are exempt from the relevant protective laws and regulations and therefore free for general use.

Printed on acid-free paper

Springer is part of Springer Science+Business Media (www.springer.com)

Preface

The aim of the series of workshops on *Coping with Uncertainty (CwU)* organized since over a decade at the International Institute for Applied Systems Analysis (IIASA), Laxenburg, Austria, has been to provide researchers and practitioners from different areas with an interdisciplinary forum for discussing various ways of effective dealing with uncertainties and risks in diverse areas, including environmental and social sciences, economics, policy making, management, and engineering. The workshops proved to be successful, especially in cross-disciplinary sharing methods, ideas, and open problems.

Science-based support for effective coping with uncertainties and risks in complex policy-making and engineering problems needs practical solutions for fundamentally new scientific problems that in turn require new concepts and tools. A key issue concerns a vast variety of practically irreducible uncertainties, including potential extreme events of high multidimensional consequences, which challenge traditional models, and thus require new concepts and analytical tools. Robust decisions for problems exposed to extreme events are essentially different from over-simplified decisions that ignore such events. Specifically, a proper treatment of extreme/rare events requires new paradigms of rational decisions, new performance indicators, and new spatio-temporal dimensions of heterogeneous interdependencies, including network externalities and risks.

Traditional scientific approaches usually rely on real observations and experiments. Yet no sufficient observations exist for new problems; "pure" experiments and "learning by doing" are dangerous, very expensive, and thus practically impossible. Moreover, the available historical observations are often contaminated by "experimentator," i.e., past actions or policies. The complexity of new problems does not allow to achieve enough certainty, e.g., by increasing the resolution of models or by bringing in more links. Such problems require explicit treatment of uncertainties using "synthetic" information derived by integration of "hard" elements, including available data, results of possible experiments, and formal representations of scientific facts, as well as "soft" elements based on diverse representations of scenarios, and opinions of public, stakeholders, and experts.

However, even a best possible integration of all these factors results in assessments having poor estimates. Therefore, the science-based support for addressing the new class of problems summarized above needs to replace the traditional "deterministic predictions" analysis by new methods and tools for designing strategies that are robust against the involved uncertainties and risks, and is also suitable for effectively coping with new challenges, such as spatiotemporal heterogeneities, interdependencies, externalities, and endogenous (i.e., caused by possible future actions) risks.

Contributions to this volume are based on selected presentations at the *CwU 2009* workshop. The workshop aimed at contributing to a better understanding between practitioners dealing with safety of complex heterogeneous systems under uncertainty, and scientists working on either corresponding modeling approaches, or on methods that can be adapted for improving the understanding and management of uncertainty. The focus of the *CwU 2009* was on novel approaches to supporting robust decision making and design, especially when uncertainty is irreducible, consequences might be enormous, and the decision process involves stakeholders with diverse interests. Presentations dealt with open problems in this field, limitations of known approaches, novel methods and techniques, or lessons from applications of various approaches.

The workshop was organized at IIASA in December 2009, jointly by:

- International Institute for Applied Systems Analysis, Laxenburg, Austria, and
- Federal Armed Forces University Munich, Germany.

The scientific Program Committee included:

- Yuri Ermoliev, IIASA, Laxenburg (A),
- Marek Makowski, IIASA, Laxenburg (A),
- Kurt Marti, Federal Armed Forces University Munich (D), and
- Gerhard I. Schuëller, University of Innsbruck (A).

This volume is composed of chapters based on selected contributions to the *CwU 2009* workshop. The first chapter summarizes key issues related to supporting decision-making under uncertainties and risks, in particular for managing safety of heterogeneous systems. The other 17 chapters are organized into the following five parts:

1. *Decisions under systemic risks and uncertainties* discusses support of robust decisions involving threats generated by intelligent agents, and under lack of imprecise probabilities, as well as decision analysis through combining second-order belief distributions with qualitative statements, and an econometric model based on the max–min expected utility concept.
2. *Modeling uncertainties of heterogeneous systems* presents effective approaches to cope with diverse types heterogeneous systems, such as technological change under increasing returns and uncertainty, an agency problems, as well as sustainable agriculture, food security, socioeconomic risks, and water management.

Preface

3. *Uncertainty and optimization* deals with novel optimization methods for analysis of uncertainties; in particular in global optimization, fuzzy linear programming, and clustering of uncertain data.
4. *Analysis and optimization of technical systems and structures under stochastic uncertainty* discusses optimal open-loop feedback control of dynamic structural systems, and deals with problems in civil engineering and large spatial trusses.
5. *Analysis and optimization of economic systems under uncertainty* presents approaches to estimation and reduction of environmental impacts for sustainable agriculture, portfolio analysis of financial and insurance instruments, and pricing catastrophe bonds.

The organizers gratefully acknowledge the generous support IIASA provided for the workshop logistics, which enabled the participation of many researchers who otherwise could not have attended this meeting.

The editors express gratitude to all referees who have helped the authors to improve their contributions by providing constructive comments, in several cases on a short notice. We thank the authors for delivering their contributions that conformed to the substantive comments by the reviewers, and the technical guidelines that were necessary to prepare this volume with limited resources for technical edition. Furthermore, we thank Ms Suchitra Subramanian of the Integrated Modeling Environment (IME) Project at IIASA for her support in the preparation of this volume.

Finally, we thank the Springer-Verlag for including the Proceedings into the Springer Lecture Notes Series in *Economics and Mathematical Systems (LNEMS)*.

Laxenburg

Laxenburg

Munich

Yuri Ermoliev

Marek Makowski

Kurt Marti

Contents

Robust Management of Heterogeneous Systems under Uncertainties 1
Yuri Ermoliev, Marek Makowski, and Kurt Marti

Part I   Decisions Under Systemic Risk and Uncertainties

Systemic Risk and Security Management 19
Yuri Ermoliev and Detlof von Winterfeldt

Robust Decisions under Risk for Imprecise Probabilities 51
Włodzimierz Ogryczak

**Combining Second-Order Belief Distributions with Qualitative
Statements in Decision Analysis** 67
Ola Caster and Love Ekenberg

**An Econometric Model Based on the Maxmin Expected Utility
Model: An Application to Earthquake Insurance** 89
Toshio Fujimi and Hirokazu Tatano

Part II   Modeling Uncertainties of Heterogeneous Systems

**Modeling Technological Change Under Increasing Returns
and Uncertainty** 109
Andrei Gritsevskyi and Yuri Ermoliev

**Stochastic Programming Perspective
on the Agency Problems Under Uncertainty** 137
Alexei A. Gaivoronski and Adrian Werner

Sustainable Agriculture, Food Security, and Socio-Economic Risks in Ukraine 169
Oleksandra Borodina, Elena Borodina, Tatiana Ermolieva,
Yuri Ermoliev, Günther Fischer, Marek Makowski,
and Harrij van Velthuizen

Multiple-Criteria Decision Support System for Siemianówka Reservoir under Uncertainties 187
Adam Kiczko and Tatiana Ermolieva

Part III   Uncertainty and Optimization

A Deterministic Algorithm for Global Optimization 205
Yury Evtushenko and Mikhail Posypkin

Robust Optimization by Fuzzy Linear Programming 219
Masahiro Inuiguchi

Various Types of Objective Functions of Clustering for Uncertain Data 241
Yasunori Endo and Sadaaki Miyamoto

Part IV   Analysis and Optimization of Technical Systems Under Uncertainty

Stochastic Optimal Open-Loop Feedback Control of Dynamic Structural Systems under Stochastic Uncertainty 263
Kurt Marti and Ina Stein

Modeling and Processing of Uncertainty in Civil Engineering by Means of Fuzzy Randomness 291
Uwe Reuter, Jan-Uwe Sickert, Wolfgang Graf,
and Michael Kaliske

Optimal Design and Sensitivity of Large Spatial Trusses Under Uncertainty 307
Simone Zier

Part V   Analysis and Optimization of Economic Systems Under Uncertainty

Sustainable Agriculture in China: Estimation and Reduction of Nitrogen Impacts 327
Günther Fischer, Wilfried Winiwarter, Tatiana Ermolieva,
Gui-Ying Cao, Harrij van Velthuizen, Zbigniew Klimont,
Wolfgang Schoepp, Wim van Veen, David Wiberg,
and Fabian Wagner

Evaluation of Portfolio of Financial and Insurance Instruments: Simulation of Uncertainty 351
Piotr Nowak, Maciej Romaniuk, and Tatiana Ermolieva

Pricing Catastrophe Bonds under Safety Constraints 367
Shuo Liu and Liyan Han

Contributors

Elena Borodina Institute of Economics and Forecasting, P. Mirnogo 26, 01011 Kiev, Ukraine, oborodina@ief.org.ua

Oleksandra Borodina Institute of Economics and Forecasting, P. Mirnogo 26, 01011 Kiev, Ukraine, oleksandra.borodina@gmail.com

Gui-Ying Cao International Institute for Applied Systems Analysis, Schlossplatz 1, A-2361, Laxenburg, Austria, cao@iiasa.ac.at

Ola Caster Uppsala Monitoring Centre, WHO Collaborating Centre for International Drug Monitoring, Box 1051, 751 40 Uppsala, Sweden

Department of Computer and Systems Sciences, Stockholm University, ola.caster@who-umc.org

Love Ekenberg Department of Computer and Systems Sciences, Stockholm University, Forum 100, 164 46 Kista, Sweden, lovek@dsv.su.se

Yasunori Endo Faculty of Systems and Information Engineering, University of Tsukuba, Tennodai 1-1-1, Tsukuba, Ibaraki 305-8573, Japan, endo@risk.tsukuba.ac.jp

Yuri Ermoliev International Institute for Applied Systems Analysis, Schlossplatz 1, A-2361, Laxenburg, Austria, ermoliev@iiasa.ac.at

Tatiana Ermolieva International Institute for Applied Systems Analysis, Schlossplatz 1, A-2361, Laxenburg, Austria, ermol@iiasa.ac.at

Yury Evtushenko Institution of Russian Academy of Sciences, Dorodnicyn Computing Centre of RAS, Vavilov st. 40, 119333 Moscow, Russia, evt@ccas.ru

Günther Fischer International Institute for Applied Systems Analysis, Schlossplatz 1, A-2361, Laxenburg, Austria, fisher@iiasa.ac.at

Toshio Fujimi Graduate School of Science and Technology, Kumamoto University, Kurokami 2-39-1, Kumamoto, Japan, fujimi@kumamoto-u.ac.jp

Alexei Gaivoronski Department of Industrial Economics and Technology Management, Norwegian University of Science and Technology, Trondheim, Norway, Alexei.Gaivoronski@iot.ntnu.no

Wolfgang Graf Institute for Structural Analysis, Technische Universität Dresden, Dresden, Germany, Wolfgang.Graf@tu-dresden.de

Andrei Gritsevskyi International Atomic Energy Agency, Vienna International Centre, POBox 100, A-1400 Vienna, Austria, A.Gritsevskyi@IAEA.org

Liyan Han School of Economics and Management, Beihang University, Xueyuan Road 37, Haidian District 100191, Beijing, hanly1@163.com

Masahiro Inuiguchi Graduate School of Engineering Science, Osaka University, 1-3 Machikaneyama, Toyonaka, Osaka 560-8531, Japan, inuiguti@sys.es.osaka-u.ac.jp

Michael Kaliske Institute for Structural Analysis, Technische Universität Dresden, Dresden, Germany, Michael.Kaliske@tu-dresden.de

Adam Kiczko Institute of Geophysics, Polish Academy of Science, ul. Ksiecia Janusza 64, Warsaw, Poland, akiczko@igf.edu.pl

Zbigniew Klimont International Institute for Applied Systems Analysis, Schlossplatz 1, A-2361, Laxenburg, Austria, klimont@iiasa.ac.at

Shuo Liu School of Economics and Management, Beihang University, Xueyuan Road 37, Haidian District 100191, Beijing, liushuo.buaa@gmail.com

Marek Makowski International Institute for Applied Systems Analysis, Schlossplatz 1, A-2361, Laxenburg, Austria, marek@iiasa.ac.at

Kurt Marti Federal Armed Forces University Munich, Aerospace Engineering and Technology, 85577 Neubiberg/Munich, Germany, kurt.marti@unibw-muenchen.de

Sadaaki Miyamoto Faculty of Systems and Information Engineering, University of Tsukuba, Tennodai 1-1-1, Tsukuba, Ibaraki 305-8573, Japan, miyamoto@risk.tsukuba.ac.jp

Piotr Nowak Systems Research Institute Polish Academy of Sciences, ul. Newelska 6, 01-447 Warszawa, Poland, pnowak@ibspan.waw.pl

Włodzimierz Ogryczak Warsaw University of Technology, 00-665 Warsaw, Poland, wogrycza@ia.pw.edu.pl

Mikhail Posypkin Institution of Russian Academy of Sciences, Institute for Systems Analysis RAS, pr-t. 60-letja Oktjabrja, Moscow, Russia, mposypkin@mail.ru

Uwe Reuter Department of Civil Engineering, Technische Universität Dresden, Dresden, Germany Uwe.Reuter@tu-dresden.de

Maciej Romaniuk Systems Research Institute Polish Academy of Sciences, ul. Newelska 6, 01-447 Warszawa, Poland, mroman@ibspan.waw.pl

Wolfgang Schöpp International Institute for Applied Systems Analysis, Schlossplatz 1, A-2361, Laxenburg, Austria, schoepp@iiasa.ac.at

Jan-Uwe Sickert Institute for Structural Analysis, Technische Universität Dresden, Dresden, Germany, Jan-Uwe.Sickert@tu-dresden.de

Ina Stein Federal Armed Forces University Munich, Aerospace Engineering and Technology, 85577 Neubiberg/Munich, Germany, ina.stein@online.de

Hirokazu Tatano Disaster Prevention Research Institute, Kyoto University, Gokasho, Uji, Kyoto, Japan, tatano@imdr.mbox.media.kyoto-u.ac.jp

Harrij van Velthuizen International Institute for Applied Systems Analysis, Schlossplatz 1, A-2361, Laxenburg, Austria, velt@iiasa.ac.at

W.C.M. van Ween International Institute for Applied Systems Analysis, Schlossplatz 1, A-2361, Laxenburg, Austria, W.C.M.vanVeen@sow.vu.nl

Fabian Wagner International Institute for Applied Systems Analysis, Schlossplatz 1, A-2361, Laxenburg, Austria, fwagner@iiasa.ac.at

Adrian Tobias Werner Department of Applied Economics and Operations Research, SINTEF Technology and Society, Trondheim, Norway, Adrian.Tobias.Werner@sintef.no

David Wiberg International Institute for Applied Systems Analysis, Schlossplatz 1, A-2361, Laxenburg, Austria, wiberg@iiasa.ac.at

Wilfried Winiwarter International Institute for Applied Systems Analysis, Schlossplatz 1, A-2361, Laxenburg, Austria, winiwarter@iiasa.ac.at

Detlof von Winterfeldt International Institute for Applied Systems Analysis, Schlossplatz 1, A-2361, Laxenburg, Austria, detlof@iiasa.ac.at

Simone Zier Federal Armed Forces University Munich, Aerospace Engineering and Technology, 85577 Neubiberg/Munich, Germany, simone.zier@unibw.de

List of Figures

Fig. 1	Results from Example 1, where the considered decision tree contains two utilities whose unconstrained belief distributions were both standard uniform. Following the constraint $\tilde{U}_a \leq \tilde{U}_b$, the resulting (constrained) belief U_b over the second utility is now Beta$(2, 1)$	74
Fig. 2	Decision tree considered in Example 2	76
Fig. 3	Decision tree considered in Example 4	84
Fig. 4	Results from the simulation of Example 4. (**a**) Histogram over the simulated belief for the difference $E(A_1) - E(A_2)$. (**b**) Two-dimensional histogram over the simulated beliefs for $E(A_1)$ and $E(A_2)$. Darker color indicates a higher density of points	85
Fig. 1	Experience curves for gas turbines, windmills and photovoltaics. Cost improvements per unit installed capacity, in USD (1990) per kW, are shown against the cumulative installed capacity, in MWe, on logarithmic scale	112
Fig. 2	Cumulative costs	113
Fig. 3	Cost reduction for on-shore (low cost) vs. off-shore (high cost) wind energy technologies in the EU	114
Fig. 4	Diminishing (*left*) and increasing (*right*) returns	117
Fig. 5	Schematic diagram illustrating network structure of the energy model	122
Fig. 6	Uncertain costs (*left*) and worst-case solution (*right*)	128
Fig. 1	Dependence of the fixed upfront payment f on the average alternative profit of the agent for the optimal contract (s^*, f^*) and the optimal contract (s', f') with averaged participation constraint	145

xvii

Fig. 2	Dependence of the principal's profit $\pi\,(s, f)$ on the average alternative profit of the agent for the optimal contract (s^*, f^*) and the optimal contract (s', f') with averaged participation constraint	146
Fig. 1	Robust and deterministic allocations of new rural activities (in number of jobs) in each region	180
Fig. 2	Total costs associated with robust solution and optimal solution of the deterministic model	182
Fig. 1	Schematic map of the study area	189
Fig. 2	The model of Upper Narew river system	190
Fig. 3	Verification of UNET model for he Upper Narew river; water levels at Suraż river gauge during a spring freshet in 1983	191
Fig. 4	An example forecast obtained with $k\text{-}NN$ method for Bondary; solid lines stands for forecasted trajectories, dashed lines for an expected ones and lines with circles for observations	194
Fig. 5	Computed trajectories of reservoir storage S (upper plot) and inundation extend at wetland areas (lower plot); CS – trajectory for stochastic formulation, CD – trajectory for deterministic formulation; CP – trajectory for the "perfect" forecast	198
Fig. 1	Estimation of fuzzy set Γ	222
Fig. 2	Possibility and necessity	223
Fig. 3	Possibility and necessity measures	224
Fig. 4	An example of possibly optimal solution	226
Fig. 5	An example of necessarily optimal solution	226
Fig. 6	An example of necessarily optimal solution with degree 0.176471	227
Fig. 7	Functions μ_{Dif} and μ_{Rat}. (a) μ_{Dif} (b) μ_{Rat}	229
Fig. 8	An example of necessarily soft-optimal solution with degree 0.481481	231
Fig. 9	An example of the best necessarily soft-optimal solution	231
Fig. 10	Illustration of necessity measure optimization and necessity fractile optimization models	236
Fig. 1	An example hyper-sphere type tolerance on the two dimensional Euclidean space: $\|\varepsilon_k\|^2 \le \kappa_k^2$	244
Fig. 2	An example of hyper-rectangle tolerance on the two dimensional Euclidean space: $|\varepsilon_{kj}| \le \kappa_{kj}$	245

	List of Figures	xix

Fig. 1   Remaining time interval 266
Fig. 2   Principle of active structural control 286

Fig. 1   Modeling of a smooth transition between two
complementary states 1 and 2 as fuzzy variable $\tilde{x}$ 293
Fig. 2   Convex fuzzy realizations $\tilde{X}(\omega)$ of a fuzzy random
variable $\tilde{X}$, e.g. as a result of uncertain measurements 294
Fig. 3   Fuzzy stochastic structural analysis – Variant I 297
Fig. 4   Fuzzy stochastic structural analysis – Variant II 298
Fig. 5   FE model of the hypar shell roof 300
Fig. 6   R-S-plot and fuzzy failure probability $\tilde{P}_f$ 302
Fig. 7   Fuzzy earth pressures $\tilde{d}_k$ ($k = 1, 2, \ldots, 8$) at eight
measurement dates 303
Fig. 8   Fuzzy failure probability $\tilde{P}_f$ 305

Fig. 1   1-storey spatial truss 311
Fig. 2 n-storey spatial truss 312
Fig. 3   Discretization of the normal distribution
$N \sim (10^5, 10^8)$ with 17 realizations 313
Fig. 4   Optimal cross-sectional areas A_1 and A_8 in dependence
on the number of storeys using the EVP 314
Fig. 5   Optimal cross-sectional areas A_4, A_5, A_{12} and A_{13} in
dependence on the number of storeys using the EVP 315
Fig. 6   Comparison of different cross-sectional areas in
dependence on the number of storeys using the EVP 315
Fig. 7   Optimal cross-sectional areas A_2, A_7, A_{11} and A_{14} in
dependence on the number of storeys using the EVP 316
Fig. 8   Expected initial, recourse and total costs using the EVP........... 316
Fig. 9   Probability of failure and expected initial (*),
recourse (x) and total (+) costs using the EVP..................... 317
Fig. 10   Optimal cross-sectional areas in dependence on the
number of storeys using the RPD 318
Fig. 11   Comparison of the optimal cross-sectional areas in
dependence on the number of storeys using the EVP
($\square$, $\triangle$, $\circ$) and the RPD ($\blacksquare$, $\blacktriangle$, $\bullet$) 319
Fig. 12   Comparison of the probability of failure and the
expected initial, recourse and total costs using the EVP
($\circ$, x, $+$) and the RPD ($\bullet$, $\blacklozenge$, $\oplus$).................................... 319
Fig. 13   Optimal cross-sectional area A_8 and expected costs in
dependence on the standard deviation considering the 3-storey 320
Fig. 14   Optimal 3-storey spatial truss 321

Fig. 15	Optimal cross-sectional area A_8, probability of failure and expected initial (*), recourse (x) and total (+) costs in dependence on the standard deviation considering the 5-storey	321
Fig. 16	Optimal cross-sectional area A_8 and probability of failure in dependence on the standard deviation considering the 3- (x), 5- (∗), 7- (△) and 10- (■) storey	322
Fig. 17	Number of variables and constraints in dependence of the number of storeys using 17 realizations	323
Fig. 1	Nitrogen cascading: Schematic structure of the model	334
Fig. 2	Nitrogen leaching fraction, in percent terms	336
Fig. 3	Leaching in kg / ha cultivated land, in 2000	336
Fig. 4	Leaching in kg / ha cultivated land, in 2030	337
Fig. 5	Leaching in severity classes by number of affected counties, in kg / ha cultivated land, in 2000 (**a**) and 2030 (**b**)	337
Fig. 6	N2O in kg / ha cultivated land, in 2000	338
Fig. 7	N2O in kg / ha cultivated land, in 2030	338
Fig. 8	N2O emissions by size classes and number of affected counties, in kg per ha of cultivated land, for 2000 (**a**) and 2030 (**b**)	339
Fig. 9	Ammonia emissions from agriculture (kg ammonia/ha cultivated land) in 2000	339
Fig. 10	Ammonia emissions from agriculture, in kg ammonia/ha cultivated land, in 2030	340
Fig. 11	Ammonia emissions by size classes and number of affected counties, in kg per ha of cultivated land, for 2000 (**a**) and 2030 (**b**)	340
Fig. 12	Absolute (million people) and relative (share of total population) distribution of population according to classes of severity of environmental pressure (measured in terms of kg nitrogen in ammonia emitted per ha cultivated land), 2000. The label on the horizontal axis indicates the regions in China: N, NE, E, C, S, SW, NW stand for North, North-East, East, Center, South, South-West, North-West, respectively, business-as-usual scenario	341
Fig. 13	Number of people by classes of severity of ammonia losses (kg nitrogen in ammonia emitted per ha total area), by economic regions and scenarios (BU = business-as-usual, RA = reallocation, OM = optimized manure; MA = minimized ammonia), in 2030	342
Fig. 1	Cash flow of catastrophe bonds	368

Fig. 2	Total (for the world) number of catastrophe bond issued in 1997-2007	368
Fig. 3	The sequential decision selection model structure	371
Fig. 4	Accepted coupon rate and issue volume combination of a simulated catastrophe bond	372
Fig. 5	Accepted coupon rate and issue volume combination of Chinese typhoon catastrophe bond	373
Fig. 6	Risk reserve dynamics with catastrophe bond (initial model)	374
Fig. 7	Risk reserve movement of insurer with catastrophe bond (revised model)	376

List of Tables

Table 1	Comparison between sample and population means	93
Table 2	Estimation results of simple models	99
Table 3	Variables of personal characteristics	100
Table 4	Estimation results of EU and MEU including variables of personal characteristics	101
Table 5	Risk premium and ambiguity premium	102
Table 1	Values of α_I and n_I coefficients used in computations	198
Table 2	Fit measures for the U, S and A trajectories obtained for the deterministic and stochastic control in respect to the "perfect" control; dU_{mean}, dS_{mean}, dA_{mean} – respectively mean deviation and dU_{max}, dS_{max}, dA_{max} – maximal deviation	199
Table 1	Running time in seconds for random polynomial unconstrained optimization problems	214
Table 2	The comparison with other works	215
Table 1	Excerpt of joint width measurements (courtesy of Staedtisches Vermessungsamt Dresden)	294
Table 2	Uncertain input variables	301
Table 3	Excerpt of the measured earth pressure data (Franke et al. 2003) (GL = ground line)	303
Table 1	Input parameters	313
Table 1	Numerical features of Portfolio II	362
Table 2	Numerical features of Portfolio III	363
Table 3	Numerical features of Portfolio IV	363
Table 4	Numerical features of Portfolio V	364

Robust Management of Heterogeneous Systems under Uncertainties

Yuri Ermoliev, Marek Makowski, and Kurt Marti

Abstract The chapter summarizes the key issues related to robust management of heterogeneous systems under uncertainties. It focuses on challenges types of decision problems under uncertainties for which standard approaches are inadequate, and builds on the related background and key concepts discussed during all workshops on *Coping with Uncertainties*. The selected key issues are summarized in a condensed manner, and illustrated by simple examples.

1 Context

Global change processes, in particular climate change, involve inherently unpredictable complex interactions between natural and human-created systems therefore proper modeling of these processes must rely on adequate treatment of uncertainties, socio-economic and environmental heterogeneities, and their interdependencies with human's decisions. Traditional natural science models are based on relations whose validity is estimated from repetitive experiments and observations. If experiments do not affect the underlying relations, and the processes are stationary, then repetitive observations allow to develop the corresponding models by using the statistical decision theory. In reality, however, human-created processes do not follow fixed relations. Such processes have typically structure, relations, and parameters that not only change over time but also depend on the decisions that affect the processes; for example, introduction of new technologies may

---

M. Makowski (✉) · Y. Ermoliev
International Institute for Applied Systems Analysis, Laxenburg, Austria
e-mail: ermoliev@iiasa.ac.at; marek@iiasa.ac.at

K. Marti
Federal Armed Forces University Munich, Aerospace Engineering and Technology,
Neubiberg/Munich, Germany
e-mail: kurt.marti@unibw-muenchen.de

Y. Ermoliev et al. (eds.), *Managing Safety of Heterogeneous Systems*, Lecture Notes
in Economics and Mathematical Systems 658, DOI 10.1007/978-3-642-22884-1_1,
© Springer-Verlag Berlin Heidelberg 2012

increase or reduce uncertainties, modify threats, interdependencies, create systemic risks, critical thresholds, and potential discontinuities. Such new type of decision problems under inherent uncertainty requires qualitatively new approaches than methods supported by the traditional statistical decision theory.

Traditional models based on statistical decision theory deal with situations in which a model of uncertainty, and the corresponding optimal solutions are defined by a sampling model characterized by a probability measure P with an unknown vector x of "true parameters" x^*. Vector $x = x^*$ defines a desirable optimal solution, the performance of which can be observed from the sampling model providing a sequence $\{\omega^1, \omega^2, \dots\}$ of random observations of x^*. Therefore, the problem is to recover x^* from these data. Potential estimates of x^* define feasible solutions x of the corresponding statistical decision problem. It is essential that x does not affect the sampling model P so that the performances of solutions x can be evaluated by a distance from x^* by using observable performance $\{\omega_1, \omega_2, \dots\}$.

Support for the new type of decision-making under uncertainty requires fundamentally new approaches. The model of uncertainty, feasible solutions, and performance of the optimal solution are not given; all of these elements have to be modeled based on analysis of the decision making situation, i.e., considering heterogeneous dimensions, such as socioeconomic, technological, environmental, and safety. Moreover, there is no information on the actual optimal performance; therefore the performance of desirable solutions cannot be characterized by a distance from an observable, actual optimal performance. Thus, the general decision problems typically have rather diversified facets (dimensions) of robust performance.

Actually, good evaluations of global change processes are unrealistic because such processes are non-stationary, have delayed responses, and experiments are dangerous or even impossible. Moreover, some human or natural actions qualitatively change the underlying processes, e.g., causing discontinuities or irreversibilities. Under inherent uncertainty of such heterogeneous processes, the role of integrated modeling rests on its ability to guide comparative analysis of rational decisions. Although exact evaluations are impossible, the preference structure among decisions can provide a stable basis for a relative ranking of alternatives, and thus enable designing robust policies, which are, in a sense, optimal against all relevant uncertainties. To illustrate this approach let us recall a commonly known observation: finding out (without exact measurements) which of two given parcels is heavier is much easier than evaluating weight of each of them.

The term *robust* was first introduced into statistics in 1953 by Box (1953); it was widely recognized after publication of a path-breaking paper by Huber (1981), who admitted that researchers had long been concerned with the sensitivity of standard estimation procedures for "bad" observations (outliers). Appeal for robustness (Hampel et al. 1986) probably dates back to prehistory of statistics. A distant outlier in observations ruins the least square analysis, therefore rejection of outliers is considered a robust statistical procedure. The mean is not robust to outliers, whereas the median is robust; therefore, switching from the mean to the median for long-tailed data increases robustness. According to Huber, . . . *any statistical procedure. . . should be robust in the sense that small deviations from the model assumptions should impair the performance only slightly.* This concept of robustness corresponds to standard mathematical ideas of continuity and stability: when disturbances become small, the performance of the perturbed model also deviates slightly. In other words, a robust procedure is in a certain sense optimal with respect to all uncertainties from a neighborhood of the model. Huber introduced rigorous notions of robustness based on probabilistic minimax approach and Choquet capacity (imprecise probabilities), which led to specific non-smooth stochastic optimization models. By using appropriate neighborhoods of probability measures (e.g., with respect to ε-contaminated probabilities, Levy distance, or Kolmogorov distance), Huber derived robust estimators optimizing the worst that can happen over the neighborhood of the model with respect to a certain performance indicator. Neighborhoods of probability measures can also be characterized by Choquet capacities, i.e., functions which define sets of probability measures by taking all probabilities which lie bellow (or above) a capacity (pointwise). These basic ideas of robust statistics, as well as the infinitesimal robustness introduced by Hampel et al. (1986) can also be used for more general decision problems under uncertainty.

2 Decisions Under Uncertainties

In general, decision problems under uncertainty, any related decision x results in multiple outcomes such as costs, benefits, damages, and risks, as well as indicators of fairness, equity, and environmental impacts. The outcomes $g_i(x, \omega)$, $i = 1, \ldots, K$ depend not only on decisions x but also on uncertainty characterized by $\omega \in \Omega$, where Ω denotes a set of admissible scenarios.

Under uncertainties, a given decision x often has qualitatively different outcomes for different scenarios ω; therefore it difficult to assess which decision is reasonably good for all considered scenarios. In 1738 mathematician Bernoulli (1977) introduced the concept of expected utility maximization as a rule for choosing decisions under multiple outcomes. This approach assumes that all outcomes are represented by a single measure of preferability, e.g., a monetary payoff denoted by: $q(x, \omega) = Q(g_1, \ldots, g_K)$. The standard expected utility model suggests to choose a decision x that maximizes an expected utility function

$$U(x) = Eu(q(x, \omega)) = \int u(q(x, \omega)) P(d\omega),$$

i.e., in a sense, for all $\omega \in \Omega$, where $u(\cdot)$ is a utility associated with an aggregate outcome $q(x, \omega)$. The shape of u defines attitudes to risks. The use of a probability measure as a degree of belief was formalized by Ramsey (1931). Savage published (Savage 1972) a thorough treatment of expected utility maximization based on subjective probability as a degree of belief. As a result of this work the use of probability measure became a standard approach for modeling uncertainty in a consistent way within a single model, by using "hard" observations, and soft public and expert opinions. Although an optimal decision maximizes, in a sense, the expected utility for all scenarios, it often still cannot be considered to be a robust solution, especially with respect to inherent heterogeneities of systems.

The shortcomings of the expected utility model are well known. Generally, it is practically impossible to find a utility function that enables satisfactory aggregation in one preferability measure of various attributes, including attitudes to different risks, distributional aspects of gains and losses, rights of future generations, and responsibilities for environmental protection. For complex heterogeneous systems, it is natural that different performance indicators should be used to evaluate its robustness in the same way as we use indicators of health, e.g., temperature and blood pressure for humans. The expected utility model is a specific case of STO[1] models that use various performance indicators $f_i(x, \omega)$, $i = 1, \ldots, m$, one of which can be the expected utility (disutility). These indicators depend on outcomes $g_k(x, \omega)$, $k = 1, \ldots, K$, on x and ω, i.e.,

$$f_i(x, \omega) := q_i(g_1, \ldots, g_K, x, \omega).$$

A rather general STO problem is formulated as optimization (maximization or minimization) of the expectation function

$$F_0(x) = E f_0(x, \omega) = \int f_0(x, \omega) P(d\theta)$$

subject to

$$F_i(x) = E f_i(x, \omega) = \int f_i(x, \omega) P(d\theta) \geq 0, i = 1, \ldots, m.$$

The choice of proper indicators $f_i(x, \omega)$ and outcomes $g_k(x, \omega)$, $k = 1, \ldots, K$, is essential for the robustness of x. Globally or regionally aggregated outcomes are less uncertain, but they may not reveal potentially dramatic heterogeneities induced by global changes on individuals, governments, and the environment. For instance, an aggregate income or growth indicators may not reveal an alarming gap between poor and rich regions, which may cause future instabilities.

By choosing appropriate outcomes $g_k(x, \omega)$ and functions $f_i(x, \omega)$, STO models enable a natural and flexible way to represent various risks, abrupt changes, spatio-temporal heterogeneities, equity constraints and the sequential resolution of uncertainty in time. Often, under proper robustness requirements, $f_i(x, \omega)$ are analytically intractable, non-smooth, and even discontinuous functions; probability measure P is chosen from a feasible set, thus is imprecise; moreover, P often depends on x, which is essential for modeling robustness.

It is also often practically impossible to uniquely identify degree of beliefs in terms of subjective probability. Most people cannot clearly distinguish between

---

[1] STO is the commonly used acronym for Stochastic Optimization.

probabilities ranging roughly from 0.3 to 0.7. Decision analysis often has to rely on imprecise statements, for example, that event e_1 is more probable than event e_2, or that the probabilities p_1, p_2 of event e_1 or of event e_2 are greater than 50% and smaller than 90%. Therefore, feasible sets of probabilities are often implicitly defined by inequalities such as $p_1 \geq p_2, 0.5 \leq p_1 + p_2 \leq 0.9$.

As in robust statistics, the robust solutions of general decision models can be derived by using worst-case (for a given decision) probability distribution from the feasible sets of distributions satisfying constraints of a STO model.

The standard expected utility maximization model suggests two types of decisions in the response to uncertainty: either risk averse (including risk neutral) or risk prone decisions. These two options also dominate, for example, the climate change policy debates, emphasizing either ex-ante anticipative emission reduction programs, or ex-post adaptation to climate changes when full information becomes available. Clearly, a robust policy should include both options, i.e., a robust strategy should be flexible enough to allow for later adjustments of earlier decisions. Two-stage and multi-stage recourse models of stochastic optimization incorporate both fundamental ideas of anticipation and adaptation within a single model, and support analysis of trade-offs between long-term anticipatory strategies with respect to some slices of risks and short-term adaptive adjustments with respect to other slices. Therefore, the adaptive capacity can be properly designed ex-ante, say, through emergency plans and insurance arrangements. Explicit incorporation of ex-ante and ex-post decisions induces risk aversion measures that cannot, in general, be imposed exogenously by a standard utility function.

Co-existence of ex-ante and ex-post decisions induces quantile-type risk measures. However, convexity (concavity) of models can be preserved by substituting mean values by median, or/and other quantiles.

STO methods can also be applied to some problems unsolvable by standard deterministic methods. For example, STO methods deal directly with the variability of $f_i(x, \omega)$ affected by variability of ω and decisions x, i.e., they deal simultaneously with uncertainty and decision analysis. Some decisions x can considerably reduce the variability of indicators $f_i(x, \omega)$, despite significant variability of ω, e.g., decisions $x_1 = 0$, $x_2 = 0$ for function $\omega_1 x_1 + \omega_2 x_2$. Therefore, STO models can significantly reduce requirements on data quality in contrast to standard approaches to data uncertainty analysis separated from decision analysis.

3 Illustrative Examples

The main challenge confronted by STO theory is that it is practically impossible to evaluate exact values of $F_i(x)$, $i = 0, 1, \ldots, m$, as the following Example 1 illustrates.

Example 1 (Safety constraints in pollution control). A common feature of most models used in designing pollution-control policies is the use of transfer coefficients

a_{ij} that link the amount of pollution x_j emitted by source j to the pollution concentrations $g_i(x, \omega)$ at the receptor location i as

$$g_i(x, \omega) = \sum_{j=1}^{n} a_{ij} x_j, i = 0, 1, \ldots, m.$$

The coefficients a are often computed with Gaussian-type diffusion equations. These equations are solved over all possible meteorological conditions, and the outputs are then weighted by the frequencies of meteorological conditions over a given time interval, yielding average transfer coefficients a_{ij}.

Deterministic models ascertain cost-effective emission strategies $x_j, j = 1, \ldots, n$ subject to achieving exogenously specified environmental goals, such as ambient average standard b_i at receptors $i = 1, \ldots, m$. Such models can be improved by the inclusion of safety constraints that account for the random nature of coefficients a_{ij} and ambient standards b_i to reduce impacts of extreme events:

$$F_i(x) = Prob[\sum_{j=1}^{n} a_{ij} x_j \leq b_i] \geq p_i, i = 1, \ldots, m;$$

namely, the probability that the deposition level in each receptor (region, grid, or country) i will not exceed uncertain critical load (threshold) b_i at a given probability (acceptable safety level) p_i.

These constraints can be written in the form of the standard STO model with discontinuous functions

$$f_j(x, \omega) = \begin{cases} p_i - 1, & \text{if } \sum_{j=1}^{n} a_{ij} x_j - b_i \leq 0, \\ p_i, & \text{otherwise.} \end{cases}$$

If there is a finite number of possible scenarios

$$\omega = (a_{ij}, b_i), \quad i = 1, 2, \ldots, m, \quad j = 1, \ldots, n$$

reflecting, say, prevailing weather conditions, then $F_i(x)$ are piecewise constant functions, i.e., gradients of $F_i(x)$ are equal to 0 almost everywhere. Hence, the straightforward conventional optimization methods cannot be used.

Ignorance of risks defined by safety constraints may cause irreversible, catastrophic events.[2] Therefore, safety constraints are important for regulation of stability

---

[2]Consider, for example, concentration of a toxicant in a lake that on average is far below a damaging level. A short-time substantial increase of the concentration may not increase long-term average in a noticeable way, but may still be lethal.

in the insurance industry, known as the insolvency constraints. The safety regulations of nuclear reactors require $p_i = 1 - 10^{-7}$, i.e., a major failure occurring on average only once in 10^7 years. Yet, such constraints (assuming $p_i < 1$) do not exclude the risk that a disaster may occur even next hour.

Deterministic decision problems are usually formulated in two steps. First, statistical procedures are used to estimate average values $\overline{\omega}$ of input data ω. Second, a deterministic problem with goal functions $f_i(x, \overline{\omega}), i = 1, \ldots, m$ is solved. For multi-mode distributions, the use of $\overline{\omega}$ may even orient analysis on decision that are actually infeasible. Moreover, for nonlinear $f_i(x, \omega)$, typically

$$\overline{f_i(x, \omega)} \neq f_i(x, \overline{\omega}).$$

For example, for ω uniformly distributed on $[-1, 1]$ $\overline{(\omega x)^2} > (\overline{\omega}x)^2$ (because $\overline{\omega} = 0$).

If one can evaluate explicitly multidimensional integrals

$$F_i(x) = E f_i(x, \omega) = \int_{\Omega} f_i(x, \omega) P(d\omega),$$

then the STO problem reduces to a standard deterministic optimization model, which, however may be difficult to solve even for simple functions. To illustrate this let us consider two random variables ω_1, ω_2 with known probability distributions; evaluation of probability distribution of their sum $(\omega_1 + \omega_2)$ is (in general) an analytically intractable task requiring the evaluation of an integral. Moreover, the distribution of $f_i(x, \omega)$, depends on x; e.g., consider $f_i(\cdot) = \omega_1 x_1 + \omega_2 x_2$, and compare two pairs of very simple functions: $\{x_1 = 0, x_2 = 1\}$ and $\{x_1 = 1, x_2 = 0\}$.

Using average or expected values of typical indicators (such as income, growth, daily pollutant concentration, losses, utility, returns, growth, incomes) is often misleading, especially for heterogeneous systems. Note that the projected global mean temperature change falls within the difference between the average temperature of cities and their surrounding rural areas. Therefore, heterogeneities of global climate change impacts can be properly evaluated only in terms of local temperature variability and related extreme events, in particular, heat waves, floods, droughts, wind-storms, diseases, and sea level rise.

A proper treatment of indicators characterized by non-normal, especially multimodal distributions requires special attention. The mean value of a multi-modal distribution can be even outside the support of the distribution (the set of admissible values). This value can be reasonably interpreted in the case of frequent repetitive observations. Subjective multi-modal probability distributions and rare extreme events call for the use of quantiles, e.g., the median. Unfortunately, this destroys the additive structure and concavity (convexity) of standard models, as (in contrast to the average value)

$$\text{median}(\sum_l v_l) \neq \sum_l \text{median}(v_l)$$

where v_l are random variables. Therefore applicability of well-known decomposition schemes and optimization methods is limited to special cases.

Example 2 (Dynamic problems). Discrete-time optimal control can be viewed as a specific case of STO models, where x is composed of state variables $z(t)$, and control variables $u(t)$, i.e.,

$$x = \{z(t), u(t)\}, t = 0, 1, \ldots, T$$

where T is a given time horizon.

Note that even if $f_i(x, \omega)$ are additive, i.e.,

$$f_i(x, \omega) = \sum_{t=1}^{T} g_i(z(t), u(t), \omega_t, t),$$

where ω_t is a stochastic disturbance at time t, the use of $median(f_i(x, \omega))$ destroys the additive structure of optimal control problems that is needed for applying the Pontriagin's Maximin Principle and Bellman's recursive equations.

Modeling heterogeneous global changes with possible dramatic interactions among humans, nature and technology call for nonsmooth stochastic models. Nonsmooth and discontinuous processes are typical for systems undergoing structural changes and potential melt-downs, collapses, bankruptcies. There are a number of methodological challenges involved in the policy analysis of nonsmooth processes. Traditional local or marginal analysis cannot be used because continuous derivatives do not exist, i.e., a nonsmooth, even deterministic, system cannot be predicted (in contrast to classical smooth systems) outside an arbitrary small neighborhood of local points. The use of average values often smoothes the problem, but leads to wrong conclusions. The following simple Example 3 of abrupt changes shows that the use of average characteristics and representing the problem by a linear deterministic model leads to misleading results from the model analysis that is not capable to detect abrupt changes that can result in environmental collapse.

Example 3 (Abrupt changes). Consider concentration of a pollutant characterized by equation

$$r_t = r_0 - x_t + \sum_{k=1}^{N(t)} e_k,$$

where $\{e_k\}$ is a sequence of emissions from episodes in interval $[0, t]$; $N(t), t \geq 0$, is a counting process for the number of episodes in $[0, t]$; x is a rate of emission reduction, and r_0 is given initial concentration.

Assume that e_k are independent, identically distributed random variables with mean value $\bar{e}$, $N(t)$ is a Poisson process with intensity α, $EN(t) = \alpha t$, and $\{e_k\}$, $N(t)$ are independent. Then, the expected concentration

$$\overline{r_t} = r_0 + (\alpha \overline{e} - x)t,$$

i.e., the complex random jumping process r_t, is replaced by simple, linear function that decreases in time for $x > \alpha \overline{e}$. Thus, the deterministic model given by $\overline{r_t}$ suggests, that if x slightly exceeds the average emission rate $\alpha \overline{e}$, then $\overline{r_t}$ decreases in time. This, however, is a wrong conclusion because the actual trajectory may exceed a critical concentration level at some t. Sensitivity scenario analysis of the linear deterministic model $\overline{r_t}$ under different scenarios of α and $\overline{e}$ produces also wrong conclusions that a robust emission reduction x needs to only slightly exceed $\alpha \overline{e}$.

The significance of "extreme events" arguments e.g., large deviations of $\overline{r_t}$ in global climate changes has been summarized by B. Clark as follows: "Impacts accrue ... not so much from slow fluctuations in the mean, but from the tails of the distributions, from extreme events". In other words, catastrophes do not occur on average with average patterns; they occur as "spikes" in space and time. Therefore, the distributional aspects, i.e., temporal and spatial heterogeneous distributions of values and threats are the key issues for capturing the main sources of vulnerability for designing robust policies.

Extreme events are usually characterized by their expected arrival time, for example, as a 1000-year flood, that is, an event that occurs on average once in every 1000 years. Accordingly, these events are often ignored as they are evaluated as improbable during a human lifetime. In fact, a 1000-year flood may occur next year. For example, floods across Europe in 2002 were classified as 1000-, 500-, 250-, and 100-year events.

Another misleading methodology is to evaluate potential extreme impacts by using so-called annualization, i.e., by spreading losses from a potential catastrophe equally over the period equal to its expected arrival time.

Example 4 (Annualization). Consider annualization approach to a potential 30-year crash of an airplane to be evaluated as a sequence of independent annual "partial" crashes: one wheel lost in the first year, another wheel in the second year, and so on, until crash of the navigation system in the 30th year. The main conclusion from this type of deterministic approach based on averaging is that catastrophes actually do not exist.

A proper approach to analysis of temporal variability of extreme events and the corresponding robust solutions can be based on the methodology of stopping-time and the related new approaches to discounting.

Another methodological pitfall comes from ignoring actual spatial interdependencies of catastrophic impacts. A typical approach is to use so-called hazard maps, i.e., maps showing catastrophe patterns that will never be observed as a result of a real episode, as a map is the average image of all possible patterns that may follow catastrophic events. Accordingly, social losses in affected regions are evaluated as the sum of individual losses computed on a location-by-location rather than pattern-by-pattern basis w.r.t. joint probability distributions. Such approach highly underestimates the real impacts of catastrophes, as illustrated by the following simple Example 5.

Example 5 (Social and individual losses). Assume that each of 100 locations has an asset of the same type. An extreme event destroys all of them at once with probability 0.01. Consider also a situation without the extreme event, but with each asset still being destroyed independently with the same probability equal to 0.01.

From an individual point of view, these two situations are identical: an asset is destroyed with probability 0.01, i.e., individual losses are the same. Collective (social) losses are dramatically different. In the first case 100 assets are destroyed with probability 0.01, whereas in the second case 100 assets are destroyed only with probability 100^{-100}, which is practically equal to 0. This example also illustrates the potential exponential growth of vulnerability from increasing network-interdependencies. Example 5 also shows that, in a sense,

$$100 \gg \overbrace{1 + 1 + \cdots + 1}^{100}.$$

Designing a catastrophe model is a multidisciplinary task requiring the joint efforts of environmentalists, physicists, economists, engineers and mathematicians. To characterize "unknown" catastrophic risks, that is, risks with the lack of historical data and large spatial and social impacts, one should at least characterize the random patterns of possible disasters, their geographical locations, and their timing. One should also design a map of values and characterize the vulnerabilities of buildings, constructions, infrastructure, and activities. Catastrophe models allow to derive histograms of mutually dependent losses for a single location, a particular hazard-prone zone, a country, or worldwide from fast Monte Carlo simulations rather than real observations.

The development of catastrophe models can be considered as a key risk management policy providing information for robust decision analysis in the absence of historical observations, in particular, on potential extreme events that have never occurred in the past. This raises new estimation problems. Traditional statistical methods are based on the ability to obtain observations from unknown true probability distributions, whereas new problems require information to be recovered from only partially observable or even unobservable variables. Rich data may exist on occurrences of natural disasters, incomes, or production values on global and national levels. The so-called downscaling and catastrophe modeling are becoming increasingly important for estimating spatio-temporal vulnerability and catastrophic impacts. Downscaling and upscaling methods in such cases support – by using all available objective and subjective information – making plausible evaluations of local processes consistent with available global data, as well as, conversely, with global implications emerging from local data and tendencies.

The above discussion illustrates that although STO models allow to represent interdependencies among decisions, uncertainties and risks, yet inappropriate treatment of the variability of indicators $f_i(x, \omega)$ can be rather misleading for achieving desirable robustness. The following stylized Example 6 motivated by food security studies illustrates further the importance of decisions which are robust in a sense against all potential scenarios of uncertainty ω.

Example 6 (Food security). Assume there are only two scenarios of weather conditions for the next agricultural season: the spring is wet (ω_1) or the spring is dry (ω_2). A farmer needs to select one[3] crop to plant: either crop A (having high profit p_A for ω_1, but total loss for ω_2) or crop B (having high p_B profit for ω_2, but total loss for ω_1) or crop C (having moderate profits $p_C(\omega)$ for both weather scenarios).

Any of these crop alternatives may become a robust decision, depending on the farmer's preferential structure and underlying conditions, such as his/her financial reserves, available hedging instruments, probability distributions of ω, the planning horizon (one- or multi-year), relations between p_A, p_B, p_C. We leave exploration of this simple but yet interesting problem to the readers.

4   Robust Decisions for Heterogeneous Systems

Let us now consider designing robust portfolios of financial assets. Assume that $\overline{\omega_j}$ is the expected value of random returns ω_j from divisible asset j, $j = 1, \dots, n$, and x_j is a fraction of this asset in the portfolio

$$\sum_{j=1}^{n} x_j = 1, \quad x_j \geq 0.$$

Maximization of expected return

$$r(x) = \sum_{j=1}^{n} \overline{\omega_j} x_j$$

from this portfolio yields a trivial (but not robust) solution: to invest all capital in the asset with the maximal expected return. Such a decision is obviously very risky, and its value is limited to be a standard example in teaching why it should be avoided.

A design of a robust portfolio requires analysis of trade-offs between expected returns and their variability. Nobel prize laureate Markowitz proposed the mean-variance approach for designing robust portfolios of financial assets, i.e., to characterize the variability by the variance of returns $Var\rho(x, \omega)$. Therefore the portfolio composition results from maximization of

$$r(x) - \mu Var\rho(x, \omega), \quad \rho(x, \omega) = \sum_{j=1}^{n} \omega_j x_j,$$

---

[3]E.g., due to technological or market constraints.

where μ is a risk parameter. This approach is based on using variance, it therefore can be properly used for analysis of assets having distributions of random returns $\rho(x, \omega)$ close to the normal distribution. However, in reality, $\rho(\cdot)$ has typically other distributions; therefore this approach should not be used.

The following non-smooth version of the portfolio selection STO problems enables dealing with non-normal distributions. Consider maximization of the utility function $U(x) = Eu(q(x, \omega))$. If the distribution of random outcome $u(q(x, \omega))$ is not normal (for example, when the policy analysis involves the polarized beliefs of different communities), then, instead of $U(x)$ one can use a quantile $U_p(x)$ of $u(q(x, \omega))$ defined as maximal v such that $Prob[u(q(x, \omega)) \leq v] \leq p$, for $0 < p < 1$ (which can be interpreted as a safety constraint).

The robust utility maximization problem can be formulated as maximization of a risk adjusted utility function

$$U_p(x) + \mu E \min\{0, u(q(x, \omega)) - U_p(x)\}.$$

This function is not concave. However, the optimization problem can be solved by converting it to the equivalent concave STO problem: maximize w.r.t. (x, z) function $\varphi(x, z)$ defined as:

$$\varphi(x, z) = z + \mu E \min\{0, \beta - z\}, \beta = u(q(x, \omega)), \mu = 1/p.$$

Let us also notice that for $\mu = 1/p$ we have

$$U_p(x) + \mu E \min\{0, u(q(x, u)) - U_p(x)\} = (1/p) \int_{u(q(x, \omega)) \leq U_p(x)} U(q(x, \omega)) dP,$$

i.e., the risk adjusted utility function is related to the so-called expected shortfall.

Similarly, a general STO model for analyzing robust solutions of heterogeneous systems can be written in the similar form: maximize w.r.t. (x, z) function:

$$z_0 + \mu_0 E \min\{0, f_0(x, \omega) - z_0\}$$

subject to:
$$z_i + \mu_i E \min\{0, f_i(x, \omega) - z_i\} \geq 0, i = 1, \ldots, m,$$

where μ_i are weights. Components $z_i^*, i = 0, 1, \ldots, m$, of optimal solution (x^*, z^*) are quantiles of $f_i(x^*, \omega)$.

Depending on the case, the robust model can also be formulated by using safety constraints (see Example 1) or constraints:

$$E f_i(x, \omega) + \mu_i E \min\{0, f_i(x, \omega)\} \geq 0, \quad i = 1, \ldots, m.$$

Standard STO models assume that $P(d\omega)$ is known exactly. However, only some of its characteristics may be known. The elicited class $\mathscr{P}$ for admissible P is often given by constraints

$$\int \varphi_k(\omega) P(d\omega) \geq 0, \quad k = 1, \ldots, K, \quad \int P(d\omega) = 1;$$

e.g., constraints on joint moments

$$c_{s_1,\ldots,s_l} \leq \int \omega_1^{s_1} \ldots \omega_l^{s_l} P(d\omega) \leq C_{s_1,\ldots,s_l},$$

where $c_{s_1,\ldots,s_l}$, $C_{s_1,\ldots,s_l}$ are given constants. The robust STO problem can be formulated as a probabilistic maximin problem: maximize

$$F_0(x) = \min_{p \in \mathscr{P}} \int f_0(x,\omega) P(d\omega)$$

subject to general constraints of STO models. This probabilistic maximin approach was first initiated in STO. For specific sets $\mathscr{P}$, the solution of the inner minimization problem has simple analytical forms. For example, it is concentrated only in a finite number of admissible scenarios from Ω.

5 Heterogeneity and Vulnerability

The vulnerability analysis of complex coupled human-environmental systems essentially relies on discounting future losses and gains to their present values. These evaluations are used to justify catastrophic risks management decisions which may turn into benefits over long and uncertain time horizons. The misperception of proper discounting rates critically affects evaluations and may be rather misleading. The lack of proper evaluations may dramatically contribute to increasing the vulnerability of our society to human-made and natural disasters. Underestimation of low probability – high consequences potentially catastrophic scenarios often leads to the growth of buildings, industrial land, and sizable value accumulation in flood prone areas without paying proper attention to flood mitigations. A challenge is that an extreme event, say a once-in-300-year flood which occurs on average only once in 300 years, may have never occurred before in a given region. Therefore, purely adaptive policies relying on historical observations provide no awareness of the risk although a 300-year flood may occur next year. For example, floods in Austria, Germany and the Czech Republic in 2002 were classified as 1000-, 500-, 250-, and 100-year events. Yet common practice is to ignore these types of events as improbable events during a human lifetime.

5.1 Discounting

Several contributions to Marti et al. (2010) analyze the implications of potentially catastrophic events on the choice of discounting for long-term vulnerability modeling and extreme events management.

Traditional approaches to using discounting for cost-benefit analysis of risk management for extreme/rare events have fundamental limitations that are commonly known, but yet often forgotten. Properly chosen discounting can be useful for integrated analysis of implications of potentially catastrophic events, especially for modeling long-term vulnerability and risk management. Recent scientific achievements support such model-based analyses. In particular, new methods enable linking arbitrary discount factors to "stopping time" ("killing") events, which define the discount-related random horizon τ ("end of the world") of evaluations. In other words, any discounting compares potential gains and losses that occur at different points in time only within a finite random discount-related time horizon τ. The expected duration of τ for standard discount rates obtained from capital markets does not exceed a few decades and, as such, these rates cannot properly match evaluations of 1000-, 500-, 250-, 100-year catastrophes. On the other hand, any "stopping time" event induces discounting. Formally it means that there are the following relations between discounted expected values $V_t = Ev_t$ with discount factors d_t and undiscounted stopping time evaluations

$$\sum_{t=0}^{\infty} d_t V_t = E \sum_{t=0}^{\tau} v_t, d_t = P[\tau \geq t].$$

Therefore, the correct discounting can be induced by explicit modeling of scenarios of potential extreme events. These induced discount rates are conditional on the degree of social commitment to mitigate risk. In general, extreme events affect discount rates, which alter the optimal mitigation efforts that, in turn, change events. Such endogenous discounting calls for the use of equivalent undiscounted random stopping time criteria and stochastic optimization methods. Combined with explicit spatio-temporal modeling of heterogeneous systems, these criteria induce the discounting which allows to properly match random horizons of potential catastrophic scenarios with necessary risk management policies. In contrast to standard time-discounting, the resulting discounting can also be viewed as induced "spatio-temporal" discounting. This allows to examine risk profiles generated by the stopping time, in particular, connections with the safety constraints and spatio-temporal versions of CVaR (Conditional Value at Risk) risk measures.

5.2 Stochastic Optimization Versus Scenario Optimization

Scenario analysis is often suggested as a straightforward approach for finding robust solutions. Monte Carlo simulations for a STO model easy generate samples of random values $f_0(x, \omega)$, $f_1(x, \omega)$, ..., $f_m(x, \omega)$, that depend on the simulation run ω and a given vector of decisions x. Therefore, for a given x, outcomes vary at random from one simulation to another. If functions $f_i(x, \omega)$, $i = 0, 1, \ldots, m$, have well defined analytical structure with respect to x for each simulated ω, then the scenario analysis has the following steps. The Monte Carlo simulations generate scenarios $\omega^1, \omega^2, \ldots, \omega^N$ for each of which the corresponding optimal solutions $x(\omega^1), x(\omega^2), \ldots, x(\omega^N)$ of the deterministic optimization model with objective function $f_0(x, \omega)$ and constraint functions $f_i(x, \omega)$, $i = 1, \ldots, m$ are calculated. Any of these solutions calculated for one scenario may not be feasible for other scenarios. The number of possible combinations of potential scenarios ω and decisions increases exponentially. Thus, e.g., with only 10 feasible decisions of emission reductions in a given region, 10 regions, and 10 possible independent scenarios for all of them, the number of "what-if" combinations is 10^{11}. The straightforward evaluation of these alternatives would require more than 100 years if a single evaluation takes only a second. Moreover, the probability of each scenario ω^l, $l = 1, \ldots, K$, is in general, equal to 0. The choice of final robust decisions is unclear and is not explicitly addressed. Therefore, the aim of STO methods is to design a directed search of an optimal solution avoiding straightforward testing of all possible combinations. The STO methods avoid also the exact estimation of the mean values $F_i(x) = E f_i(x, \omega)$.

6 Summary

This chapter summarizes diverse facets of managing safety of heterogeneous systems, and supporting robust decision-making under uncertainties and risks. Different decision situations and the corresponding characteristics of the system require, of course, different approaches. However, methodologies and experience can and should be shared across application areas. Very different processes often have the same (or very similar) mathematical representation, therefore it is rational to re-use (or adapt) methods and tools, instead of reinventing them, or – even worse – to try to apply methods which are known (in other areas or research communities) as inadequate.

Each, except this, chapter of this book focuses on a specific either domain problem or methodology, and provides numerous references on, and other examples of, one of the issues belonging to the wide and diversified area of managing safety of heterogeneous systems, and the corresponding methods and tools for supporting robust decision-making under uncertainties and risks. This chapter provides overview of the selected key problems and methods applicable to different types of substantive problems, especially in economics, industry, environmental and social policy-making.

A comprehensive overview of the state-of-the-art would be beyond the scope of this type of chapter. Therefore we had to select a small number of issues, and focused on two types of them. First, modern methods applicable to a wide range of different problems, and still not commonly used. Second, on methods that are widely used although they are clearly inappropriate, and there exist appropriate (often not commonly known yet) methods for the corresponding class of problems.

Finally, we have refrained from providing comprehensive bibliography in this chapter: this would require a rather large set of publications. We have included only several fundamental/classical references. Numerous bibliography is included in almost all chapters of this volume, as well as in Marti et al. (2010, 2006), i.e., the publications resulting from the previous *CwU* workshops.

References

Bernoulli, D. (1777). Dijudicatio maxime probabilis plurium observationum discrepantium atque verisimililima inductio inde formanda. acta acad. sci. petropolit 1, 3-33. *Biometrika*, English translation by C.G. Allen (1961). *48*, 3–13.

Box, G. (1953). Non-normality and tests on variances. *Biometrika, 40*, 318–335.

Hampel, F., Ronchetti, E., Rousseeuw, P., & Stahel, W. (1986). *Robust Statistics: The Approach Based on Influence Functions*. New York: Wiley.

Huber, P. (1981). *Robust Statistics*. New York: Wiley.

Marti, K., Ermoliev, Y., & Makowski, M. (Eds.) (2010). *Coping with Uncertainty: Robust Decisions. Lecture Notes in Economics and Mathematical Systems*. Berlin, Heidelberg, New York: Springer. ISBN 978-3-642-03734-4.

Marti, K., Ermoliev, Y., Makowski, & M., Pflug, G. (Eds.) (2006). *Coping with Uncertainty: Modeling and Policy Issues, Lecture Notes in Economics and Mathematical Systems*, vol. 581. Springer, Berlin, Heidelberg, New York ISBN 978-3-540-35258-7.

Ramsey, F. (1931). Truth and probability. In R. Braithwaite (Ed.) *The Foundations of Mathematics and other Logical Essays* (pp. 156–198). London: Kegan, Paul, Trench, Trubner & Co. Originally published in 1926.

Savage, L. (1972). *The Foundations of Statistics*. New York: Dover Publications. 2nd edn.; originally published in 1954.

Part I
Decisions Under Systemic Risk and Uncertainties

Systemic Risk and Security Management

Yuri Ermoliev and Detlof von Winterfeldt

Abstract The aim of this paper is to develop a decision-theoretic approach to security management of uncertain multi-agent systems. Security is defined as the ability to deal with intentional and unintentional threats generated by agents. The main concern of the paper is the protection of public goods from these threats allowing explicit treatment of inherent uncertainties and robust security management solutions. The paper shows that robust solutions can be properly designed by new stochastic optimization tools applicable for multicriteria problems with uncertain probability distributions and multivariate extreme events.

1 Introduction

Standard risk management deals with threats generated by exogenous events. Typically, such situations allow to separate risk assessment from risk management. Repetitive observations are used to characterize risk by a probability distribution that can be used in risk management. Statistical decision theory, expected utility theory and more general stochastic optimization (STO) theory provide common approaches for this purpose.

Security management includes threats generated (intentionally or unintentionally) by intelligent agents. Obvious examples are threats to public goods and homeland security from terrorists (Ezell and von Winterfeldt 2009). Less evident examples are floods which are often triggered by rains, hurricanes, and earthquakes in combination with inappropriate land use planning, maintenance of flood protection systems and behavior of various agents. The construction of levees, dikes, and dams which may break on average, say, once in 100 years, create an illusion of safety

---

Y. Ermoliev (✉) · D. von Winterfeldt
International Institute for Applied Systems Analysis, Schlossplatz 1, A-2361, Laxenburg, Austria,
e-mail: ermoliev@iiasa.ac.at; detlof@iiasa.ac.at

Y. Ermoliev et al. (eds.), *Managing Safety of Heterogeneous Systems*, Lecture Notes
in Economics and Mathematical Systems 658, DOI 10.1007/978-3-642-22884-1_2,
© Springer-Verlag Berlin Heidelberg 2012

and in the absence of proper regulations developments close to these constructions may create catastrophic floods of high consequences.

Other examples include social, financial, economic, energy, food and water security issues. Water and food security deals with the robust functioning of complex multi-agent water and food supply networks. Threats associated with such systems depend on decisions of different agents. For example, an increase of bio-fuel production may change market prices, induce threats of environmental degradation, destabilize supplies of food and water, and disturb economic developments.

These examples illustrate threats that cannot be characterized by a single probability distribution. Inherent uncertainties of related decision problems with the lack and even absence of repetitive observations restrict exact evaluations and predictions. The main issue in this case is the design of robust solutions. Although exact evaluations are impossible, the preference structure among feasible alternatives provides a stable basis for relative ranking of them in order to find solutions robust with respect to all potential scenarios of uncertainties. As we know, the heavier parcel can be easily found without exact measuring of the weight.

The goal of this paper is to develop a decision-theoretic approach to security management. It shows that robustness of solutions in security management can be achieved by developing new stochastic optimization tools for models with uncertain multi-dimensional probability distributions which may implicitly depend on decisions. The common approach, using the concept of two-stage Stackelberg game is built on strong assumptions of perfect information about preference structures of agents which lead to unstable solutions and discontinuous models even with respect to slight variations of initial data in linear criteria functions. Our proposed decision-theoretic approach explicitly deals with uncertainties. It does not destroy convexities but still preserves the two-stage structure of the Stackelberg "leader-follower" decisions.

In order to develop robust approaches, Sects. 2, 3, and 4 analyze similarities and fundamental differences between frequent standard risks, multivariate multi-agent catastrophic risks generated by natural disasters with the lack and even absence of repetitive observations, and risks generated by intelligent agents.

In the case of standard risks, the term "robust" was introduced in statistics (Huber 1981) in connections with irrelevant "bad" observations (outliers) which ruin the standard mean values, least square analysis, regression and variance/covariance analysis. Section 2 shows, that switching from quadratic (least square) smooth optimization principles in statistics to non-smooth stochastic minimax optimization principles leads to robust statistical decisions. This idea is generalized in the following sections.

In general decision problems (Sect. 3) under inherent uncertainty the robustness of decisions is achieved by a proper representation of uncertainty, adequate sets of feasible decisions and performance indicators allowing to characterize main socio-economic, technological, environmental concerns and security requirements. This leads to specific STO problems. In particular, a key issue is the singularity of robust solutions with respect to low-probability extreme events.

Section 3 introduces new robust STO models applicable for managing systemic risks involving multivariate extreme events. Section 4 analyses security management problems with several agents where a principal agent (PA) uses PA's perception of uncertainties to regulate behavior of other agents to secure the overall performance of a system. This section and Sect. 5 demonstrate that although these problems have features of a two-stage Stackelberg game, the applicability of this game is problematic because of the assumption about exactly known decisions of agents. Implicitly, similar assumptions are also used in bi-level mathematical programs (Dempe 2002; Kocvara and Outrata 2004; Luo et al. 1996). Section 4.1 discusses serious limitations of Bayesian games. In particular, the use of Nash games destroys essential two-stage structure of the PA problems.

Sections 4 and 5 introduce concepts of robust PA's solutions. Section 5 analyses systemic security management problems, in particular, preventive solutions in randomized strategies, defensive allocation of resources, and modeling of systemic failures. Section 6 discusses security of electricity networks. Section 7 analyses computational methods. Applications of these methods to socio-economic and homeland security management can be found in (Borodina et al. 2011; Wang 2010). Section 8 provides conclusions.

Since the focus of CwU workshop is on broad audience, this paper avoids mathematical technicalities. In particular, it pays specific attention to motivations and clarifications.

2 Standard Risks

Standard risk analysis relies on observations from an assumed true model specified by a probability distribution P. Repetitive observations allow deriving the probability distribution P and its characteristics required for related decision support models. A key issue in this case is concerned with "bad" observations or "outliers", which may easily ruin standard mean values, variance, least-square analysis, regressions and covariances (Ermoliev and Hordijk 2006; Huber 1981; Koenker and Bassett 1978). Therefore, traditional deterministic models using mean values may produce wrong results. The main approach in such cases is to use robust models which are not sensitive to irrelevant bad observations and at the same time, which are able to cope with relevant rare extreme events of high consequences.

The term "robust" was introduced into statistics in 1953 by Box and received recognition after the path-breaking publication by Huber (1981), although the discussion about rejection of bad observations is at least as old as the 1777 publication of Daniel Bernoulli. The straightforward rejection of outliers is practically impossible in the case of massive data sets, because it may also delete important and relevant observations. Huber introduced rigorous notions of robustness based on probabilistic minimax approach. Its main idea can be developed for general decision problems emerging in security management (Sect. 4). By using appropriate neighborhoods of probability distributions (e.g. $\varepsilon-$ contaminated probabilities, neighborhoods of imprecise probabilities) Huber derived robust estimates optimizing the worst that can happen in a specific probabilistic sense over the neighborhood of the model. In other words, robust statistical analysis is equivalent to switching from smooth least square optimization principles to non-smooth minimax STO principles. The mean is not robust to outliers, whereas the median is robust. The mean value of a random variable θ minimizes the quadratic function

$$M(x) = E(x - \theta)^2 = \int (x - \theta)^2 P(d\theta), \tag{1}$$

whereas the median and more generally a quantile minimizes function

$$Q(x) = E \max\{\alpha(x - \theta), \beta(\theta - x)\} = \int \max\{\alpha(x - \theta), \beta(\theta - x)\} P(d\theta), \tag{2}$$

i.e., it solves a stochastic minimax problem with non-smooth random function $\max\{\alpha(x - \theta), \beta(\theta - x)\}$, where P is a probability distribution function, and $\alpha, \beta > 0$. This follows from convexity of functions $M(x)$, $Q(x)$. For example assume that P has a continuous density, i.e., M(x), Q(x) are continuously differentiable functions. Then intuitively we have

$$Q'(x) = \alpha \operatorname{Prob}[\theta < x] - \beta \operatorname{Prob}[\theta \ge x] = 0$$

i.e., a solution x of stochastic minimax problem (2) satisfies the equation (Ermoliev and Leonardi 1982; Ermoliev and Yastremskii 1979; Koenker and Bassett 1978; Rockafellar and Uryasev 2000):

$$\operatorname{Prob}[\theta \ge x] = q, q = \frac{\alpha}{\alpha + \beta}. \tag{3}$$

Remark 1 (Uniqueness of quantile). If $Q(x)$ is not a continuously differentiable function, then optimality conditions satisfy analogue of (3) equations using subgradients (Ermoliev and Leonardi 1982) of function (2). In this case, (3) has a set of solutions. Quantile x_q is defined as minimal x satisfying equation $\operatorname{Prob}[\theta \ge x] \le q$. A slight contamination of θ in (2), say by normal random variable, $(1 - \varepsilon)\theta + \varepsilon N(0, 1)$, makes $Q(x)$ strongly convex and continuously differentiable function (Ermoliev and Norkin 2003). The convergence of resulting quantile x_q^ε to x_q follows from the monotonicity of x_q^ε, that is $x_q^{\varepsilon_2} < x_q^{\varepsilon_1}$ for $\varepsilon_2 < \varepsilon_1$. Therefore, in the following we avoid using subgradients by assuming that (3) has a unique solution. For $\alpha = \beta$ (3) defines the median.

Remark 2 (Equivalent calculations of quantiles). It is easy to see that $Q(x) = \alpha x + (\alpha + \beta)E \max\{0, \theta - x\} - \alpha E\theta$. Therefore, x_q minimizes also function

$$x + (1/q)E \max\{0, \theta - x\}, q = \frac{\alpha}{\alpha + \beta}. \tag{4}$$

This simple rearrangement is used in Sect. 3 to formulate robust STO decision support models applicable for security management. Formula (3) connects quantiles with a simple convex STO model (2). This became a key approach (Rockafellar and Uryasev 2000) in risk management. Note, that a direct use of quantiles destroys continuity of even linear performance indicators (Ermoliev and Wets 1988). Minimal value of function (4) defines the important risk measure called Conditional Value at Risk (CVaR).

Problems (1), (2) are simplest examples of STO models. Model (2) is an example of important stochastic minimax problems arising in the security analysis (Sect. 4). Equation (3) shows that even the simplest case of such problems generates robust solutions characterized by quantiles. In general decision models under uncertainty, any relevant decision x results in multiple outcomes dependent on x and uncertainty characterized by a scenario (event, random vector) $\omega \in \Omega$, where Ω denotes a set of admissible states ω. For complex systems it is natural that different performance indicators should be used (see, e.g., Ermolieva and Ermoliev (2005); Ermoliev and Hordijk (2006); Huber (1981)) to evaluate robustness of x similar to the use of different indicators of health (e.g., temperature and blood pressure) for humans. This leads to STO models formulated as optimization (maximization or minimization) of an expectation function.

$$F_0(x) = E f_0(x, \omega) = \int_\Omega f_0(x, \omega) P(d\omega) \tag{5}$$

subject to constraints

$$F_i(x) = E f_i(x, \omega) = \int_\Omega f_i(x, \omega) P(d\omega) \geq 0, i = 1, \ldots, m, \tag{6}$$

where vector $x \in X \subseteq R^n$ and ω in general represent decisions and uncertainties in time $t = 0, 1, \ldots$, i.e., $x = (x(0), x(1), \ldots)$, $\omega = (\omega(0), \omega(1), \ldots)$. Models with ex-ante and ex-post time dependent decisions can be always formulated, see (Ermoliev and Wets 1988), in terms of the first stage solutions x as in (5), (6). Therefore, model of type (5), (6) allows to assess multi-stage dynamic trade-offs between anticipative ex-ante and adaptive ex-post decisions arising in security management (Sect. 6). Random performance indicators $f_i(x, \omega)$, $i = \overline{0, m}$, are often non-smooth functions as in (2). In the case of discontinuous functions $f_i(x, \omega)$, expected values $F_i(x)$ of constraints (6) characterize often risks of different parts $\overline{1, m}$ of the system (see Birge and Louveaux (1997); Ermoliev and Norkin (2003); Ermoliev and Wets (1988); Marti (2005); Shapiro et al. (2009)) in the form of the chance constraints: Prob $[f_i(x, \omega) \geq 0] \geq p_i$, $i = \overline{1, m}$, where p_i is a desirable level of safety. Say, an insolvency of insurers is regulated with $1 - p_i \approx 8 * 10^{-2}$, meaning that balances (risk reserves) may be violated only once in 800 years. In models presented in (Ermolieva and Ermoliev 2005) these type constraints characterize a dynamic systemic risk of systems composed of individuals, insurers, governments, and investors.

Remark 3 (Scenario analysis). It is often used as a straightforward attempt to find a decision x that is robust with respect to all scenarios ω by maximizing $f_0(x, \omega)$, s.t. $f_i(x, \omega) \geq 0, i = 1, \ldots, m$, for each possible scenarios ω. Unfortunately, a given decision x for different scenarios ω may have rather contradictory outcomes, which do not really tell us which decision is reasonably good (robust) for all of them. For example, models (1), (2) show that for any scenario ω the optimal solution is $x(\omega) = \omega$, i.e., the scenario-by-scenario analysis will not suggest solutions in the form of quantile (3). This straightforward scenario analysis faces computational limits even for very small number of examined decisions and scenarios, e.g., analysis of all combinations of 10 scenarios and 10 different decisions may easy require 10^{10} sec. > 100 years.

Models (1), (2) illustrate the main specifics of STO problems of the following sections. Objective functions (1), (2) are analytically intractable because in statistics the probability distribution P is unknown. Instead only observations of ω are available. Analytical intractability of functions $F_i(x)$ is a common feature of STO models. For example, even a sum of two random variables commonly has analytically intractable probability distribution although distributions of both variables are given analytically. Therefore, the main issue of this paper is the development of effective "distribution-free" methods applicable for different type of distributions (see Birge and Louveaux (1997); Ermoliev and Wets (1988); Marti (2005); Mulvey et al. (1995); Shapiro et al. (2009)) and large numbers of decision variables and uncertainties (Sect. 7).

Remark 4 (Uncertain probabilities, Bayesian and non-Bayesian models). The standard stochastic optimization model (5), (6) is characterized by a single probability distribution P, therefore can be defined as Bayesian STO model. When observations are extremely sparse or not available distribution P is elicited from experts (Keeney and von Winterfeldt 1991, 1994; Otway and von Winterfeldt 1992). Yet, often it is difficult to identify uniquely probability P. Most people cannot clearly distinguish between probability ranging roughly from 0.3 to 0.5. Decision analysis then has to rely on imprecise statements, for example, that event e_1 is more probable than event e_2 or that the probability of event e_1 or of event e_2 is greater than 50% and less than 90%. Therefore only feasible sets of probabilities are identified by inequalities such as $p_1 > p_2$, $0.5 \leq p_1 + p_2 \leq 0.9$. It is typical for models arising in security management (Sects. 4 and 5). In such cases we may speak of non-Bayesian STO models., i.e. STO models which are not defined by a single probability distribution, but by a family of distributions with uncertain parameters or, more generally, by an uncertain distribution. Probability distributions depending on decisions are discussed in Sect. 4.

3 Catastrophic and Systemic Risks

Standard "known" risks are characterized by a single probability distribution that can be derived from repetitive observations of ω. The essential new feature of catastrophic risks is the lack and even absence of real repetitive observations. Experiments may be expensive, dangerous, or impossible. The same catastrophe never strikes twice the same place. In addition, catastrophes affect different location

Systemic Risk and Security Management

and agents generating multivariate risks and needs for developing new STO models integrating risk reductions, risk transfers and risk sharing (Ermolieva and Ermoliev 2005).

As a substitute of real observations, so-called catastrophe modeling (catastrophe generators) is becoming increasingly important for estimating spatio-temporal hazard exposures and potential catastrophic impacts. The designing of a catastrophe model is a multidisciplinary task. To characterize "unknown" catastrophic risks, that is, risks with the lack of repetitive real observations we should at least characterize the random patterns of possible disasters, their geographical locations, and their timing. We should also design a map of values and characterize the vulnerabilities of buildings, constructions, infrastructure, and activities. The resulting catastrophe model allows deriving histograms of mutually dependent losses for a single location, a particular zone, a country, or worldwide from fast Monte-Carlo simulations rather than real observations (Ermolieva and Ermoliev 2005; Walker 1997).

3.1 Applicability of Mean Values, Systemic Risk

The use of different sources of information, including often rather contradictory expert opinions usually leads to multimodal distributions of ω and random indicators $f_i(x, \omega)$. The mean value of such indicator can be even outside the set of admissible values requiring the use of quantile, e.g., the median of $f_i(x, \omega)$. Unfortunately, the straightforward use of quantiles destroys the additive structure and concavity (convexity) of model (5), (6), even for linear functions $f_i(\cdot, \omega)$ because, in contrast to the mean value

$$\text{quantile} \sum_i f_i \neq \sum_i \text{quantile} f_i.$$

This lack of additivity makes it practically impossible to use many computational methods relying on additive structure of models, e.g., dynamic programming equations and Pontryagin's maximum principle.

Equations (3), (4) enable the following promising quantile-related decision-theoretic approach. Let us denote a quantile of $f_i(x, \omega)$ by $Q_i(x), i = 0, 1, \ldots, m$. Then we can formulate the following robust version of STO model (5)–(6): maximize

$$Q_0(x) + \mu_0 E \min\{0, f_0(x, \omega) - Q_0(x)\}$$

subject to

$$Q_i(x) + \mu_i E \min\{0, f_i(x, \omega) - Q_i(x)\} \geq 0, i = 1, \ldots, m,$$

where $\mu_i > 1$ are risk parameters regulating potential variability of $f_i(x, \omega)$ below $Q_i(x), i = 0, 1, \ldots, m$. Unfortunately the direct use of $Q_i(x)$ destroys concavity of functions $F_i(x)$. This can be avoided by the following reformulation of the problem.

According to model (2), (3) and Remark 2, the formulated above robust version of STO model (5), (6) can be equivalently rewritten in a similar to (4) form: maximize w.r.t. (z, x) function

$$z_0 + \mu_0 E \min\{0, f_0(x, \omega) - z_0\}, \tag{7}$$

subject to

$$z_i + \mu_i E \min\{0, f_i(x, \omega) - z_i\} \geq 0, i = 1, \ldots, m. \tag{8}$$

For concave functions $f_i(\cdot, \omega)$ this is a concave STO model. The following Proposition 1 shows, that components $z_i^*(x), i = 0, 1, \ldots, m$, solving (7), (8) w.r.t. $z = (z_0, z_1, \ldots, z_m)$ are quantiles $Q_i(x)$. Therefore, (7), (8) is a robust version of model (5), (6) where mean values $E f_i$ are substituted by quantiles of indicators f_i with a safety levels μ_i controlling their variability. In a sense, the model (7), (8) can also be viewed as a concave version of STO models with probabilistic safety constraints (see (Birge and Louveaux 1997; Ermoliev and Hordijk 2006; Ermoliev and Wets 1988; Marti 2005)) outlined in Sect. 2. Equation (9) shows that model (7), (8) is defined by multicriteria versions of VaR and CVaR risk measures (Rockafellar and Uryasev 2000) controlling safety/security of overall system, i.e., a systemic risk. An alternative formulation of quantile optimization problems (subject to quantile constraints) and a corresponding mixed-integer programming solution technique is considered in (O'Neill et al. 2006).

Let us also note that the variability of outcomes $f_i(x, \omega)$ can be controlled by using a vector of quantiles $z^i = (z_{i0}, z_{i1}, \ldots, z_{il})$ generated as in (7)–(8) by performance indicators $\sum_l (z_{il} + \mu_{il} \min\{0, f_i(x, \omega) - z_{il}\}), i = \overline{0, m}$, where $1 < \mu_{i1} < \mu_{i2} < \cdots < \mu_{il}$.

Proposition 1. *(Quantiles of* $f_i(x, \omega)$*): Assume* $f_i(x, \cdot), i = 0, 1, \ldots, m$, *have continuous densities (Remark 1);* $\mu_i > 1$ *,* (z^*, x^*) *is a solution of model (7), (8) and* $\lambda^* = (\lambda_1^*, \ldots, \lambda_m^*) \geq 0$ *is a dual solution. Then for i = 0 and active constraints* $i = \overline{1, m}$,

$$\text{Prob}\left[f_i(x^*, \omega) \leq z_i^*\right] = 1/\mu_i, i = 0, 1, \ldots, m. \tag{9}$$

Proof. Let $\varphi_i(z_i, x, \omega) := z_i + \mu_i \min\{0, f_i(x, \omega) - z_i\}$. From the duality theory follows that z_i^* maximizes

$$E \varphi_0(z_0, x^*, \omega) + \sum_{i=1}^{m} \lambda_i^* E \varphi_i(z_i, x^*, \omega).$$

Thus, if $\lambda_i^* > 0, i = 1, \ldots, m$ then z_i^* maximizes $E \varphi_i(z_i, x^*, \omega)$. Therefore, from Remark 2 follows (9) for $i = 1, \ldots, m$. Equation (9) for $i = 0$ follows from the complementary condition $\sum_{i=1}^{m} \lambda_i^* E \varphi_i(z_i^*, x, \omega) = 0$ and formula (3). $\qquad\square$

3.2 Extreme Events and Unknown Risks

The following simple examples illustrate critical importance of quantiles to represent distributional characteristics of performance indicators.

Systemic Risk and Security Management

Example 1 (Annualization, temporal heterogeneity). Extreme events are usually characterized by their expected arrival time say as a 200-year flood, that is, an event that occurs on average once in 200 years. Methodologically, this view is supported by so-called annualization, i.e., by spreading losses from a potential, say, 30-year crash of airplane, equally over 30 years. In this case, roughly speaking, the crash risk is evaluated as a sequence of independent annual crashes: one wheel in the first year, another wheel in the second year, and so on, until the final crash of the navigation system in the 30th year. The main conclusion from this type of deterministic mean value analysis is that catastrophes are not a matter although they occur as random "explosions" in time and space that may destabilize a system for a long time.

Example 2 (Collective losses). A key issue is the use of proper indicators for collective losses. In a sense, we often have to show that $100 >> \overbrace{1 + 1 + \ldots + 1}^{100}$. Assume that each of 100 locations has an asset of the same type. An extreme event destroys all of them at once with probability 1/100. Consider also a situation without the extreme event, but with each asset still being destroyed independently with the same probability 1/100. From an individual point of view, these two situations are identical: an asset is destroyed with probability 1/100, i.e., individual losses are the same. Collective (social) losses are dramatically different. In the first case 100 assets are destroyed with probability 1/100, whereas in the second case 100 assets are destroyed with probability 100^{-100}, which is practically 0. This example also bears on individual versus systemic (collective) risk, risk sharing and the possibility to establish a mutuality.

Model (7), (8) allows to analyze properly risk sharing portfolios involving both type of situations. In Example 2 the standard worst case scenario is identical for both situations, that is losses of 100 assets. Stochastic worst case scenario as in stochastic maximin problems (16) of Sect. 4.2 is determined only by extreme events, i.e., losses of 100 assets with probability 1/100.

A fundamental methodological challenge in dealing with systemic risks is their endogenous character. Catastrophic losses occur often due to inappropriate land use planning and maintenance of engineering standards. In these cases functions $F_i(x)$ in (5)- (6) have the following structure:

$$F_i(x) = \int f_i(x, \omega) P(x, d\theta), i = 0, 1, \ldots, m.$$

In other words, there is no single probability distribution defining the structure of functions $F_i(x)$ for all x. Instead, there are probability distributions $P(x, d\theta)$, which are different for different decisions x. Therefore, this is a non-Bayesian STO model (Remark 4). Usually probability distribution $P(x, d\theta)$ is given implicitly by a Monte Carlo type simulations, which allow to observe in general only values of random functions $f_i(x, \omega)$ for a given x (Sect. 7.1). The decision dependent measure $P(x, d\omega)$ may easily overthrow convexity. Fortunately, this is not the case with decision dependent measure defined as in (15) of Sect. 4.2.

4 Security Management, Principal Agent Problem

Security management typically involves multi-agent. The main source of uncertainty and risks is associated with behavioral patterns of agents motivated often and shaped by other uncertainties. In contrast to "unknown" risks of Sect. 3 which can be characterized by catastrophe models, security management deals in a sense with "unknowable" risks dependent on decisions of agents. This section analyzes two ways to represent behavioral uncertainties: game theoretic and decision theoretic approaches.

4.1 Game Theoretic Approach

The search for proper regulations protecting public goods is often formulated as the principal-agent problem (Audestad et al. 2006; Gaivoronski and Werner 2007; Mirrlees 1999) or Stackelberg game (Paruchuri et al. 2009, 2008). Important issues concerns non-market institutions (Arnott and Stiglitz 1991). In rather general terms the problem is summarized as the following. The principal agent (PA) introduces a regulatory policy characterized by a vector of decision variables $x = (x_1, \ldots, x_n)$. Other agents, which are often called adversaries, know x and they commit to a unique response characterized by a vector function $y(x)$. The PA knows $y(x)$ and he knows that agents commit to $y(x)$. Therefore his main problem is formulated as to find a decision x^* maximizing an objective function

$$R(x, y(x)) \tag{10}$$

subject to some constraints given by a vector-function $r(x, y)$,

$$r(x, y(x)) \geq 0. \tag{11}$$

The game theoretic approach assumes that components of the vector-function $y(x)$ maximize individual objective functions of agents

$$A(x, y) \tag{12}$$

subject to their individual feasibility constraints

$$a(x, y) \geq 0, \tag{13}$$

where A, a are in general vector-functions, i.e., in general, there may be many principals and agents. For the sake of notational simplicity, we will view them as single-valued functions. Since PA knows functions A, a, he can derive responses $y(x)$ by solving agents individual optimization problems. Since $y(x)$ is assumed to

Systemic Risk and Security Management

be a unique solution, then agents have strong incentive to choose $y(x^*)$ afterwards, i.e., x^* is the Stackelberg equilibrium.

This approach relies on the strong assumptions of perfect information that the PA has about the preference structure of agents and their commitments to a unique response $y(x)$. Section 5 shows that $R(x, y(x))$, $r(x, y(x))$ are non-convex and discontinuous functions even for linear functions $R(x, y)$, $r(x, y)$, $A(x, y)$, $a(x, y)$. This leads to degenerated solutions and sensitivity of solutions to small variations of data.

Remark 5 (bi-level mathematical programming). A solution procedure for PA can be defined by solving bi-level mathematical programs (Dempe 2002): maximize

$$R(x, y), \tag{14}$$

subject to constraints $r(x, y) \geq 0$ and optimality conditions (for a given x) for all individual models (12), (13).

Example 3 (Bayesian games: Cournot duopoly). These games deal with situations in which some agents have private information. Therefore, agents make decisions relying on their beliefs about each other under certain consistency assumptions. The following example illustrates these assumptions restricting the applicability of Bayesian games for PA models.

The profit function of two firms are given as

$$\pi_i = (x_i + x_j - \omega_i)x_i, i \neq j, i, j = 1, 2.$$

Firm 1 has $\omega_1 = 1$, but firm 2 has private information about ω_2. Firm 1 believes that $\omega_2 = \alpha$ with probability p and $\omega_2 = \beta$ with probability $1 - p$. Decision problem of firm 2 is to

$$\max_{x_2}(x_2 + x_1 - \omega_2)x_2,$$

which has solution $x_2^*(x_1, \omega_2) = \frac{1}{2}(\omega_2 - x_1)$. Assume that firm 1 knows response function $x_2^*(x_1, \omega_2)$, then its decision problem is to

$$\max_{x_1}[p(x_1 + x_2^*(x_1, \alpha) - 1)x_1 + (1 - p)(x_1 + x_2^*(x_1, \beta) - 1)x_2],$$

which has solution $x_1^*(\alpha, \beta, p)$ dependent on α, β, p. Assume that the private information of firm 2 is consistent with the believe of firm 1: firm 2 is type $\omega_2 = \alpha$ (observes $\omega_2 = \alpha$ before making decisions) with probability p and $\omega_2 = \beta$ with probability $1 - p$. Only then firm 2 (agent) has incentives to use decisions $x_2^*(x_1^*, \alpha)$, $x_2^*(x_1^*, \beta)$. Therefore, the Bayesian games are applicable in the cases when firm 1 (PA) exactly knows the unique response function $x_2^*(x_1, \omega_2)$ of firm 2 (agent) and the exact distribution of agent's uncertainties ω_2. For general model (10)–(13) Bayesian games require exact information about dependencies of functions A, a on uncertainties ω (say, functions $A(x, y, \omega)$, $a(x, y, \omega)$) and probability distribution of ω, assuming also a unique response function $y(x, \omega)$ solving problem (12), (13).

4.2 Decision-Theoretic Approach

The game theoretic approach introduces behavioral scenarios of agents by uniquely defined known response functions $y(x)$. This raises a key issue regarding actual outcomes of derived solutions in the presence of uncertainty. The decision-theoretic approach explicitly addresses uncertainty based on PA's perceptions of agents behavioral scenarios. These scenarios can be represented (see examples in Sects. 5 and 6) either by a set Π of mixed strategies $\pi \in \Pi$ defined on a set of pure strategies Y, or by a set Y of pure strategies $y \in Y$. This leads then to two classes of STO models.

Probabilistic maximin models associate robust solutions with distributions characterizing desirable indicators (say, social welfare function) over the worst that may happen from $\pi \in \Pi$, i.e., of the form:

$$F(x) = \min_{\pi \in \Pi} E \int f(x, y, \omega) \pi(dy), \tag{15}$$

for some random function $f(x, y, \omega)$, where ω is an exogenous uncertainty.

Stochastic maximin models of the type (2) associate the robustness with respect to the worst-case random events generated by $y \in Y$:

$$F(x) = E \min_{y \in Y} f(x, y, \omega). \tag{16}$$

where Y may depend on x, ω.

Remark 6 (Extreme events and robust statistics). Extreme values (events) theory analyses distributions of minimum (maximum) $M_n = \min(\xi_1, \dots, \xi_n)$, where $\xi_1, \dots, \xi_n$ is a sequence of identically distributed independent random variables (Embrechts et al. 2000). The model (16) has connections with this theory: it focuses on random events generated by extreme values $\min_{y \in Y} f(x, y, \omega)$ with respect to scenarios $y \in Y$. In other words, (16) can be viewed as a decision oriented analogue of the extreme events models with mutually dependent multivariate endogenous (dependent on decision variables x) extreme events. The use of expected values in (16) may not be appropriate, i.e., (16) has to be modified as (7)–(8). Probabilistic maximin model (15) corresponds to minimax approaches introduced by Huber in robust statistics. The integral (15) with respect to an extreme measure $\Pi(x, dy)$ indicates links to Choquet integrals used also by Huber for simple sets Π of imprecise probabilities. The key issue is a proper representation of Π that is discussed in Sect. 7.

Decision theoretic approaches aim to address uncertainties of agents responses $y(x)$. Namely, assumptions of game theoretic approach:

1. Agents commit to a unique $y(x)$,
2. PA knows $y(x)$ and the commitments of agents and, hence, chooses x maximizing function (10)

are substituted by assumptions about the PA perception of agents scenarios. For example, the PA may use his perceptions $A(x, y, \omega)$, $a(x, y, \omega)$ of real functions $A(x, y)$, $a(x, y)$ "contaminated" by uncertain parameters ω. In this case random sets of agents scenarios $Y(x, \omega)$ can be defined as

$$Y(x, \omega) = \{y : a(x, y, \omega) \geq 0\}.$$

In other cases (Wang 2010) these sets can be characterized by experts opinions combined with probabilistic inversions. The overall decision problem is formulated as multicriteria (multi-objective) STO problem with random functions $R(x, y, \omega)$, $A(x, y, \omega)$, $r(x, y, \omega)$, $a(x, y, \omega)$. For example, it can be formulated as the maximization of function

$$F(x) = E \min_{y \in Y(x, \omega)} R(x, y, \omega)/A(x, y, \omega)$$

or

$$F(x) = E \min_{y \in Y(x, \omega)} [R(x, y, \omega) - A(x, y, \omega)]$$

under constraints defined by functions $r(x, y, \omega)$. This leads to stochastic maximin models (16). In general, function $F(x)$ may have the form

$$F(x) = E \min_{y \in Y(x, \omega)} \varphi(A, R, x, y, \omega)$$

for some function φ, e.g., a welfare function $\varphi = \delta A + (1 - \delta) R$, $0 < \delta < 1$ with economic perspectives of welfare analysis regarding possible transferable utilities, side payments, contracts, contingent claims. Definitely, in these cases insurance and finance supplement the safety measures and may mitigate many related problems besides prevention.

5 Systemic Security

Under increasing interdependencies of globalization processes the protection of public goods is becoming a critical topic, especially against uncertain threats generated by agents. In rather general terms such problems can be formulated by using "defender-attacker" terminology. The agents can be intentional attackers such as terrorists, or agents generating extreme events such as electricity outage, oil spills, or floods by the lack of proper regulations, e.g., land use planning. The main issues in these cases concern coping with extreme events generated by agents directly and indirectly through cascading systemic failures. As a result, the security of the whole system can be achieved only by coordinated security management of all its interconnected subsystems, i.e., the systemic security management. In general, arising complex interdependent problems require developing new specific models and methods. This section and Sect. 6 discuss some related issues.

5.1 Preventive Randomized Solutions

This section analyzes situations requiring solutions in randomized strategies as in probabilistic maximin model (15). The simplicity of selected model allows easy to illustrate specifics of both game theoretic and decision theoretic approaches.

The following model is a simplified version of the model analyzed in (Paruchuri et al. 2009). Consider a PA (defender) providing civil security say to houses $i = \overline{1,n}$ to prevent an attack (robbery). A pure strategy i is to visit a house i, whereas x_i is portion of times the pure strategy i is used in overall security control policy $x = (x_1, \dots, x_n)$, $\sum_i x_i = 1$, $x_i \geq 0$. It is assumed that the agent (attacker) knows randomized strategy x and commits to a randomized strategy $y(x) = (y_1(x), \dots, y_n(x))$ maximizing his expected rewards:

$$A(x, y) = \sum_{i,j} r_{ij} x_i y_j, \sum_j y_j = 1, y_j \geq 0, j = \overline{1,n}, \tag{17}$$

assuming that the response $y(x)$ is a unique vector-function. Since PA knows the agent's commitment to $y(x)$, the PA maximizes his expected rewards

$$R(x, y(x)) = \sum_{i,j} R_{ij} x_i y_j(x), \sum_i x_i = 1, x_i \geq 0, i = \overline{1,n}. \tag{18}$$

The randomized strategy x definitely increases the security of the PA. At the same time, the randomized strategy y increases uncertainty about the agent.

A discontinuity of $R(x)$ can be easily seen for $n = 2, i = 1, 2$. The response function $y(x) = (y_1(x), y_2(x))$ maximizes $(r_{11}x_1 + r_{21}x_2)y_1 + (r_{12}x_1 + r_{22}x_2)y_2$, $y_1 + y_2 = 1, y_1, y_2 \geq 0$, and it has the following simple structure. Let $\alpha = (r_{22} - r_{21})/(r_{11} - r_{12})$, then

$$\left. \begin{array}{l} y_1(x) = 1, y_2(x) = 0, \text{ for } x_1 < \alpha x_2 \\ y_1(x) = 0, y_2(x) = 1, \quad \text{otherwise.} \end{array} \right\}, \tag{19}$$

i.e., $R(x, y(x))$ is a discontinuous function on the line $x_1 = \alpha x_2$:

$$R(x, y(x)) = \begin{cases} R_{11}x_1 + R_{21}x_2, & \text{for } x_1 > \alpha x_2, \\ R_{12}x_1 + R_{22}x_2, & \text{for } x_1 < \alpha x_2. \end{cases}$$

The deterministic game theoretic model (17), (18) relies strongly on perfect information about randomized strategies x, y. As a result $y(x)$ attains degenerated 0-1 values. It is natural to expect that formulations which take into account uncertainties will lead to more reasonable solutions. Consider first a straightforward generalization of model (17), (18). Instead of deterministic r_{ij}, let us assume that the PA perceives agent's rewards as random variables $r_{ij}(\omega)$ defined on a set Ω of admissible probabilistic scenarios ω. In general, $\{r_{ij}(\omega)\}$ is a random matrix of interdependent variables. The PA uses now his perception of the agent model and can derive agent's random response function $y(x, \omega)$ by maximizing with respect to y.

$$A(x, y, \omega) = \sum_{i,j} r_{ij}(\omega) x_i y_j, \sum_j y_j = 1, y_j \geq 0, j = \overline{1, n}. \tag{20}$$

Assuming that the PA still follows exactly the logic of model (10), (11), i.e., PA maximizes now the expected value

$$\overline{R}(x) = E \sum_{i,j} R_{ij} x_i y_j (x, \omega), \tag{21}$$

where for the simplicity of illustration we assume that $\{R_{ij}\}$ is a deterministic matrix. It is easy to see that this formal introduction of uncertainty into the game-theoretic model already smoothes function $R(x, y(x))$. Consider random variable $\alpha(\omega) = (r_{22}(\omega) - r_{21}(\omega))/(r_{11}(\omega) - r_{12}(\omega))$. Then similar to (19):

$$y_1(x, \omega) = 1, y_2(x, \omega) = 0 \quad \text{with} \quad \text{Prob } [\alpha(\omega) > x_1/x_2],$$

$$y_1(x, \omega) = 0, y_2(x, \omega) = 1 \quad \text{with} \quad \text{Prob } [\alpha(\omega) \leq x_1/x_2].$$

Therefore,

$$\overline{R}(x) = (R_{11} x_1 + R_{21} x_2) \text{Prob } [\alpha(\omega) > x_1/x_2]$$
$$+ (R_{12} x_1 + R_{22} x_2) \text{Prob } [\alpha(\omega) \leq x_1/x_2].$$

Remark 7 (Non-concave and discontinuous models). If the distribution of $\alpha(\omega)$ has a continuous density, then $\overline{R}(x)$ is a continuous but, in general, non-concave function. Otherwise, $\overline{R}(x)$ is again a discontinuous function purely due to the structure of the Stackelberg models, i.e., in fact, meaningful only under perfect information about commitments of agents to $y(x, \omega)$.

Thus, the game theoretic approach orients PA decisions on unique best-case scenarios $y(x)$ or $y(x, \omega)$ from agents' perspectives, whereas the decision theoretic approach orients decisions on extreme random scenarios of agents from PA perspectives. In particular, the PA can take position to oppose the agent's interests, i.e., to view perceived rewards $A(x, y, \omega)$ as his losses. Therefore, the PA decision model can be formulated as the following stochastic maximin model: maximize

$$F(x) = E \min_{y \in Y} f(x, y, \omega), x \in X,$$

where

$$f(x, y, \omega) = R(x, y, \omega) - A(x, y, \omega),$$

$$X = \{x \geq 0 : \sum_i x_i = 1\}, \qquad Y = \{y \geq 0 : \sum_i y_i = 1\}.$$

In general cases X and Y may reflect various additional feasibility constraints of agents. For example, Y may represent prior information in the form of such comparative statements as the following: the agent plans to visit i more probably then j, $y_i \geq y_j$; or the probability to visit objects i, k, l is higher then objects k, m, n, s, t, i.e., $y_i + y_k + y_l \geq y_m + y_n + y_s + y_t$, etc. Sets X, Y may include also budget constraints. In particular, if c_i is the cost per visit of location i, then the total costs should not exceed a given budget C, $\sum_i c_i x_i \leq C$.

Example 4 (uncertain distributions). It is essential that decision theoretic models can be formulated in a different case-dependent manner. Consider an important situation. Practically, the PA observes results of random trials i, j from randomized strategies x, y and he can see whether $i = j$, or not. If the information about rewards is not available, then the PA problem can be formulated as finding randomized strategy $x = (x_1, \ldots, x_n)$ that "matches" feasible randomized strategy $y = (y_1, \ldots, y_n)$ of the agent as much as it is possible. In this case, a rather natural way to derive optimal randomized strategy x is by minimizing the function

$$\max_{y \in Y} \sum_i x_i \ln \frac{x_i}{y_i}, x \in X,$$

where $\sum_i x_i \ln \frac{x_i}{y_i}$ defines the Kullback-Leibler distance between distributions x and y. This distance is a concave in x and a convex in y function. A simple effective solution procedures similar in spirit to sequential downscaling methods (Fischer et al. 2006) can be developed in the case of sets X, Y defined by linear constraints.

5.2 Defensive Resource Allocation

A problem of resource allocation for protecting public goods against attackers is demonstrated in (Wang 2010) as an application of the stochastic minimax model (16). A typical setting is that the PA (defender) wants to minimize the perceived payoffs to the agents (attackers). In the following we shortly summarize this study advanced during 2010 IIASA[1] Young Scientists Summer Program.

Suppose the defender is faced with potential attacks on a collection of targets (e.g., cities, critical infrastructures, public transportation systems, financial systems, energy or food supply systems, and etc). The defender's objective is to minimize the consequences from attacker choices. A Stackelberg game is usually used to model this situation when there is no uncertainty about the attacker preferences. In reality, the attacker's preferences are not fully known to the defender. In the face of such uncertainty, the defender cannot predict the attacker's best response for sure; therefore, a STO model is needed to minimize the perceived total consequences.

For simplicity, suppose the defender is faced with one attacker, whose decision is to choose a target i among n targets with the highest payoff to attack. The defender objective is to minimize

$$E \max_i g_i(x, \omega)$$

---

[1]International Institute for Applied System Analysis.

Systemic Risk and Security Management

where $x \in X$ is the defensive resource allocation decision among targets under a budget constraint

$$X = \{x \in R^n : \sum_{i=1}^{n} x_i \leq B, x_i \geq 0, i = 1, \ldots, n\}$$

for some $B > 0$; $g_i(x, \omega)$ is the perception of attacker utility function on each target. Therefore, this model focuses on extreme attacks (events) maximizing perceived utility of attackers (see also Remark 6). In general, this model also considers the interdependencies between multiple targets and agents if the agent's utility functions depend on all components of x, ω. In particular, $g_i(x, \omega) = p(x_i)u_i(\omega)$ is a product of target vulnerability (success probability)

$$p(x_i) = e^{-\lambda_i x_i}$$

and the attack consequence

$$u_i(\omega) = \sum_{j=1}^{m-1} w_j A_{ij} + w_m \varepsilon_i.$$

Note that in this model $\omega = (w_1, \ldots, w_m, \varepsilon_1, \ldots, \varepsilon_n)$ is a random vector representing all uncertain parameters in the attacker's utility function, λ_i is the cost effectiveness of defensive investment on target i. For example, at the cost effectiveness level of 0.02, if the investment is measured in millions of dollars, then every million dollars of defensive investment will reduce the success probability of an attack by about 2%.

It is assumed that consequences are valued by the attacker according to a multi-attribute utility function with m attributes (of which $m - 1$ are assumed to be observable by the defender). A_{ij} is attacker utility of target i on the j-th attribute, where A_{ij} takes values in $[0,1]$, with 1 representing the best possible value and 0 the worst, ε_i is utility of the unknown (by the defender) mth attribute of target i, $(w_1, \ldots, w_m)$ are weights on the m attributes, where $\sum_{j=1}^{m} w_j = 1$ and $w_j \geq 0, j = 1, \ldots, m$.

The inherent and deep uncertainty about agent behaviors is critical to models of protecting public goods. Solutions obtained in a deterministic model are usually unstable to even a subtle change in the agent parameters. The STO models are developed for robust solutions against such uncertainties. Therefore, quantifying uncertainty becomes an important task to provide input for the STO models. When direct judgments on the uncertain parameters ω are available, the uncertainties can be quantified directly through probability distributions or simulated scenarios.

However, in some cases direct judgments are not available. For example, in the case study of defensive resource allocations against intentional attacks (Wang 2010) available are only expert opinions about attacker's ranking of cities (targets i). Therefore, so-called probabilistic inversion is used to simulate scenarios about attribute weights and unobserved attributes $\omega = (w_1, \ldots, w_m, \varepsilon_1, \ldots, \varepsilon_n)$. In other words, if there are expert opinions on attacker rankings of potential targets, it is possible to probabilistically invert subjective distributions (as simulated scenarios) on the relative importance of targets attributes (e.g., expected loss or profits from attacks, population, national icon, difficulty of launching an attack, and etc), and even the characteristics of unknown attributes.

5.3   Systemic Failures and Damages

The model of this section can be used as a module of systemic security management model. The main issues concerns an "attack" involving different agents such as a catastrophic flood, financial melt-down, oil spill, or terrorists strikes which may have direct and indirect long-term consequences with cascading failures and damages. Example 2 illustrates a vital importance of systemic damages distributions of which may significantly exceed the sum of isolated damages of related subsystems.

The development of an appropriate model reflecting dependencies among failures, damages, and decisions of different subsystems requires special attention. An attack may produce a chain of indirect damages. For example, a rain affects simultaneously different locations of a region and may cause landslides and formation of damps, lakes; overfilling and breakdowns of dams may further cause floods, fires, and destruction of buildings, communication networks, and transportation systems. Fires may affect computer networks and destroy important information, etc. A failure in a peripheral power grid and financial organization may trigger cascading failures with catastrophic systemic outages and global financial crisis. The indirect losses can even significantly exceed direct impacts. Therefore it is important to develop a model capable of analyzing the propagation of failures through the system and their total direct and indirect impacts. In the following a simple model is described, which is related to notions such as random fields and Bayesian nets. Versions of this model have been used in studies of catastrophic risks at IIASA. The model distinguishes N subsystems or elements (buildings, infrastructures, locations, agents, etc.) $l = 1, \ldots, N$ of a system (region). Possible damage at each l is characterized by random variable ς_l assuming M levels: for sake of simplicity $1, 2, \ldots, M$. Hence damages of the system are described by the random vector $\varsigma = (\varsigma_1, \ldots, \varsigma_N)$. A fixed value of this vector is denoted by z and the set of all possible damages by Z. Let us denote by p_k^{lt} the probability that the damage at l is equal k at time t, $\sum_{k=1}^M p_k^{lt} = 1$, $p_k^{lt} \geq 0$. Dependencies between subsystems are represented as a graph, where elements $i = 1, \ldots, N$ are nodes of the graph and links between them are represented by arrows between nodes. The dependency graph $G = (V, U)$ is characterized then by the set of nodes $V = \{1, 2, \ldots, N\}$ and the set of arrows (directed arcs) U. If nodes l, s belong to V, $l, s \in V$, and there is an arrow from l to s, then l is an adjacent to s node. Define as V_s the set of all adjacent to s nodes and z_{V_l} is sub-vector of the vector of damages indexed by V_l. For example, $z_{V_l} = (z_2, z_5)$ for $V_l = (2, 5)$.

Damages z_l are described by a conditional probability $H^l(z_l | z_{V_l}, x)$, i.e., damages at l depend on current values of damages at l and adjacent nodes as a function

Systemic Risk and Security Management

of available mitigation measures x. Let this function be known for each l. This is a common assumption of catastrophe modeling (Sect. 3). Say, the probability of a dam break is conditional on probabilities of potential discharge curves; the probability of inundation is conditional on a dike break; damages of buildings and other constructions are conditional on inundation patterns, and so on. In the same manner we can model, say, financial crisis spreading through regions. Functions H^l define the propagation of indirect events and related damages through the system according to the following relation

$$p_k^{l,t+1} = \sum_{z_{V_l} \in Z} H^l(\varsigma_l^t = k \,|\, \varsigma_{V_l}^{t-1} = z_{V_l}, x) P(\varsigma_{V_l}^{t-1} = z_{V_l}),$$

where $p_k^{l,t} = P(\varsigma_l^t = k)$. To define completely the propagation of failures and damages it is necessary to fix an initial distribution of ς_l^t for $t = 0$, i.e., at the moment when the attack occurred. This equation together with initial distribution allow the exact calculation (under certain assumptions on the structure of graph G) of $p_k^{l,t}$ for any $t \geq 0$. Of course, for complex graphs it is practically impossible to derive analytical formulas for $p_k^{l,t}$ as functions of decision variables x. Hence the damages may have rather complex dynamic implicit dependencies on decision vector x requiring developments of specific decision support tools. The most important approaches have to rely on STO in combination with fast Monte Carlo simulation as in Sect. 7.1. The paper by (Wang 2010) reports on computational effectiveness of these methods for realistic problems of security management with very large number of simulated scenarios and two-stage decision variables required for coping with extreme events. The values $p_k^{l,t}$ reflect the dynamic of propagation of initial (direct) damages through the system after the occurrence of an attack. Scenarios of damages can be simulated at any $t \geq 0$. For example, $t = 0$ corresponds to the distribution of direct damages.

6 Security of Electricity Networks

This section presents a decision theoretic model for regulating electricity markets (networks). The State California energy crises in 2001 and the collapse of ENRON raised serious concerns about proper regulation of the market power, that is, the ability of electricity suppliers to raise prices above competitive levels for a significant period of time. This is considered as a major obstacle to successful reforms of centralized electricity sectors to competitive markets (Cardell et al. 1996; Yao 2006). Leader-follower type models are being used to support policy decisions on design of electricity markets and various regulatory tasks. Unfortunately, these models are usually inherently non-convex and sensitive to assumptions on their parameters. It is recognized (Cardell et al. 1996) that no modeling approach can predict prices in oligopolistic markets, therefore the value of models is considered in their ability to provide robust results on relative differences of feasible market structure and regulations. Let us consider a model (Yao 2006) where the independent system operator (ISO) controls the transmission system and generator outputs so as to maximize social welfare of consumers while meeting all the network and security constraints.

An electricity market can be represented by a set N of nodes and a set of transmission lines. The strategic decision variables of the ISO are import/export quantities $r_i, i \in N$, which must be balanced

$$\sum_i r_i = 0, \tag{22}$$

and such that the resulting power flows don't exceed secure thermal limits of the transmission lines in both directions

$$-K_l \leq \sum_i D_{il} r_i \leq K_l, l \in L, \tag{23}$$

where D_{il} are distribution factors (parameters) which specify the flow on a line l from a unite of flow increase at a node i.

Given the ISO's (leader) decisions r_i, each producer (follower) i maximizes the profit function

$$P_i(q_i + r_i)q_i - C_i(q_i), 0 \leq q_i \leq \overline{q}_i, i \in N, \tag{24}$$

where $P_i(\cdot), C_i(\cdot)$ are the inverse demand function (wiliness-to-pay) and generation cost function; $\overline{q}_i$ are upper capacity bounds.

The leader-follower models rely on the following perfect information assumptions: the ISO knows the response functions $q_i(r_i)$ and chooses $r_i, i \in N$, maximizing the welfare function of consumers

$$\sum_i \left[\int_0^{r_i + q_i(r_i)} P_i(v)dv - C_i(q_i(r_i)) \right]. \tag{25}$$

The resulting model is one-leader multi-follower Stackelberg game. This type of model may have none or multiple degenerated solutions. There might exist also no-equilibrium in pure strategies due to non-convexity and even discontinuities in the welfare function (25). Slight deviations in $q_i(r_i)$, say due to volatility of price/demand functions $P_i(\cdot)$ may have significant consequences (Cardell et al. 1996) on the market power mitigation and equilibrium.

Another approach is to assume that the ISO is a Nash player that acts simultaneously with producers. This unilateral approach for regulation of network interdependencies ignores dependence of agents' decisions q_i on regulations r_i, that removes the non-convexity from ISO's optimization problem (22), (23), (25). Yet again, this approach requires perfect information about all producers, demand and system contingencies. It ignores uncertainties of fluctuations that stem from unforeseen events, such as demand uncertainty and transmission and generation outranges. The critical shortcoming of the Nash equilibrium is that it ignores the

Systemic Risk and Security Management

two-stage character of the ISO and producers decisions. In particular, it excludes proper modeling of forward markets allowing participants to secure more stable prices reducing opportunities to manipulate the market.

We shall now describe a two-stage STO model (see also Sect. 7.1) in which the ISO determines its forward decisions under uncertainties at stage 1 (comprising possibly many random time interval), and producers act at stage 2 after a scenario of uncertainty is revealed. Let us consider a network affected by a set of random events (shocks) $\omega \in \Omega$ which are assumed to be elements of a probability space. These events lead to variability of functions $P_i(\cdot, \omega)$, bounds $K_i(\omega), \overline{q}_i(\omega)$, and cost functions $C_i(\cdot, \omega)$. In general, these random functions and parameters can be viewed as ISO's perception of producer's model.

In the presence of uncertainties the best ISO strategy would be a collective risk sharing maximizing the social welfare function of consumers and producers:

$$F(r) = \sum_i E f_i(r_i, \omega), \tag{26}$$

$$f_i(r_i, \omega) = \max_{q_i} \int_0^{q_i + r_i} P_i(\upsilon, \omega) d\upsilon + P_i(q_i + r_i, \omega) q_i - C_i(q_i, \omega), 0 \le q_i \le \overline{q}_i(\omega),$$

under constraints (22), (23). This is a two-stage STO problem as in Sect. 7.1. Function $F(r)$ orients regulatory decisions on achieving best possible outcomes with respect to all potential behavioral scenarios of agents (see also Remark 6).

7 Computational Methods

A discussion of computational methods and applications of Stackelberg games can be found in (Gaivoronski and Werner 2007; Kocvara and Outrata 2004; Luo et al. 1996; Paruchuri et al. 2009, 2008). The concept of Nash equilibrium smoothes the problem but it ignores the essential two-stage structure of leader-follower decisions (Sect. 6). Explicit treatment of uncertainty in PA models with bi-level structure is considered in (Audestad et al. 2006; Gaivoronski and Werner 2007). The paper by (Wang 2010) advances the decision-theoretic approach to homeland security models.

The development of effective decision-theoretic computational methods essentially depends on specifics of arising STO models. The main issue is analytical intractability of performance indicators $F_i(x) = E f_i(x, \omega) = \int f_i(x, \omega) P(x, d\omega)$, where $P(x.d\omega)$ may implicitly depend on x as in problem (15) and Remark 4. If functions $F_i(x)$ are analytically tractable, then the problem can be solved by using standard deterministic methods. Unfortunately, this is rarely the case. In fact, standard deterministic models are formulated usually by switching from $F_i(x) = E f_i(x, \omega)$ to deterministic functions $f_i(x, E\omega)$. Simple examples show that this substitution may result in wrong conclusions as in Sect. 3.2. More specifically, outcomes $\exp(\omega x)$ for $\omega = \pm 100$ with probability $1/2$ have considerable variability, whereas $\exp(E\omega x) = 1$.

Instead straightforward evaluations of integrals $F_i(x)$, STO methods (Birge and Louveaux 1997; Borodina et al. 2011; Ermoliev and Wets 1988; Marti 2005; Shapiro et al. 2009) use only random values $f_i(x, \omega)$ available from Monte Carlo simulations. The following section outlines the main idea of these powerful methods which avoid the "curse of dimensionality" (Remark 3) and allow to solve problems which cannot be solved by other existing methods (Ermoliev 2009; Ermolieva and Ermoliev 2005; Gaivoronski 2004). An important application of these methods for security management can be found in (Wang 2010).

7.1 Adaptive Monte Carlo Optimization for Two-Stage STO Models

For simplicity of illustration, let us consider the minimization of function $F(x) = \int f(x, \omega) P(x, d\omega)$ without constraints.

Computations evolve from an initial solution x^0. Instead of computing values of integral $F(x)$, what is practically impossible, the procedure uses only observable (simulated) random values $f(x, \omega)$.

For a solution x^k calculated after k-th step, simulate two independent observations $\omega^{k,1}$, $\omega^{k,2}$ of ω correspondingly from $P(x^k + \gamma_k \eta^k, d\omega^{k,1})$, $P(x^k, pd\omega^{k,2})$, and calculate values $f(x^k + \gamma_k \eta^k, \omega^{k,1})$, $f(x^k, \omega^{k,2})$, where γ_k is a positive number, $\eta^k = (\eta_1^k, \ldots, \eta_n^k)$ is a random vector with, say, independent identically uniformly distributed in interval $[-1, 1]$ components. New approximate solution x^{k+1} is computed by moving from x^k in direction of so-called stochastic quasi-gradient (Ermoliev 2009; Gaivoronski 2004):

$$\xi^k = \frac{f(x^k + \gamma_k \eta^k, \omega^{k,1}) - f(x^k, \omega^{k,2})}{\gamma_k} \eta^k$$

with a step size $\rho_k > 0$. The convergence of x^k to the set of optimal solutions with probability 1 follows from the fact that random vector ξ^k is a stochastic quasigradient (SQG) of $F(x)$, i.e., $E[\xi^k | x^k] \approx F_x(x^k)$. In other words, ξ^k is an estimate of the gradient $F_x(x^k)$ or its analogs for nondifferentiable and discontinuous functions (Ermoliev 2009; Ermoliev and Norkin 2003; Gaivoronski 2004). Step-sizes ρ_k, γ_k have to satisfy some simple requirements, e.g., $\rho_k = const/k$, $\sum_k \rho_k \gamma_k < \infty$.

This method simulates realistic adaptive processes. Only two random observations of function $F(x)$ are used at each step to identify the direction of transition from x^k to x^{k+1} and its size, whereas values of $F(x)$ and its derivatives remain unknown. In fact, changes of $F(x^k)$ can be relatively tracked, in a sense, by $F^k = \frac{1}{k} \sum_{s=1}^{k} f(x^s, \omega^s)$ due to the convergence of $F(x^k)$, $k \to \infty$. This allows designing adaptive regulations of ρ_k to speed up the convergence. Applications of fast Monte Carlo simulations usually require nontrivial analytical analysis (Ermoliev and Norkin 2003) of involved stochastic processes. The next section outlines a version of the method that utilizes the analytical structure of $f(x, \omega)$ to achieve faster simulations.

Systemic Risk and Security Management 41

The robustness is achieved by the two-stage structure of decisions combining both for coping with uncertainty mechanisms of anticipation and adaptation. Forward-looking anticipative decisions are made before new information about uncertainty become available, whereas other options are created and remain open for adaptive adjustments to potential new information when it becomes available. The two-stage STO problem (Birge and Louveaux 1997; Ermoliev and Wets 1988; Marti 2005; Shapiro et al. 2009) seems to be the most suitable for framing principle-agent decision models under uncertainty. An example of two-stage model is given in Sect. 6, (26). Problem (2) has also two-stage formulation important for modeling the climate change dilemma. A rather general two-stage STO model is formulated as follows. A long-term decision x must be made at stage 1 before the observation of uncertainty ω is available. At stage 2, for given $x \in X$ and observed ω, the adaptive short-term decision $y(x, \omega)$ is chosen so as to solve the problem: find $y \in Y$, such that

$$g_i(x, y, \omega) \geq 0, i = \overline{1, l}, \tag{27}$$

and

$$g_0(x, y, \omega), \tag{28}$$

is maximized (minimized) for some functions $g_i, i = \overline{1, l}$. Then the main problem is to find decision x, such that

$$F(x) = Ef(x, \omega), f(x, \omega) = g_0(x, y(x, \omega), \omega), x \in X \tag{29}$$

is maximized.

We can see that (27), (28) correspond to (12), (13) of the agents' models, and (29) to the goal (10) of the principle-agent model. Computational methods for this general model are discussed in (Ermoliev 2009). If $y(x, \omega)$ maximizes (28), this problem corresponds to the (two-stage) recourse STO model; otherwise – stochastic maximin problem (16). In general, (29) may be (Sect. 2) a dynamic two-stage STO model. Multistage STO models with more then two stages arise in cases when ω remains unknown after new information become available.

In general, two-stage STO models (29) are solved by using adaptive Monte Carlo optimization (Ermoliev 2009) based on stochastic quasigradient (SQG) methods. The main idea can be easy illustrated by using the simplest stochastic minimax model (2). As in Sect. 7.1, instead of integral $F(x) = Ef(x, \omega)$, $f(x, \omega) = \max\{\alpha(x - \omega), \beta(\omega - x)\}$, the method sequentially updates an initial solution x^0 by using only available on-line independent observations or simulations $\omega_0, \omega_1, \ldots$ of random variable ω. Let x^k is an approximate solution computed at step $k = 0, 1, \ldots$. Observe (simulate) ω_k and change x^k by the rule

$$x^{k+1} = x^k - \rho_k \xi^k, k = 0, 1, \ldots, \xi^k = \begin{cases} \alpha, & if \ x^k \geq \omega_k, \\ \beta, & otherwise. \end{cases}$$

Again, ξ^k is a SQG of non-smooth function $F(x)$ at $x = x^k$. The step-size multiplier ρ_k satisfies the same type conditions as in Sect. 7.1.

7.2 Uncertain Distributions

In this and next sections we assume that ω is characterized by a vector $v = (v_1, \ldots, v_m)$ of random parameters $v \in V \subset R^m$. To avoid technicalities, STO models which can be formulated as maximization of the function

$$F(x) = Ef(x, v) = \int_V f(x, v) dH(v), x \in X \subset R^n, \tag{30}$$

where $v \in V \subset R^m$ is a vector of random parameters, $H(y)$ is a cumulative distribution function, i.e., $P(dv) = dH(v)$, and $f(x, \cdot)$ is a random function possessing all the properties necessary for expression (30) to be meaningful.

As previous sections show, we often do not have full information on $H(v)$. For new decision problems, in particular, arising in security analysis, we often have large a number of unknown interdependent variables v, x and only very restricted samples of real observations which don't allow to derive the distribution P.

Experiments to generate new real observations may be extremely expensive, dangerous or simply impossible. Instead, the natural approach for dealing with new problems can be based on using all additional information on P to derive a set of feasible distributions.

Let us denote by K the set of distributions consistent with available information on P. The robust solution can be defined as $x \in X$ maximizing

$$F(x) = \min_{P \in K} \int f(x, v) P(dv) = \int f(x, v) P(x, dv), \tag{31}$$

where $P(x, dv)$ denotes the extreme distribution as in Sect. 4. Thus, we have a general case of STO models with probability measure affected by decision x as in Sect. 3.2.

Assume that in accordance with available sample and our beliefs we can split the set V into disjoint subsets$\{C_s, s = 1, \ldots, S\}$. Some of them may correspond to clusters of available observations whereas others may reflect expert opinions on the degree of uncertainty and its heterogeneity across the admissible set V. For instance, we can distinguish some critical zones ("catalogues of earthquakes") which may cause significant reductions of performance indicators. Accordingly, the additional beliefs can be given in terms of a "quantile" class

$$K = \left\{ P : \int_{C_s} P(dv) = \alpha_s, s = 1, \ldots, S \right\}, \tag{32}$$

where $\sum_{s=1}^{S} \alpha_s = 1$; more generally – in terms of ranges of probabilities

$$K = \left\{ P : \alpha_s \leq \int_{C_s} P(dv) \leq \beta_s, s = 1, \ldots, S \right\}, \tag{33}$$

Systemic Risk and Security Management

where α_s, β_s are given numbers such that $\sum_{s=1}^{S} \alpha_s \leq 1 \leq \sum_{s=1}^{S} \beta_s$. This class is considered as the most natural elicitation mechanism.

Let us denote $\gamma_s = \int_{C_s} P(dv)$, $\gamma = (\gamma_1, \ldots, \gamma_S)$. In general, additional beliefs can be represented in a form of various inequalities among components of vector γ (see Remark 4), of the type

$$K = \{P : A\gamma \leq b, \gamma \geq 0\} \tag{34}$$

for some matrix A and vector b.

Proposition 2. *For any function $f(x, v)$ assumed to be integrable w.r.t. all P in K defined by (34)*

$$\min_{P \in K} \int f(x, v) P(dv) = \min_{\gamma} \left\{ \sum_{s=1}^{S} \gamma_s \min_{v \in C_s} f(x, v) : A\gamma \leq b \right\}.$$

In the case of K defined by (32), (33)

$$\min_{P \in K} \int f(x, v) P(dv) = \sum_{s=1}^{S} \alpha_s \min_{v \in C_s} f(x, v),$$

$$\min_{P \in K} \int f(x, v) P(dv) = \min_{\{\gamma_s\}} \left\{ \sum_{s=1}^{S} \gamma_s \min_{v \in C_s} f(x, v) : \alpha_s \leq \gamma_s \leq \beta_s, \sum_{s=1}^{S} \gamma_s = 1 \right\}.$$

Proof. We can choose a distribution P concentrated at any collection of points $v_s \in C_s$, $s = \overline{1, S}$, therefore

$$\min_{P \in K} \int f(x, v) P(dv) \leq \min_{\gamma} \left\{ \sum_{s=1}^{S} \gamma_s \min_{v \in C_s} f(x, v) : A\gamma \leq b \right\}.$$

On the other hand,

$$\int f(x, v) P(dv) = \sum_{s} \int_{C_s} f(x, v) P(dv) \geq \min_{\gamma} \left\{ \sum_{s=1}^{S} \gamma_s \min_{v \in C_s} f(x, v) : A\gamma \leq b \right\}.$$

Remark 8 (general STO models) Proposition 2 is also true for general STO problems (5)–(6) with uncertain probability distributions $P \in K$. An interesting specific case occurs for imprecise probabilities characterized by constraints (33). It is important that Proposition 2 reduces the maximization problem (31) to a deterministic maximin problems which can be solved by linear or nonlinear programming methods. There are important connections between dual relations (Sect. 7.4) of these problems and CVaR measures discussed in Sects. 2 and 3. $\square$

7.3 Generalized Moment Problem

Often we know bounds for the mean value or other moments of H in (30). Such information can often be written in terms of constraints

$$Q^k(H) = Eq^k(v) = \int_V q^k(v)dH(v) \le 0, k = \overline{1,l},$$

$$\int_V dH(v) = 1,$$

where the $q^k(v)$, $k = \overline{1,l}$, are known functions. Let K is the set of functions H satisfying these constraints.

Consider again the problem (31). Methods of maximizing $F(x)$ in (31) depend on solution procedures for the following "inner" minimization problem: find a distribution function H that minimizes

$$Q^0(H) = Eq^0(v) = \int_V q^0(v)dH(v)$$

subject to $H \in K$ for some function q^0. This is a generalization of the known moments problem (see, e.g., (Ermoliev et al. 1985)). It can also be regarded as a generalization of the nonlinear programming problem

$$\min \left\{ q^0(v) : q^k(v) \le 0, v \in V, k = \overline{1,l} \right\}$$

to an optimization problem involving randomized strategies as in Sect. 5.

There are two main approaches (Ermoliev et al. 1985) for minimizing $Q^0(H)$ in K: generalized linear programming (GLP) methods and dual maximin approach.

Minimization of $Q^0(H)$ in K is equivalent to the following GLP problem (Ermoliev et al. 1985): find points $v^j \in V$, $k = \overline{1,l}$, $t \le l + 1$ and real numbers $p_j, k = \overline{1,l}, j = \overline{1,t}$, minimizing

$$\sum_{j=1}^{t} q^0(v^j)p_j \tag{35}$$

subject to

$$\sum_{j=1}^{t} q^k(v^j)p_j \le 0, k = \overline{1,l}, \tag{36}$$

$$\sum_{j=1}^{t} p_j = 1, p_j \ge 0, j = \overline{1,t}. \tag{37}$$

Consider arbitrary points v^j, $j = \overline{1,l+1}$ (setting $t = l + 1$), and for the fixed set $\{v^1, v^2, \ldots, v^{l+1}\}$ find a solution $\overline{p} = (\overline{p}_1, \overline{p}_2, \ldots, \overline{p}_{l+1})$ of problem

Systemic Risk and Security Management 45

(35)–(37) with respect to p. Assume that $\overline{p}$ exists and that $(\overline{u}_1, \overline{u}_2, \ldots, \overline{u}_{l+1})$ are the corresponding dual variables. We know that if there exists a point v^* such that $q^0(v^*) - \sum_{k=1}^{l} \overline{u}_k q^k(v^*) - \overline{u}_{l+1} < 0$, then the solution $\overline{p}$ could be improved by dropping one of the columns $(q^0(v^j), q^1(v^j), \ldots, q^l(v^j), 1)$, $j = \overline{1, l+1}$ from the basis and replacing it by the column $(q^0(v^*), q^1(v^*), \ldots, q^l(v^*), 1)$, following the revised simplex method. Point v^* could be defined by minimizing $q^0(v) - \sum_{k=1}^{l} \overline{u}_k q^k(v)$, $v \in V$.

This conceptual framework leads to various methods (Gaivoronski 1986; Golodnikov and Stoikova 1978) for solving not only (35)–(37) but also some more general classes of nonlinear (in probability) problems. The interesting important issue is the combination of the described procedure with simultaneous gradient type adjustments of x ensuring optimal solution of compound problem (31).

7.4 Duality Relations and Stochastic Optimization

The duality relations for minimization of $Q^0(H)$ in K provide a more general approach. It can be shown that if V is compact, $q^k(v)$, $k = \overline{0, l}$, are continuous and $0 \in intco\{z : z = (q^0(v), q^1(v), \ldots, q^l(v)\}, v \in V$, then

$$\min_{H \in K} \int f(x, v) dH(v) = \max_{u \in U^+} \min_{v \in V} \left[f(x, v) - \sum_{k=1}^{m} u_k q^k(v) \right] \qquad (38)$$

for each $x \in X$, where $f(x, \cdot)$ is a continuous function. Hence, original infinite dimensional STO problem can then be reduced to finite-dimensional a maximin type problem as follows: maximize the function

$$\gamma(x, u) = \min_{v \in V} \left[f(x, v) - \sum_{k=1}^{m} u_k q^k(v) \right] \qquad (39)$$

with respect to $x \in X$, $u \geq 0$. This allows developing a number of algorithms using GLP approach and algorithms based on solving directly maximin problem (39). A general scheme of such an algorithms is the following.

According to (38), (39) the STO model with uncertain distribution is reduced to a finite-dimensional maximin problem with a possibly non-convex inner problem of minimization and a concave final problem of maximization. A vast amount of work has been done on maximin problems but virtually all of the existing methods fail if the inner problem is non-convex. The following approach allows to overcome this difficulty.

Consider a general maximin problem

$$\max_{x \in X} \min_{v \in V} g(x, v),$$

where $g(x, v)$ is a continuous function of (x, v) and a concave function of x for each $v \in V$, $X \subset R^n$, $V \subset R^m$. Although $G(x) = \min_{v \in V} g(x, v)$ is a concave function, to compute its value requires a solution $v(x)$ of non-convex problem. In order to avoid the difficulties involved in computing $v(x)$ one could try to approximate V by an ε-set.

But, in general, this would require a set containing a very large number of elements. An alternative is to use the following ideas (Ermoliev et al. 1985). Consider a sequence of sets V_s, $s = 0, 1, \ldots$ and the sequence of functions $G^s(x) = \min_{v \in V_s} g(x, v)$. It can be proven that, under natural assumptions concerning the behavior of sequence G^s, the sequence of points generated by the rule

$$x^{s+1} = x^s - \rho_s G_x^s(x^s), G_x^s(x^s) = g_x(x^s, v^s), s = 0, 1, \ldots,$$

where the step size ρ_s satisfies assumptions such as $\rho_s \geq 0$, $\rho_s \to 0$, $\sum_{s=0}^{\infty} \rho_s = \infty$, tends to follow the time-path of optimal solutions: for $s \to \infty$

$$\lim [G^s(x^s) - \max G^s(x)] = 0.$$

It was shown (see discussion in (Ermoliev et al. 1985)) how V_s (which depends on x^s) can be chosen so that we obtain the convergence

$$\min G^s(x) \to \min G(x),$$

where V_s contains only a finite number $N_s \geq 2$ of random elements. The main idea is the following.

We start by choosing initial points x^0, v^0, a probability distribution μ on set V and an integer $N_0 \geq 1$. Suppose that after the s-th iteration we have arrived at points x^s, v^s. The next approximations x^{s+1}, v^{s+1} are then constructed in the following way. Choose $N_s \geq 1$ points $v^{s,1}, v^{s,2}, \ldots, v^{s,N_s}$, which are sampled from the distribution μ, and determine the set

$$V_s = \{v^{s,1}, v^{s,2}, \ldots, v^{s,N_s}\} \bigcup v^{s,0},$$

where $v^{s,0} = v^s$. Take $v^{s+1} = Arg \max_{v \in V_s} g(x^s, v)$ and compute

$$x^{s+1} = \pi_X \left[x^s - \rho^s g_x(x^s, v^{s+1}) \right], s = 0, 1, \ldots,$$

where ρ_s is the step size and π_X is the result of the projection operation on X. The convergence analysis of this method can be found in (Ermoliev et al. 1985).

8   Concluding Remarks

In the case of perfect information the best response of the follower to a decision x of the leader is the decision $y(x)$ maximizing his reward $A(x, y)$. Therefore, the best decision of the leader is $x = x^*$ maximizing his rewards $R(x, y(x))$. Since

Systemic Risk and Security Management

$y(x)$ is assumed to be a unique solution, then the follower has strong incentive to choose $y(x^*)$ afterwards, i.e., x^* is the Stackelberg equilibrium.

In the case of uncertainty the situation is different. The leader may again use the best response $y(x,\omega)$ of the follower according to his perception of uncertainty ω and rewards $A(x,y,\omega)$ of the follower. However, the decision $y(x,\omega)$ is no longer rational for the follower to choose afterwards. In addition, as Remark 7 indicates, it may produce degenerated solutions $y(x,\omega)$ resulting in discontinuities and instabilities. Therefore, in the case of uncertainty the proposed decision-theoretic approach relies on random extreme scenarios for the leader rather than random best case scenarios for the follower. This preserves convexities of models and it allows the introduction of concepts of robust solutions based on new type of (in general) non-Bayesian multicriteria STO models with uncertain probability distributions and multivariate extreme events. As Sect. 7 shows, specific classes of such models can be solved by linear and nonlinear programming methods in the case of price-wise linear random functions. An important food security case study in (Borodina et al. 2011) was analyzed by linear programming methods in an extended space of proposed two-stage multi-agent STO model. In general, adaptive fast Monte Carlo and SQG optimization methods can be used (Gaivoronski and Werner 2007; Wang 2010) to solve arising STO models. Developments of tools for solving STO problems involving implicit dynamic dependencies of probabilities on decisions in Sect. 5.3 demand special attention. Truly multiagent security management is required for coping with systemic failures and extreme events generating disruptions of financial and economic systems, communication and information systems, food-water-energy supply networks.

Acknowledgements We are grateful to the anonymous referees for constructive suggestions resulting in important improvements of the chapter, which helped us to improve it. Fruitful discussions during the Coping with Uncertainties 2009 conference also helped us to shape the structure of the chapter and its content.

References

Arnott, R., & Stiglitz, J. E. (1991). Moral hazard and nonmarket institutions: Dysfunctional crowding out of peer monitoring? *The American Economic Review, 81*(1), 170–190.

Audestad, J. A., Gaivoronski, A. A., & Werner, A. S. (2006). Extending the stochastic programming framework for the modeling of several decision makers: pricing and competition in the telecommunication sector. *Annals of Operations Research, 142*(1): 19–39.

Birge, J., Louveaux, F., (1997). *Introduction to Stochastic Programming*. New York: Springer.

Borodina, A., Borodina, E., Ermolieva, T., Ermoliev, Y., Fischer, G., Makowski, M., et al. (2011). Sustainable agriculture, food security, and socio-economic risks in Ukraine. In Y. Ermoliev, M. Makowski, K. Marti (Eds.), *Managing safety of heterogeneous systems*. Lecture notes in economics and mathematical systems. Heidelberg: Springer.

Cardell, J. B., Hitt, C. C., & Hogan, W. W. (1996). Market power and strategic interaction in electricity networks. *Resource and Energy Economics, 19*, 109–137.

Dempe, S. (2002). Foundation of Bilevel Programming. Dordrecht: Kluwer Academic Publishers.

Embrechts, P., Klueppelberg, C., & Mikosch, T., (2000). Modeling extremal events for insurance and finance. *Applications of Mathematics, Stochastic Modeling and Applied Probability*. Heidelberg: Springer.

Ermoliev, Y. (2009). Stochastic quasigradient methods: Applications. In C. Floudas & P. Pardalos (Eds.), *Encyclopedia of Optimization* (pp. 3801–3807). New York: Springer.

Ermolieva, T., & Ermoliev, Y. (2005). Catastrophic risk management: flood and seismic risks case studies. In S. W. Wallace & W. T. Ziemba (Eds.), *Applications of Stochastic Programming*, MPS-SIAM Series on Optimization, Philadelphia, PA, USA, 2005.

Ermoliev, Y. E., Gaivoronski, A., & Nedeva, C. (1985). Stochastic Optimization Problems with Incomplete Information on Distribution Functions. *SIAM Journal Control and Optimization, 23*(5), 697–708.

Ermoliev, Y., & Hordijk, L. (2006). Global Changes: Facets of Robust Decisions. In K. Marti, Y. Ermoliev, M. Makowski, & G. Pflug, (Eds.), *Coping with Uncertainty, Modeling and Policy Issues* (pp. 4–28). Berlin, Germany: Springer.

Ermoliev, Y., & Leonardi, G. (1982). Some proposals for stochastic facility location models. *Mathematical Modeling, 3*, 407–420.

Ermoliev, Y., & Norkin, V. (2003). Stochastic optimization of risk functions via parametric smoothing. In K. Marti, Y. Ermoliev, & G. Pflug (Eds.), *Dynamic Stochastic Optimization* (pp. 225–249). Berlin, Heidelberg, New York: Springer.

Ermoliev, Y., & Wets, R. (Eds.) (1988). *Numerical Techniques for Stochastic Optimization, Computational Mathematics*. Berlin: Springer.

Ermoliev, Y., & Yastremskii, A. (1979). *Stochastic Modeling and Methods in Economic Planning*. Moscow: Nauka [in Russian].

Ezell, B., & von Winterfeldt, D. (2009). Probabilistic risk analysis and terrorism. *Biosecurity and Bioterrorism, 7*, 108–110.

Fischer, G., Ermoliev, T., Ermoliev, Y., & van Velthuizen, H. (2006). Sequential downscaling methods for estimation from aggregate data. In K. Marti, Y. Ermoliev, M. Makowski, & G. Pflug (Eds.), *Coping with Uncertainty: Modeling and Policy Issues*. Berlin, Heidelberg, New York: Springer.

Gaivoronski, A. A. (1986). Linearization methods for optimization of functionals which depend on probability measures. *Mathematical Programming Study, 28*, 157–181.

Gaivoronski, A. A. (2004). SQG: stochastic programming software environment. In S. W. Wallace, & W. T. Ziemba (Eds.), *Applications of Stochastic Programming*, MPS-SIAM Series in Optimization, 637–670.

Gaivoronski, A. A., & Werner, A. S. (2007). A solution method for stochastic programming problems with recourse and bilevel structure.

Golodnikov, A. N., & Stoikova, L. S. (1978). Numerical methods of estimating certain functionals characterizing reliability. *Cybernetics, 2*, 73–77.

Huber, P. (1981). *Robust Statistics*. New York: Wiley.

Keeney, R. L., & von Winterfeldt, D. (1991). Eliciting probabilities from experts in complex technical problems. *IEEE Transactions on Engineering Management, 38*, 191–201.

Keeney, R. L., & von Winterfeldt, D. (1994). Managing nuclear waste from power plants. *Risk analysis, 14*, 107–130.

Kocvara, M., & Outrata, J. V. (2004). Optimization problems with equilibrium constraints and their numerical solution. *Mathematical Programming Series B, 101*, 119–149.

Koenker, R., & Bassett, G. (1978). Regression quantiles. *Econometrica, 46*, 33–50.

Luo, Z. Q., Pang, J.-S., & Ralph, D. (1996). *Mathematical Programs with Equilibrium Constraints*. Cambridge: Cambridge University Press.

Marti, K. (2005). *Stochastic Optimization Methods*, (Second edition, 2008). Berlin, Haidelberg: Springer.

Mirrlees (1999). The theory of moral hazard and unobservable behavior: Part I. *The Review of Economic Studies, 66*(1), 3–21.

Mulvey, J. M., Vanderbei, R. J., & Zenios, S. A. (1995). Robust Optimization of Large Scale Systems. *Operations Research, 43*, 264–281.

O'Neill, B., Ermoliev, Y., & Ermolieva, T. (2006). Endogenous risks and learning in climate change decision analysis. In K. Marti, Y. Ermoliev, M. Makowski, & G. Pflug (Eds.), *Coping with Uncertainty: Modeling and Policy Issues*. Berlin, Heidelberg, New York: Springer.

Otway, H. J., & von Winterfeldt, D. (1992). Expert judgment in risk analysis and management: Process, context, and pitfalls. *Risk Analysis, 12*, 83–93.

Paruchuri, P., Pearce, J. P., Marecki, J., Tambe, M., Ordonez, F., & Kraus, S. (2009). Coordinating randomized policies for increasing security of agent systems. *Information Technology and Management, 10*, 67–79.

Paruchuri, P., Pearce, J. P., Marecki, J., Tambe, M., Ordonez, F., & Kraus, S. (2008). *Efficient Algorithms to Solve Bayesian Stackelberg Games for Security Applications. Proceedings of the Twenty-Third AAAI Conference on Artificial Inteligence*.

Rockafellar, T., & Uryasev, S. (2000). Optimization of conditional-value-at-risk. *The Journal of Risk, 2*, 21–41.

Shapiro, A., Dentcheva, D., & Ruszczynski, A. (2009). *Lectures on Stochastic Programming: Modeling and Theory*. Philadelphia: SIAM.

Walker, G. R. (1997). Current Developments in Catastrophe Modeling. In N. R. Britton, & J. Oliver. (Eds.), *Financial Risk Management for Natural Catastrophes* (pp. 17–35). Brisbane: Aon Group Australia Limited, Griffith University.

Wang, C. (2010). Allocation of resources for protecting public goods against uncertain threats generated by agents. IIASA Interim Report IR-10-012, Int. Inst. For Applied Systems Analysis, Laxenburg, Austria.

Yao, J. (2006). *Cournot Equilibrium in Two-settlement Electricity Markets: Formulation and Computation. A dissertation of Doctor Philosophy*. Berkeley: Graduate Division of University California.

Robust Decisions under Risk for Imprecise Probabilities

Włodzimierz Ogryczak

Abstract In this paper we analyze robust approaches to decision making under uncertainty where the expected outcome is maximized but the probabilities are known imprecisely. A conservative robust approach takes into account any probability distribution thus leading to the notion of robustness focusing on the worst case scenario and resulting in the max-min optimization. We consider softer robust models allowing the probabilities to vary only within given intervals. We show that the robust solution for only upper bounded probabilities becomes the tail mean, known also as the conditional value-at-risk (CVaR), with an appropriate tolerance level. For proportional upper and lower probability limits the corresponding robust solution may be expressed by the optimization of appropriately combined the mean and the tail mean criteria. Finally, a general robust solution for any arbitrary intervals of probabilities can be expressed with the optimization problem very similar to the tail mean and thereby easily implementable with auxiliary linear inequalities.

1 Introduction

Several approaches have been developed to deal with uncertain or imprecise data in optimization problems. In the standard stochastic programming models, we assume that the probability distribution of the data is known (or can be estimated) (Ruszczyński and Shapiro 2003). The approaches focused on the quality of the solution for some data domains (bounded regions) are considered robust (Ben-Tal et al. 2009; Bertsimas and Thiele 2006). Notion of robust solutions was first introduced for statistical decisions in 1964 by Huber (1964). Stochastic programming models with uncertain probability distributions first had been introduced in (Dupacova 1987; Ermoliev et al. 1985). Practical importance of

---

W. Ogryczak (✉)
Warsaw University of Technology, Institute of Control & Computation Engineering,
Warsaw, Poland
e-mail: wogrycza@ia.pw.edu.pl

Y. Ermoliev et al. (eds.), *Managing Safety of Heterogeneous Systems*, Lecture Notes
in Economics and Mathematical Systems 658, DOI 10.1007/978-3-642-22884-1_3,
© Springer-Verlag Berlin Heidelberg 2012

the performance sensitivity against data uncertainty and errors has later attracted considerable attention to the search for robust solutions (see (Hampel et al. 1986)). In general decision theory under uncertainty the notion of robustness may have rather broad set of definitions (Ermoliev and Hordijk 2006). The precise concept of robustness depends on the way uncertain data domains and the quality or stability characteristics are introduced.

A conservative notion of robustness focusing on worst case scenario results is widely accepted and the max-min optimization is commonly used to seek robust solutions. Although shortcomings of the worst case approaches are known (Ermoliev and Wets 1988). Recently, a more advanced concept of ordered weighted averaging was introduced into robust optimization (Perny et al. 2006), thus allowing to optimize combined performances under the worst case scenario together with the performances under the second worst scenario, the third worst and so on. Such an approach exploits better the entire distribution of objective vectors in search for robust solutions and, more importantly, it introduces some tools for modeling robust preferences.

In this paper we focus on robust approaches where the probabilities are unknown or imprecise. Having assumed that the probabilities may vary within given intervals, we optimize the worst case expected outcome with respect to the probabilities perturbation set. For the case of unlimited perturbations the worst case expectation becomes the worst outcome (max-min solution). In general case, the worst case expectation is a generalization of the tail mean. Nevertheless, it can be effectively reformulated as a Linear Programming (LP) expansion of the original problem.

The paper is organized as follows. In the next section we recall the tail mean (Conditional Value at Risk, CVaR) solution concept providing a new proof of the LP computational model which remains applicable for more general problems related to the robust solution concepts. Section 3 contains the main results. We show that the robust solution for only upper bounded probabilities is the tail β-mean solution for an appropriate β value. For proportional upper and lower limits on probability perturbation the robust solution may be expressed as optimization of appropriately combined the mean and the tail mean criteria. Finally, a general robust solution for any arbitrary intervals of probabilities or probabilities perturbations can be expressed with optimization problem very similar to the tail β-mean and thereby easily implementable with auxiliary linear inequalities. In Sect. 4 we show how for the specific case of LP problems, alternative dual models of robust solutions may be built to overcome high dimensionality caused by the number of scenarios. The computational advantages of the dual models are demonstrated on the portfolio optimization problem in Sect. 5.

2 Robust Solution Concept

Consider a decision problem under uncertainty where the decision is based on the maximization of a scalar (real valued) outcome. The simplest representation of uncertainty depends on a finite set Ω ($|\Omega| = m$) of predefined scenarios. The final outcome is uncertain and only its realizations under various scenarios $\omega \in \Omega$ are known. Exactly, for each scenario ω the corresponding outcome realization is given as a function of the decision variables $y_\omega = f_\omega(\mathbf{x})$ where $\mathbf{x}$ denotes a vector of decision variables to be selected from the feasible set $Q \subset R^n$ of constraints under consideration. Let us define the set of attainable outcomes $A = \{\mathbf{y} = (y)_{\omega \in \Omega} : y_\omega = f_\omega(\mathbf{x}) \ \forall \ \omega \in \Omega, \quad \mathbf{x} \in Q\}$. We are interested in larger outcomes under each scenario. Hence, the decision under uncertainty can be considered a multiple criteria optimization problem (Haimes 1993; Ogryczak 2002)

$$\max \ \{ \ (y_\omega)_{\omega \in \Omega} \ : \ \mathbf{y} \in A \ \}. \tag{1}$$

From the perspective of decision making under uncertainty, the model (1) only specifies that we are interested in maximization of outcomes under all scenarios $\omega \in \Omega$. In order to make the multiple objective model operational for the decision support process, one needs to assume some solution concept well adjusted to the decision maker's preferences.

Within the decision problems under risk it is assumed that the exact values of the underlying scenario probabilities p_ω $(\omega \in \Omega)$ are given or can be estimated. This is a basis for the stochastic programming approaches where the solution concept depends on the maximization of the expected value (the mean outcome)

$$\mu(\mathbf{y}) = \sum_{\omega \in \Omega} y_\omega p_\omega \tag{2}$$

or some risk function. In particular, the risk functions $\mu_{\delta^k}(\mathbf{y}) = \mu(\mathbf{y}) - \delta^k(\mathbf{y})$ based on the downside semideviations

$$\delta^k(\mathbf{y}) = \left[\sum_{\omega \in \Omega} \max\{\mu(\mathbf{y}) - y_\omega p_\omega, 0\}^k \right]^{1/k} \tag{3}$$

are consistent with the second degree stochastic dominance (Ogryczak and Ruszczyński 2001) and thereby coherent (Artzner et al. 1999). Among them, the Mean Absolute Deviation (δ^1) related risk function can be expressed as the mean of downside distribution $\mu_{\delta^1}(\mathbf{y}) = \sum_{\omega \in \Omega} \min\{\mu(\mathbf{y}), y_\omega\} p_\omega$.

Recently, the second order quantile risk measures have been introduced in different ways by many authors (Artzner et al. 1999; Embrechts et al. 1997; Ermoliev and Leonardi 1982; Ogryczak 1999; Rockafellar and Uryasev 2000). They generally represent the (worst) tail mean defined as the mean within the specified tolerance level (quantile) of the worst outcomes. Within the decision under risk literature, and especially related to finance application, the tail mean quantity is usually called Tail VaR, Average VaR or Conditional VaR (where VaR reads after Value-at-Risk) (Pflug 2000). Actually, the name CVaR after (Rockafellar and Uryasev 2000) is now the most commonly used. Although, since we will consider the measure with respect to distributions without a formally defined probabilistic space we will refer to it as the tail mean. The tail mean maximization is consistent with the second degree stochastic dominance (Ogryczak and Ruszczyński 2002) and it meets the requirements of coherent risk measurement (Pflug 2000).

For any probabilities p_ω and tolerance level β the corresponding tail mean can be mathematically formalized as follows (Ogryczak 2002; Ogryczak and Ruszczyński 2002). Having defined the right-continuous cumulative distribution function (cdf): $F_y(\eta) = \text{Prob}[y_w \leq \eta]$, we introduce the quantile function $F_y^{(-1)}$ as the left-continuous inverse of the cumulative distribution function F_y:

$$F_y^{(-1)}(\beta) = \inf \{\eta : F_y(\eta) \geq \beta\} \quad \text{for } 0 < \beta \leq 1.$$

By integrating $F_y^{(-1)}$ one gets the (worst) tail mean

$$\mu_\beta(\mathbf{y}) = \frac{1}{\beta} \int_0^\beta F_y^{(-1)}(\alpha)d\alpha \quad \text{for } 0 < \beta \leq 1. \tag{4}$$

the point value of the absolute Lorenz curve (Ogryczak 2000). The latter makes the tail means directly related to the dual theory of choice under risk (Quiggin 1982; Roell 1987; Yaari 1987).

Maximization of the tail β-mean

$$\max_{\mathbf{y} \in A} \mu_\beta(\mathbf{y}) \tag{5}$$

defines the tail β-mean solution concept. When parameter β approaches 0, the tail β-mean tends to the smallest outcome

$$M(\mathbf{y}) = \min \{y_\omega : \omega \in \Omega\} = \lim_{\beta \to 0_+} \mu_\beta(\mathbf{y}).$$

On the other hand, for $\beta = 1$ the corresponding tail mean becomes the standard mean ($\mu_1(\mathbf{y}) = \mu(\mathbf{y})$).

Note that, due to the finite number of scenarios, the tail β-mean is well defined by the following optimization

$$\mu_\beta(\mathbf{y}) = \min_{u_\omega} \left\{ \frac{1}{\beta} \sum_{\omega \in \Omega} y_\omega u_\omega : \sum_{\omega \in \Omega} u_\omega = \beta, \ 0 \leq u_\omega \leq p_\omega \ \forall \ \omega \in \Omega \right\}. \tag{6}$$

Problem (6) is a Linear Program for a given outcome vector $\mathbf{y}$ while it becomes nonlinear for $\mathbf{y}$ being a vector of variables as in the tail β-mean problem (5). It turns out that this difficulty can be overcome by an equivalent LP formulation of the β-mean that allows one to implement the β-mean problem (5) with auxiliary linear inequalities. Namely, the following theorem recalls Rockafellar and Uryasev (2000) LP model for continuous distributions which remains valid for a general distribution (Ogryczak and Ruszczyński 2002). Although we introduce a new proof which can be further generalized for a family of robust solution concepts we consider.

Robust Decisions under Risk for Imprecise Probabilities

Theorem 1. *For any outcome vector* $\mathbf{y}$ *with the corresponding probabilities* p_ω, *and for any real value* $0 < \beta \le 1$, *the tail* β-*mean outcome is given by the following linear program:*

$$\mu_\beta(\mathbf{y}) = \max_{t,d_\omega} \left\{ t - \frac{1}{\beta} \sum_{\omega \in \Omega} p_\omega d_\omega : y_\omega \ge t - d_\omega,\ d_\omega \ge 0\ \forall\ \omega \in \Omega \right\}. \quad (7)$$

Proof. The theorem can be proven by taking advantage of the LP dual to (6). Introducing dual variable t corresponding to the equation $\sum_{\omega \in \Omega} u_\omega = \beta$ and variables d_ω corresponding to upper bounds on u_ω one gets the LP dual (7). Due to the duality theory, for any given vector $\mathbf{y}$ the tail β-mean $\mu_\beta(\mathbf{y})$ can be found as the optimal value of the LP problem (7). □

Frequently, scenario probabilities are unknown or imprecise. Uncertainty is then represented by limits (intervals) on possible values of probabilities varying independently (Thiele 2008). We focus on such representation to define robust solution concept. Generally, we consider the case of unknown probabilities belonging to the hypercube:

$$\mathbf{u} \in U = \left\{ (u_1, u_2, \dots, u_m) : \sum_{\omega \in \Omega} u_\omega = 1,\ \Delta_\omega^l \le u_\omega \le \Delta_\omega^u\ \forall\ \omega \in \Omega \right\} \quad (8)$$

where obviously

$$\sum_{\omega \in \Omega} \Delta_\omega^l \le 1 \le \sum_{\omega \in \Omega} \Delta_\omega^u.$$

Focusing on the mean outcome as the primary system efficiency measure to be optimized we get the robust mean solution concept

$$\max_{\mathbf{y}} \min_{\mathbf{u}} \left\{ \sum_{\omega \in \Omega} u_\omega y_\omega : \mathbf{u} \in U,\ \mathbf{y} \in A \right\}. \quad (9)$$

Further, taking into account that all the constraints of attainable set A remain unchanged while the probabilities are perturbed, the robust mean solution can be rewritten as

$$\max_{\mathbf{y} \in A} \min_{\mathbf{u} \in U} \sum_{\omega \in \Omega} u_\omega y_\omega = \max_{\mathbf{y} \in A} \left\{ \min_{\mathbf{u} \in U} \sum_{\omega \in \Omega} u_\omega y_\omega \right\} = \max_{\mathbf{y} \in A} \mu^U(\mathbf{y}) \quad (10)$$

where

$$\begin{aligned}
\mu^U(\mathbf{y}) &= \min_{\mathbf{u} \in U} \sum_{\omega \in \Omega} u_\omega y_\omega \\
&= \min_{u_\omega} \left\{ \sum_{\omega \in \Omega} y_\omega u_\omega : \sum_{\omega \in \Omega} u_\omega = 1,\ \Delta_\omega^l \le u_\omega \le \Delta_\omega^u\ \forall\ \omega \in \Omega \right\}
\end{aligned} \quad (11)$$

represent the worst case mean outcomes for given outcome vector $\mathbf{y} \in A$ with respect to the probabilities set U.

Similar robust solution concepts can be built for various risk functions used instead of the mean. For the tail mean (CVaR) optimization, the corresponding robust tail β-mean solution can be expressed as

$$\max_{\mathbf{y} \in A} \mu_\beta^U(\mathbf{y}) \tag{12}$$

where

$$\mu_\beta^U(\mathbf{y}) = \min_{\mathbf{u} \in U} \min_{u_\omega'} \left\{ \frac{1}{\beta} \sum_{\omega \in \Omega} y_\omega u_\omega' : \sum_{\omega \in \Omega} u_\omega' = \beta, \ 0 \le u_\omega' \le u_\omega \ \forall \ \omega \in \Omega \right\}. \tag{13}$$

represents the worst case tail β-mean outcome for given outcome vector $\mathbf{y} \in A$ with respect to the probabilities set U.

3 Tail Mean and Related Robust Solution Concepts

Let us consider first the robust mean solution (10) in the case of unlimited probability perturbations ($\Delta_\omega^l = 0$ and $\Delta_\omega^u = 1$). One may easily notice that the worst case mean outcome (11) becomes the worst outcome

$$\mu^U(\mathbf{y}) = \min_{u_\omega} \left\{ \sum_{\omega \in \Omega} y_\omega u_\omega : \sum_{\omega \in \Omega} u_\omega = 1, \ 0 \le u_\omega \le 1 \ \forall \ \omega \in \Omega \right\} = \min_{\omega \in \Omega} y_\omega$$

thus leading to the conservative robust solution concept represented by the max-min approach.

For the case of probabilities lying in a given box with relaxed lower limits ($\Delta_\omega^l = 0 \ \forall \ \omega \in \Omega$) the worst case mean outcome (11) becomes the classical tail mean outcome. Hence, the robust solution (10) may be represented as the tail β-mean with respect to appropriately rescaled probabilities.

Theorem 2. *The robust solution the worst case mean outcome (9)–(11) with relaxed lower bounds may be represented as the tail β-mean with respect to probabilities*

$$p_\omega = \Delta_\omega^u \bigg/ \sum_{\omega \in \Omega} \Delta_\omega^u \quad and \quad \beta = 1 \bigg/ \sum_{\omega \in \Omega} \Delta_\omega^u,$$

and it can be found by simple expansion of the optimization problem with auxiliary linear constraints and variables to the following:

$$\max_{\mathbf{y}, \mathbf{d}, t} \left\{ t - \sum_{\omega \in \Omega} \Delta_\omega^u d_\omega : \ \mathbf{y} \in A; \ y_\omega \ge t - d_\omega, \ d_\omega \ge 0 \ \forall \ \omega \in \Omega \right\}. \tag{14}$$

Proof. Note that by simple rescaling of variables with $s^u = \sum_{\omega \in \Omega} \Delta^u_\omega$ one gets

$$\mu^U(\mathbf{y}) = \min_{u_\omega} \left\{ \sum_{\omega \in \Omega} y_\omega u_\omega : \sum_{\omega \in \Omega} u_\omega = 1, \ 0 \le u_\omega \le \Delta^u_\omega \ \forall \ \omega \in \Omega \right\}$$

$$= \min_{u'_\omega} \left\{ s^u \sum_{\omega \in \Omega} y_\omega u'_\omega : \sum_{\omega \in \Omega} u'_\omega = \frac{1}{s^u}, \ 0 \le u'_\omega \le \frac{\Delta^u_\omega}{s^u} \ \forall \ \omega \in \Omega \right\}.$$

Hence, the robust solution may be represented as the tail $(1/s^u)$-mean with respect to probabilities $p_\omega = \Delta^u_\omega / s^u$. Following Theorem 1, it can searched by solving (14). $\qquad \square$

Note that with $\Delta^u_\omega = 1$ for $\omega \in \Omega$ we represent the robust solution (11) as the tail β-mean with $p_\omega = 1/m$ and $\beta = 1/m$ thus representing the max-min model. In the case of $\Delta^u_\omega = k/m$ for $\omega \in \Omega$ we get $p_\omega = 1/m$ and $\beta = 1/k$. For the specific case of given probabilities $\bar{\mathbf{p}}$ with possible perturbations bounded proportionally it is possible to express the corresponding robust solution (11) as the tail mean based on the original probabilities. Indeed, in the case of $\Delta^u_\omega = (1 + \delta^+)\bar{p}_\omega$ we get in Theorem 2

$$p_\omega = \Delta^u_\omega \Big/ \sum_{\omega \in \Omega} \Delta^u_\omega = \bar{p}_\omega.$$

In the general case of possible lower limits, the robust mean solution concept (9)–(11) cannot be directly expressed as an appropriate tail β-mean. It turns out, however, that it can be expressed by the optimization with combined criteria of the tail β-mean and the mean.

Theorem 3. *The robust mean solution concept (9)–(11) is equivalent to the convex combination of the mean and the tail β-mean criteria maximization*

$$\max_{\mathbf{y} \in A} \mu^U(\mathbf{y}) = \max_{\mathbf{y} \in A} \left[\lambda \mu(\mathbf{y}) + (1 - \lambda) \mu_\beta(\mathbf{y}) \right] \qquad (15)$$

with

$$\beta = \left(1 - \sum_{\omega \in \Omega} \Delta^l_\omega \right) \Big/ \sum_{\omega \in \Omega} \left(\Delta^u_\omega - \Delta^l_\omega \right) \quad and \quad \lambda = \sum_{\omega \in \Omega} \Delta^l_\omega,$$

where the tail mean $\mu_\beta(\mathbf{y})$ is defined according to probabilities p'_ω while the mean $\mu(\mathbf{y})$ is considered with respect to probabilities p''_ω:

$$p'_\omega = \left(\Delta^u_\omega - \Delta^l_\omega \right) \Big/ \sum_{\omega \in \Omega} \left(\Delta^u_\omega - \Delta^l_\omega \right) \quad and \quad p''_\omega = \Delta^l_\omega \Big/ \sum_{\omega \in \Omega} \Delta^l_\omega \quad for \ \omega \in \Omega.$$

Proof. When introducing scaling factors $s^u = \sum_{\omega \in \Omega} \Delta^u_\omega$ and $s^l = \sum_{\omega \in \Omega} \Delta^l_\omega$, the worst case mean outcome (11) can be expressed as follows

$$\mu^U(\mathbf{y}) = \min_{u_\omega} \left\{ \sum_{\omega \in \Omega} y_\omega u_\omega : \sum_{\omega \in \Omega} u_\omega = 1, \ \Delta_\omega^l \le u_\omega \le \Delta_\omega^u \ \forall \ \omega \in \Omega \right\}$$

$$= \min_{u'_\omega} \left\{ \sum_{\omega \in \Omega} y_\omega u'_\omega : \sum_{\omega \in \Omega} u'_\omega = 1 - s^l, 0 \le u'_\omega \le \Delta_\omega^u - \Delta_\omega^l \ \forall \ \omega \in \Omega \right\}$$
$$+ \sum_{\omega \in \Omega} y_\omega \Delta_\omega^u$$

$$= (1 - s^l) \min_{u''_\omega} \left\{ \frac{s^u - s^l}{1 - s^l} \sum_{\omega \in \Omega} y_\omega u''_\omega : \sum_{\omega \in \Omega} u''_\omega = \frac{1 - s^l}{s^u - s^l}, \right.$$

$$\left. 0 \le u''_\omega \le \frac{\Delta_\omega^u - \Delta_\omega^l}{s^u - s^l} \ \forall \ \omega \in \Omega \right\} + s^l \sum_{\omega \in \Omega} y_\omega \frac{\Delta_\omega^l}{s^l}$$

$$= (1 - \lambda)\mu_\beta(\mathbf{y}) + \lambda\mu(\mathbf{y})$$

which completes the proof. $\qquad\square$

Corollary 1. *The robust mean solution concept (10)–(11) for the specific case of given probabilities $\bar{\mathbf{p}}$ with possible perturbations bounded proportionally $\Delta_\omega^l = (1 - \delta^-)\bar{p}_\omega$ and $\Delta_\omega^u = (1 + \delta^+)\bar{p}_\omega$ for all $\omega \in \Omega$ is equivalent to the convex combination of the mean and tail β-mean criteria maximization*

$$\max_{\mathbf{y} \in A} \mu^U(\mathbf{y}) = \max_{\mathbf{y} \in A} \left[\lambda\mu(\mathbf{y}) + (1 - \lambda)\mu_\beta(\mathbf{y}) \right] \tag{16}$$

with $\beta = \delta^-/(\delta^+ + \delta^-)$ and $\lambda = 1 - \delta^-$ where both the mean $\mu(\mathbf{y})$ and the tail mean $\mu_\beta(\mathbf{y})$ are calculated with respect to the original probabilities $\bar{p}_\omega$.

Proof. For proportionally bounded perturbations

$$\Delta_\omega^l = (1 - \delta^-)\bar{p}_\omega \quad \text{and} \quad \Delta_\omega^u = (1 + \delta^+)\bar{p}_\omega$$

formula 15 of Theorem 3 is fulfilled with

$$\beta = \frac{1 - \sum_{\omega \in \Omega} \Delta_\omega^l}{\sum_{\omega \in \Omega} \left(\Delta_\omega^u - \Delta_\omega^l \right)} = \frac{\delta^-}{\delta^+ + \delta^-}$$

and

$$\lambda = \sum_{\omega \in \Omega} \Delta_\omega^l = 1 - \delta^-.$$

Further, where the tail mean is defined according to probabilities

$$p'_\omega = \frac{\Delta_\omega^u - \Delta_\omega^l}{\sum_{\omega \in \Omega} \left(\Delta_\omega^u - \Delta_\omega^l \right)} = \frac{\left(\delta^+ + \delta^- \right) \bar{p}_\omega}{\delta^+ \sum_{\omega \in \Omega} \bar{p}_\omega + \delta^- \sum_{\omega \in \Omega} \bar{p}_\omega} = \bar{p}_\omega$$

as well as the mean is also considered with respect to probabilities

$$p''_\omega = \frac{\Delta^l_\omega}{\sum_{\omega \in \Omega} \Delta^l_\omega} = \frac{\left(1 - \delta^l_\omega\right) \bar{p}_\omega}{\left(1 - \delta^l_\omega\right) \sum_{\omega \in \Omega} \bar{p}_\omega} = \bar{p}_\omega$$

which completes the proof. $\qquad\square$

Alternatively, one can take advantages of the fact that the structure of optimization problem (11) remains very similar to that of the tail β-mean (6). Note that problem (11) is an LP for a given outcome vector $\mathbf{y}$ while it becomes nonlinear for $\mathbf{y}$ being a vector of variables. This difficulty can be overcome similar to Theorem 1 for the tail β-mean.

Theorem 4. *For any arbitrary intervals $[\Delta^l_\omega, \Delta^u_\omega]$ (for all $\omega \in \Omega$) of probabilities, the corresponding robust mean solution (10)–(11) can be given by the following optimization problem*

$$\max_{\mathbf{y}, t, d^u_\omega, d^l_\omega} \left\{ t - \sum_{\omega \in \Omega} \Delta^u_\omega d^u_\omega + \sum_{\omega \in \Omega} \Delta^l_\omega d^l_\omega : \right. \tag{17}$$
$$\left. \mathbf{y} \in A; \quad t - d^u_\omega + d^l_\omega \leq y_\omega, \ d^u_\omega, d^l_\omega \geq 0 \quad \forall \omega \in \Omega \right\}.$$

Proof. The theorem can be proven by taking advantages of the LP dual to (11). Introducing dual variable t corresponding to the equation $\sum_{\omega \in \Omega} u_\omega = 1$ and variables d^u_ω and d^l_ω corresponding to upper and lower bounds on u_ω, respectively, one gets the following LP dual to problem (11)

$$\mu^U(\mathbf{y}) = \max_{t, d^u_\omega, d^l_\omega} \left\{ t - \sum_{\omega \in \Omega} \Delta^u_\omega d^u_\omega + \sum_{\omega \in \Omega} \Delta^l_\omega d^l_\omega : \right.$$
$$\left. t - d^u_\omega + d^l_\omega \leq y_\omega, \ d^u_\omega, d^l_\omega \geq 0 \quad \forall \omega \in \Omega \right\}$$

which completes the proof. $\qquad\square$

While considering the tail mean as the basic optimization criterion (CVaR optimization) we have to deal with the robust tail mean solution concepts (12)–(13) to allow for imprecise probabilities. It turns out that this robust solution concept for any arbitrary perturbation set U (8) may be expressed as the standard tail mean with appropriately defined tolerance level and rescaled probabilities.

Theorem 5. *The robust tail β-mean solution (12)–(13) with arbitrary set U (8) may be represented as the tail β'-mean with respect to probabilities*

$$p'_\omega = \Delta^u_\omega \left/ \sum_{\omega \in \Omega} \Delta^u_\omega \right. \quad \text{and} \quad \beta' = \beta \left/ \sum_{\omega \in \Omega} \Delta^u_\omega \right.,$$

and it can be found by simple expansion of the optimization problem with auxiliary linear constraints and variables to the following:

$$\max_{\mathbf{y}, \mathbf{d}, t} \left\{ t - \frac{1}{\beta} \sum_{\omega \in \Omega} \Delta^u_\omega d_\omega : \quad \mathbf{y} \in A; \quad y_\omega \geq t - d_\omega, \ d_\omega \geq 0 \ \forall \omega \in \Omega \right\}. \tag{18}$$

Proof. Note that

$$\mu_\beta^U(\mathbf{y}) = \min_{u \in U} \min_{u'_\omega} \left\{ \frac{1}{\beta} \sum_{\omega \in \Omega} y_\omega u'_\omega : \sum_{\omega \in \Omega} u'_\omega = \beta, \ 0 \le u'_\omega \le u_\omega \ \forall \ \omega \in \Omega \right\}$$

$$= \min_{u'_\omega} \left\{ \frac{1}{\beta} \sum_{\omega \in \Omega} y_\omega u'_\omega : \sum_{\omega \in \Omega} u'_\omega = \beta, \ 0 \le u'_\omega \le \Delta_\omega^u \ \forall \ \omega \in \Omega \right\}$$

Thus by simple rescaling of variables with $s^u = \sum_{\omega \in \Omega} \Delta_\omega^u$ one gets

$$\mu_\beta^U(\mathbf{y}) = \min_{u''_\omega} \left\{ \frac{s^u}{\beta} \sum_{\omega \in \Omega} y_\omega u''_\omega : \sum_{\omega \in \Omega} u''_\omega = \frac{\beta}{s^u}, 0 \le u''_\omega \le \frac{\Delta_\omega^u}{s^u} \ \forall \ \omega \in \Omega \right\}.$$

Hence, the robust solution may be represented as the tail (β/s^u)-mean with respect to probabilities $p_\omega = \Delta_\omega^u/s^u$. Following Theorem 1, it can searched by solving (18). $\square$

Corollary 2. *The robust tail β-mean solution concept (12)–(13) for the specific case of given probabilities $\bar{\mathbf{p}}$ with possible perturbations upper bounded proportionally $\Delta_\omega^u = (1 + \delta^+) \bar{p}_\omega$ and arbitrary lower bounded (any $\Delta_\omega^l \le \bar{p}_\omega$) for all $\omega \in \Omega$ is equivalent to the tail β'-mean with respect to probabilities $\bar{\mathbf{p}}$ and $\beta' = \beta/(1+\delta^+)$, and it can be found by simple expansion of the optimization problem with auxiliary linear constraints and variables to the following:*

$$\max_{\mathbf{y}, \mathbf{d}, t} \left\{ t - \frac{1 + \delta^+}{\beta} \sum_{\omega \in \Omega} \bar{p}_\omega d_\omega : \quad \mathbf{y} \in A; \quad y_\omega \ge t - d_\omega, \ d_\omega \ge 0 \ \forall \ \omega \in \Omega \right\}.$$
$$(19)$$

4 Dual LP Models

Following (10), the robust mean solution concept is given as

$$\max_{\mathbf{y} \in A} \mu^U(\mathbf{y}) = \max_{\mathbf{y} \in A} \left\{ \min_{u \in U} \sum_{\omega \in \Omega} y_\omega u_\omega \right\} = \max_{\mathbf{y} \in A} \min_{u \in U} \sum_{\omega \in \Omega} u_\omega y_\omega$$

where the inner optimization problem (11) represents the worst case mean outcome for given outcome vector $\mathbf{y} \in A$ with respect to the probabilities set U. It is an LP for a given vector $\mathbf{y}$ but it turns into nonlinear within the entire robust optimization problem (5), due to the quadratic objective function $\sum_{\omega \in \Omega} y_\omega u_\omega$. This difficulty is overcome by an equivalent dual LP formulation of problem (6). Indeed, introducing dual variable t corresponding to the equation $\sum_{\omega \in \Omega} u_\omega = 1$ and variables d_ω^u and d_ω^l corresponding to upper and lower bounds on u_ω, respectively, we get the following LP dual to problem (11)

Robust Decisions under Risk for Imprecise Probabilities

$$\mu^U(\mathbf{y}) = \max_{t,d_\omega^u,d_\omega^l} \left\{ t - \sum_{\omega \in \Omega} \Delta_\omega^u d_\omega^u + \sum_{\omega \in \Omega} \Delta_\omega^l d_\omega^l : \atop t - d_\omega^u + d_\omega^l \leq y_\omega, \; d_\omega^u, d_\omega^l \geq 0 \quad \forall \, \omega \in \Omega \right\} \tag{20}$$

This leads us to the standard LP model (17) of Theorem 4 for the robust optimization. The model dimensionality is strongly affected by the number of scenarios under consideration. The latter may be huge in the case of more advanced simulation models employed for scenario generation (Pflug 2001).

An alternative robust optimization models can be built for LP problems by taking advantages of the minimax theorem. Note that both sets A and U are convex polyhedra. Hence, formula (5) can be rewritten into a dual form

$$\max_{\mathbf{y} \in A} \min_{\mathbf{u} \in U} \sum_{\omega \in \Omega} u_\omega y_\omega = \min_{\mathbf{u} \in U} \max_{\mathbf{y} \in A} \sum_{\omega \in \Omega} u_\omega y_\omega = \min_{\mathbf{u} \in U} D(\mathbf{u}) \tag{21}$$

with the inner optimization problem

$$D(\mathbf{u}) = \max_{\mathbf{y}} \left\{ \sum_{\omega \in \Omega} u_\omega y_\omega : \mathbf{y} \in A \right\}. \tag{22}$$

The inner optimization problem although being an LP for a given vector $\mathbf{u}$ has the quadratic objective function $\sum_{\omega \in \Omega} u_\omega y_\omega$ within the entire robust optimization problem (21) where $\mathbf{u}$ is also a vector of variables. Again, this difficulty can be resolved by taking advantages of the LP dual $D^*(\mathbf{u})$ to the inner problem $D^*(\mathbf{u})$. Indeed:

$$\min_{\mathbf{u} \in U} D(\mathbf{u}) = \min_{\mathbf{u} \in U} D^*(\mathbf{u}) \tag{23}$$

but solving the latter problem allows us to use the LP methodology. Moreover, set U has only one equation (structural constraint) which makes the problem $\min_{\mathbf{u} \in U} D^*(\mathbf{u})$ much simpler than those of (20). In the next section we illustrate potential advantages of the alternative (dual) model with the portfolio optimization problem.

5 Portfolio Optimization

The portfolio optimization problem we consider follows the original Markowitz' formulation and is based on a single period model of investment. At the beginning of a period, an investor allocates the capital among various securities, thus assigning a nonnegative weight (share of the capital) to each security. Let $J = \{1, 2, \ldots, n\}$ denote a set of securities considered for an investment. For each security $j \in J$, its rate of return is represented by a random variable R_j with a given mean $\mu_j = \mathbb{E}\{R_j\}$. Further, let $\mathbf{x} = (x_j)_{j=1,\ldots,n}$ denote a vector of decision variables x_j expressing the weights defining a portfolio. The weights must satisfy a set of constraints to represent a portfolio. The simplest way of defining a feasible set Q is by a requirement that the weights must sum to one and they are nonnegative (short sales are not allowed), i.e.

$$Q = \left\{ \mathbf{x} : \sum_{j \in J} x_j = 1, \quad x_j \geq 0 \quad \forall \, j \in J \right\}. \tag{24}$$

Hereafter, we perform detailed analysis for the set Q given with constraints (24). Nevertheless, the presented results can easily be adapted to a general LP feasible set given as a system of linear equations and inequalities, thus allowing one to include short sales, upper bounds on single shares or portfolio structure restrictions which may be faced by a real-life investor.

Each portfolio $\mathbf{x}$ defines a corresponding random variable $R_{\mathbf{x}} = \sum_{j \in J} R_j x_j$ that represents the portfolio rate of return while the expected value can be computed as $\mu(\mathbf{x}) = \sum_{j \in J} \mu_j x_j$. We consider m scenarios $\omega \in \Omega$ with probabilities p_ω. We assume that for each random variable R_j its realization r_j^ω under the scenario ω is known. Typically, the realizations are derived from historical data treating m historical periods as equally probable scenarios ($p_\omega = 1/m$). Although the models we analyze do not take advantages of this simplification. The realizations of the portfolio return $R_{\mathbf{x}}$ are given as

$$y_\omega = \sum_{j \in J} r_j^\omega x_j. \tag{25}$$

Following Theorem 4 and taking into account (25), for any arbitrary intervals $[\Delta_\omega^l, \Delta_\omega^u]$ (for all $\omega \in \Omega$) of probabilities, the corresponding robust portfolio optimization problem (10) can be given by the following LP problem:

$$
\begin{aligned}
\max_{\mathbf{x}, y, t, d_\omega^u, d_\omega^l} \quad & t - \sum_{\omega \in \Omega} \Delta_\omega^u d_\omega^u + \sum_{\omega \in \Omega} \Delta_\omega^l d_\omega^l : \\
\text{s.t.} \quad & \sum_{j \in J} x_j = 1, \quad x_j \geq 0 \qquad\qquad \text{for } j \in J \\
& d_\omega^u - d_\omega^l - t + \sum_{j \in J}^{n} r_j^\omega x_j \geq 0, \ d_\omega^u, d_\omega^l \geq 0 \text{ for } \omega \in \Omega
\end{aligned}
\tag{26}
$$

where t is an unbounded variable.

As a particular case of relaxed lower bounds on scenario probabilities ($\Delta_\omega^l = 0$ $\forall \omega \in \Omega$), following Corollary 2 one gets the classical CVaR portfolio optimization model (Mansini et al. 2003):

$$
\begin{aligned}
\max_{\mathbf{x}, y, t, d_\omega} \quad & t - \frac{1}{\beta} \sum_{\omega \in \Omega} p_\omega d_\omega \\
\text{s.t.} \quad & \sum_{j \in J} x_j = 1, \quad x_j \geq 0 \qquad\qquad \text{for } j \in J \\
& d_\omega - t + \sum_{j \in J} r_j^\omega x_j \geq 0, \ d_\omega \geq 0 \text{ for } \omega \in \Omega
\end{aligned}
\tag{27}
$$

with probabilities $p_\omega = \Delta_\omega^u / \sum_{\omega \in \Omega} \Delta_\omega^u$ and the tolerance level $\beta = 1/\sum_{\omega \in \Omega} \Delta_\omega^u$.

Except from the corresponding portfolio constraints (24), model (27) contains m nonnegative variables d_ω plus single variable t and m corresponding linear inequalities. Hence, its dimensionality is proportional to the number of scenarios m. Exactly, the LP model contains $m + n + 1$ variables and $m + 1$ constraints. It does not cause any computational difficulties for a few hundreds scenarios as in several computational analysis based on historical data (Mansini et al. 2007), However, in the case of more advanced simulation models employed for scenario generation one may get several thousands scenarios (Pflug 2001). This may lead to the LP model (27) with huge number of variables and constraints thus decreasing the computational efficiency of the model.

The dual model (23) allows us to formulate the corresponding robust portfolio optimization problem (10), for any arbitrary intervals of probabilities (8), as the following LP problem:

$$
\begin{aligned}
&\min_{\mathbf{u},q} \; q \\
&\text{s.t.} \;\; q - \sum_{\omega \in \Omega} r_j^\omega u_\omega \geq 0 \text{ for } j \in J \\
&\quad\quad \sum_{\omega \in \Omega} u_\omega = 1 \\
&\quad\quad \Delta_\omega^l \leq u_\omega \leq \Delta_\omega^u \quad \text{for } \omega \in \Omega.
\end{aligned}
\tag{28}
$$

For the specific case of the CVaR model (27) representing the case of relaxed lower bounds, the dual model takes the following form:

$$
\begin{aligned}
&\min_{\mathbf{u},q} \; q \\
&\text{s.t.} \;\; q - \sum_{\omega \in \Omega} r_j^\omega u_\omega \geq 0 \quad \text{for } j = 1, \ldots, n \\
&\quad\quad \sum_{\omega \in \Omega} u_\omega = 1 \\
&\quad\quad 0 \leq u_\omega \leq \frac{p_\omega}{\beta} \quad \text{for } \omega \in \Omega.
\end{aligned}
\tag{29}
$$

The dual LP model contains m variables u_ω, but only $n+1$ constraints (n inequalities and one equation) excluding the simple bounds on u_ω not affecting the problem complexity. Actually, the number of constraints in (29) is proportional to the portfolio size n, thus it is independent from the number of scenarios. Exactly, there are $m + 1$ variables and $n + 1$ constraints. This guarantees a high computational efficiency of the dual model even for very large number of scenarios. Note that possible additional portfolio structure requirements are usually modeled with rather small number of linear constraints thus generating small number of additional variables in the dual model. Certainly, the optimal portfolio shares x_j are not directly represented within the solution vector of problem (29) but they are easily available as the dual variables (shadow prices) for inequalities $q - \sum_{\omega \in \Omega} r_j^\omega u_\omega \geq 0$. Moreover, the dual model (29) may be considered a special case within the general theory of dual representations of coherent measures of risk, following from conjugate duality (Sect. 5 in (Miller and Ruszczyński 2008)).

We have run computational tests (Ogryczak and Śliwiński 2010) on the large scale CVaR portfolio optimization instances developed by Lim et al. (2010). The instances were originally generated from a multivariate normal distribution for 50, 100 or 200 securities with the number of scenarios 50,000. All computations were performed on a PC with the Intel Core i7 2.66GHz processor and 6GB RAM employing the simplex code of the CPLEX 12.1 package. An attempt to solve the primal model (27) with $\beta = 0.05$ resulted in 580, 1443 and 5006 seconds of computation on average, for problems with 50, 100 and 200 securities, respectively. Solving the dual models (29) directly by the primal method (standard CPLEX settings) results in computation times 5.3, 13.6 and 38.9 CPU seconds, respectively. Moreover, the computation times remain very low for various confidence levels (Ogryczak and Śliwiński 2010).

6 Conclusions

We have analyzed the robust mean solution concept where uncertainty is represented by limits (intervals) on possible values of scenario probabilities varying independently. Such an approach, in general, leads to complex optimization models with variable coefficients (probabilities). We have shown, however, that the robust mean solution concepts can be expressed with auxiliary linear inequalities, similar to the tail β-mean solution concept based on maximization of the mean in β portion of the worst outcomes. Actually, the robust mean solution for upper limits on probabilities turns out to be the tail β-mean for an appropriate β value. For upper and lower limits the robust mean solution may be sought by optimization of appropriately combined the mean and the tail mean criteria. Thus, a general robust mean solution for any arbitrary intervals of probabilities can be expressed with optimization problem very similar to the tail β-mean and thereby easily implementable with auxiliary linear inequalities. While considering the tail mean as the basic optimization criterion (CVaR optimization) the corresponding robust solution concept for any arbitrary perturbation set may be expressed as the standard tail mean with appropriately defined tolerance level and rescaled probabilities.

Our analysis has shown that the robust mean solution concept is closely related with the tail mean which is the basic equitable solution concept (Kostreva et al. 2004). It corresponds to recent approaches to the robust optimization based on the equitable optimization (Miettinen et al. 2008; Perny et al. 2006; Takeda and Kanamori 2009). Further study on equitable solution concepts and their relations to robust solutions seems to be a promising research direction. In particular, more complex robust preferences can be modeled by combining with various weights the tail means for larger and smaller perturbations thus leading to the combinations of multiple CVaR measures (Mansini et al. 2007).

Robust Decisions under Risk for Imprecise Probabilities

Acknowledgements The research was partially supported by the Polish National Budget Funds 2009–2011 for science under the grant N N516 3757 36.

References

Artzner, P., Delbaen, F., Eber, J.-M., & Heath, D. (1999). Coherent masures of risk. *Mathematical Finance, 9*, 203–228.

Ben-Tal, A., El Ghaoui, L., & Nemirovski, A. (2009). *Robust Optimization*. Princeton: Princeton University Press.

Bertsimas, D., & Thiele, A. (2006). Robust and data-driven optimization: modern decision making under uncertainty. *Tutorials on Operations Research, INFORMS, Chap. 4*, 195–122.

Dupacova, J. (1987). Stochastic programming with incomplete information: A survey of results on postoptimization and sensitivity analysis. *Optimization, 18*, 507–532.

Embrechts, P., Klüppelberg, C., & Mikosch T. (1997). *Modelling Extremal Events for Insurance and Finance*. New York: Springer.

Ermoliev, Y., Gaivoronski, A., & Nedeva, C. (1985). Stochastic optimization problems with incomplete information on distribution functions. *SIAM Journal on Control and Optimization, 23*, 697–716.

Ermoliev, Y., & Hordijk, L. (2006). Facets of robust decisions. In K. Marti, Y. Ermoliev, M. Makowski, & G. Pflug (Eds.), *Coping with Uncertainty, Modeling and Policy Issues* (pp. 4–28). Berlin: Springer.

Ermoliev, Y., & Leonardi, G. (1982). Some proposals for stochastic facility location models. *Mathematical Modelling, 3*, 407–420.

Ermoliev, Y., & Wets, R.J.-B. (1988). Stochastic programming, an introduction. Numerical techniques for stochastic optimization. *Springer Series in Computational Mathematics, 10*, 1–32.

Haimes, Y. Y. (1993). Risk of extreme events and the fallacy of the expected value. *Control and Cybernetics, 22*, 7–31.

Hampel, F., Ronchetti, E., Rousseeuw, P., & Stahel, W. (1986). *Robust Statistics: The Approach Based on Influence Function*. New York: Wiley.

Huber, P. J. (1964). Robust estimation of a location parameter. *The Annals of Mathematical Statistics, 35*, 73–101.

Kostreva, M. M., Ogryczak, W., & Wierzbicki, A. (2004). Equitable aggregations and multiple criteria analysis. *European Journal of Operational Research, 158*, 362–367.

Lim, C., Sherali, H. D., & Uryasev, S. (2010). Portfolio optimization by minimizing conditional value-at-risk via nondifferentiable optimization. *Computational Optimization and Applications, 46*, 391–415.

Mansini, R., Ogryczak, W., Speranza, M. G. (2003). On LP solvable models for portfolio selection. *Informatica, 14*, 37–62.

Mansini, R., Ogryczak, W., & Speranza, M. G. (2007). Conditional value at risk and related linear programming models for portfolio optimization. *Annals of Operations Research, 152*, 227–256.

Miettinen, K., Deb, K., Jahn, J., Ogryczak, W., Shimoyama, K., & Vetchera, R. (2008). Future challenges (Chap. 16). In *Multi-Objective Optimization – Evolutionary and Interactive Approaches, Lecture Notes in Computer Science*, vol. 5252 (pp. 435–461). New York: Springer.

Miller, N., & Ruszczyński, A. (2008). Risk-adjusted probability measures in portfolio optimization with coherent measures of risk. *European Journal of Operational Research, 191*, 193–206.

Ogryczak, W. (1999). Stochastic dominance relation and linear risk measures. In A. M. J. Skulimowski (Ed.), *Financial Modelling – Proc. 23rd Meeting EURO WG Financial Modelling, Cracow, 1998* (pp. 191–212). Cracow: Progress & Business Publisher.

Ogryczak, W. (2000). Multiple criteria linear programming model for portfolio selection. *Annals of Operations Research, 97*, 143–162.

Ogryczak, W. (2002). Multiple criteria optimization and decisions under risk. *Control and Cybernetics, 31*, 975–1003.

Ogryczak, W., & Ruszczyński, A. (2001). On consistency of stochastic dominance and mean-semideviation models. *Mathematical Programming, 89*, 217–232.

Ogryczak, W., & Ruszczyński, A. (2002). Dual stochastic dominance and quantile risk measures. *International Transactions on Operational Research, 9*, 661–680.

Ogryczak, W., & Śliwiński, T. (2010). On solving the dual for portfolio selection by optimizing conditional value at risk. *Computational Optimization and Applications*, DOI: 10.1007/s10589-010-9321-y.

Perny, P., Spanjaard, O., & Storme, L.-X. (2006). A decision-theoretic approach to robust optimization in multivalued graphs. *Annals of Operations Research, 147*, 317–341.

Pflug, G.Ch. (2001). Scenario tree generation for multiperiod financial optimization by optimal discretization. *Mathematical Programming, 89*, 251–271.

Pflug, G.Ch. (2000). Some remarks on the value-at-risk and the conditional value-at-risk. In S. Uryasev (Ed.), *Probabilistic Constrained Optimization: Methodology and Applications*. Dordrecht: Kluwer.

Quiggin, J. (1982). A theory of anticipated utility. *Journal of Economic Behavior and Organization, 3*, 323–343.

Rockafellar, R. T., & Uryasev, S. (2000). Optimization of conditional value-at-risk. *Journal of Risk, 2*, 21–41.

Roell, A. (1987). Risk aversion in Quiggin and Yaari's rank-order model of choice under uncertainty. *Economic Journal, 97*, 143–159.

Ruszczyński, A., & Shapiro, A. (Eds.) (2003). *Stochastic Programming, Handbooks in Operations Research and Management Science*, vol. 10 (pp. 272–281). Elsevier

Takeda, A., & Kanamori, T. (2009). A robust approach based on conditional value-at-risk measure to statistical learning problems. *European Journal of Operational Research, 198*, 287–296.

Thiele, A. (2008). Robust stochastic programming with uncertain probabilities. *IMA Journal of Management Mathematics, 19*, 289–321.

Yaari, M. E. (1987). The dual theory of choice under risk. *Econometrica 55*, 95–115.

Combining Second-Order Belief Distributions with Qualitative Statements in Decision Analysis

Ola Caster and Love Ekenberg

Abstract There is often a need to allow for imprecise statements in real-world decision analysis. Joint modeling of intervals and qualitative statements as constraint sets is one important approach to solving this problem, with the advantage that both probabilities and utilities can be handled. However, a major limitation with interval-based approaches is that aggregated quantities such as expected utilities also become intervals, which often hinders efficient discrimination. The discriminative power can be increased by utilizing second-order information in the form of belief distributions, and this paper demonstrates how qualitative relations between variables can be incorporated into such a framework. The general case with arbitrary distributions is described first, and then a computationally efficient simulation algorithm is presented for a relevant sub-class of analyses. By allowing qualitative relations, our approach preserves the ability of interval-based methods to be deliberately imprecise. At the same time, the use of belief distributions allows more efficient discrimination, and it provides a semantically clear interpretation of the resulting beliefs within a probabilistic framework.

1 Introduction

It is questionable whether people are capable of providing the inputs that utility theory requires, when most people cannot clearly distinguish between widely separated probabilities Shapira (1995). This indicates that precise numerical information does

O. Caster (✉)
Uppsala Monitoring Centre, WHO Collaborating Centre for International Drug Monitoring, Uppsala, Sweden

Department of Computer and Systems Sciences, Stockholm University
e-mail: ola.caster@who-umc.org

L. Ekenberg
Department of Computer and Systems Sciences, Stockholm University, Kista, Sweden
e-mail: lovek@dsv.su.se

Y. Ermoliev et al. (eds.), *Managing Safety of Heterogeneous Systems*, Lecture Notes in Economics and Mathematical Systems 658, DOI 10.1007/978-3-642-22884-1_4,
© Springer-Verlag Berlin Heidelberg 2012

not make much sense in real-life decision making. Furthermore, even if a decision maker is able to discriminate between different probabilities, very often complete, adequate, and precise information is missing. Hence, decision problems frequently contain far less information than classical utility theory requires. In particular, quite often we might, at best, have access to some vague probability beliefs and qualitative preferences among the consequences, and very little more than that. This is the class of decision problems we aim at in this article.

It has since long been recognized that decision theory needs to accommodate imprecise probabilities (and utilities) and a vast amount of models with representations allowing imprecise probability statements have been suggested, including possibility theory, capacity theory, evidence theory and belief functions in the Dempster-Shafer sense, various kinds of logic, upper and lower probabilities, hierarchical models and sets of probability measures. A multitude of articles have been presented on various methods. For some early examples of these, see e.g. (Choquet 1954; Dempster 1967; Dubois and Prade 1988; Ellsberg 1961; Good 1962; Shafer 1976; Smith 1961). It is interesting to note that, during recent years, the activities within the area of imprecise probabilities have increased substantially and special conferences are now dedicated to contributions on this theme. An example of this is Jaffray (1999) from the first International Symposium on Imprecise Probabilities and Their Applications (ISIPTA).

Some general approaches to evaluating imprecise decision situations include both imprecise probabilities and utilities. We have earlier discussed various aspects on these issues in a sequence of articles and argued that there are strong arguments for modeling quantitative impreciseness as intervals (and similar constraints), enabling representation and modeling of qualitative information as constraint sets of relations. Cf. Danielson and Ekenberg (1998); Danielson et al. (2009); Ding et al. (2010); Ekenberg and Thorbiörnson (2001).

An obvious advantage of approaches using upper and lower probabilities is that they do not require taking particular probability distributions into consideration. On the other hand, the expected utility range resulting from an evaluation is then also an interval. To our experience, in real-life decision situations, it is hard to discriminate between the alternatives in a pure interval approach, even if various relations are added. In effect, an interval-based decision procedure preserves all alternatives with overlapping expected utility intervals, even if the overlap is quite small. Consequently, there is a need to extend the representation of the decision situation using more information, but keeping the requirement that the decision maker does not have to be more precise than what is possible. In pursuit of more discriminative power, we have also developed methods for handling belief distributions over sub-parts of the probabilities as well as the utilities involved, see e.g. Danielson et al. (2007); Ekenberg (2000). Furthermore, we have developed a calculus for aggregating these in various ways, predominantly using a generalization of the expected utility function Ekenberg et al. (2005).

Thus, we have developed computationally meaningful methods – that have also been implemented as software – for solving multi-linear expressions with respect to constraints sets. We have also developed – and implemented – methods for handling

Combining Second-Order Belief Distributions with Qualitative Statements 69

distributions over independent variables. Various aspects of these latter methods are provided in Ekenberg et al. (2007, 2006); Sundgren et al. (2009). However, a great and embarrassing dilemma has been to find the combination of these two approaches, i.e. to use qualitative statements and distributions at the same time. Dependencies of the kind that qualitative statements give rise to are not particularly straightforward to handle in the context of belief distributions. Nevertheless, such statements arise naturally in many decision situations. For instance, usually a decision maker has access only to local information and qualitative statements of relations between different parameters, in terms of constraints, and, consequently, has no explicit idea about the overall distribution.

This article presents a computationally meaningful method for solving this dilemma. We present how to solve expected utilities of quite complex structure, considering general decision trees and belief distributions over all the probabilities and utilities involved. Furthermore, and most importantly herein, we demonstrate a method for including qualitative statements, while still preserving the possibility to use these distributions efficiently. Starting from decision trees, we first solve the general case, followed by a computationally feasible method for handling this practically in a relevant sub-class of analyses. Compared to other approaches, this merging of second-order belief distributions and qualitative statements not only improves discrimination but also provides an easier interpretation within a probabilistic framework.

2 Decision Trees

A *decision tree* represents a decision problem, collecting all information necessary for the model into one structure.

Definition 1. A graph is a structure $\langle V, E \rangle$ where V is a set of nodes and E is a set of node pairs (edges).

Definition 2. A tree is a connected graph without cycles. A rooted tree is a tree containing a finite set of nodes and that has a dedicated node at level 0. The adjacent nodes to a node at level i, except the nodes at level $i - 1$, are at level $i + 1$. A node at level $i + 1$ that is adjacent to a node at level i is a child of the latter. A node at level 1 is an *alternative*. A node at level i is a leaf or *consequence* if it has no adjacent nodes at level $i + 1$. A node that is at level 2 or more and has children is an *event* (an intermediary node). The depth of a rooted tree is $\max(n|\text{there exists a node at level n})$.

For convenience we can, for instance, use the notation that the n children of a node c_i are denoted $c_{i1}, c_{i2}, \ldots, c_{in}$ and the m children of the node c_{ij} are denoted $c_{ij1}, c_{ij2}, \ldots, c_{ijm}$, etc.

Definition 3. Given a rooted tree, a decision tree T is formed by assigning a p symbol to each edge not starting in the root node, and a u symbol to each consequence node.

Generally, the p and u symbols can be given any meaning, but here they will represent probabilities and utilities, respectively. As such they are all constrained to $[0, 1]$, and further the probabilities on edges from a common parent node (not the root) must sum to 1. Such a set of probabilities will henceforth be called a *probability group*.

Primary evaluation rules of a decision tree model are based on the expected utility.

Definition 4. Given a decision tree T and an alternative A_i, the expression

$$E(A_i) = \sum_{i_1=1}^{n_{i_0}} p_{i i_1} \sum_{i_2=1}^{n_{i_1}} p_{i i_1 i_2} \cdots \sum_{i_{m-1}=1}^{n_{i_{m-2}}} p_{i i_1 i_2 \ldots i_{m-2} i_{m-1}}$$
$$\times \sum_{i_m=1}^{n_{i_{m-1}}} p_{i i_1 i_2 \ldots i_{m-2} i_{m-1} i_m} u_{i i_1 i_2 \ldots i_{m-2} i_{m-1} i_m}$$

where m is the depth of the tree corresponding to A_i, n_{i_k} is the number of possible outcomes following the event with probability p_{i_k}, $p_{\ldots i_j \ldots}$, $j \in [1, \ldots, m]$, denote probability variables and $u_{\ldots i_j \ldots}$ denote utility variables as above, is the expected utility of alternative A_i in T.

This is a general representation and one option is thus to define probability distributions and utility functions in the classical way. Another option that also covers impreciseness is to define sets of possible probability distributions and utility functions. The possible functions are then conveniently expressed as vectors in polytopes that are solution sets to the constraints involved.

A number of evaluation procedures, earlier suggested by us, then yield first-order interval estimates of the evaluations, i.e. upper and lower bounds for the expected utilities of the alternatives Danielson and Ekenberg (2007). However, the expected utility range resulting from an evaluation now also becomes an interval. In real-life decision situations, it is then often hard to discriminate between the alternatives, i.e. an interval-based decision procedure will not separate out alternatives with overlapping expected utility intervals, even if the overlap is quite small. Furthermore, a decision maker does not necessarily believe with equal faith in all the epistemologically possible probability distributions, represented by a set of interval statements. Therefore, it is interesting to extend the representation of the decision situation using more information, such as distributions over classes of probability and utility measures, in pursuit of more discriminative power.

3 Belief Distributions

The idea is now that distributions can be used for expressing various beliefs over multi-dimensional spaces where each dimension corresponds to, for instance, possible probabilities or utilities of consequences. The distributions can consequently be used to express strengths of beliefs in different vectors in the solution sets.

Combining Second-Order Belief Distributions with Qualitative Statements

Approaches for extending the interval representation using distributions over classes of probability and value measures in this way have been developed into various hierarchical models, such as second-order probability theory, cf. Ekenberg et al. (2006).

In such an approach, it is possible to make use of distributions rather than intervals for expressing beliefs regarding the probabilities and utilities involved. However, since general distributions over the entire solution sets are very hard to imagine, already when handling just a few dimensions, the marginal distributions, and the relations between these, are of high importance. A more comprehensible distribution in the latter sense can straightforwardly be defined.

Definition 5. For a utility or probability variable x in a decision tree T, the continuous random variable $\tilde{X}$ is the belief distribution over x. $\tilde{X}$ is defined on $[a, b]$ with $a, b \in [0, 1]$ and $a < b$.

We will frequently use the density function $f_{\tilde{X}}(x)$ of $\tilde{X}$ to represent and visualize the belief distribution. In some cases we will be interested in the joint density function for the belief regarding a set of utilities or probabilities. Exemplifying with the joint belief distribution over the utilities $u_1, \ldots, u_n$, this density function will be denoted $f_{\tilde{U}_1, \ldots, \tilde{U}_n}(u_1, \ldots, u_n)$, or more compactly as $f_{\tilde{\mathbf{U}}}(\mathbf{u})$. [1]

3.1 Constrained Belief Distributions

Our main objective here is to extend the earlier approach to allow for comparative constraints. The type of constraints considered are linear relations between two variables, i.e. of the type $u_i \leq u_j$, which here translate to constraints for the corresponding belief distributions. These constraints, together with the specified belief distributions, make up the decision maker's perception; they are his or her statements about the decision situation.

We will allow comparative constraints between any two utilities and between any two probabilities in the same probability group[2].

Definition 6. Given a decision tree T, the total set of constraints for $\tilde{U}_1, \ldots, \tilde{U}_n$ is denoted by $A_{\mathbf{U}}$. Furthermore $B_{\mathbf{U}}$ is the corresponding subspace of $[0, 1]^n$ implied by $A_{\mathbf{U}}$. Analogously, the total set of constraints for the belief distributions $\tilde{P}_{k1}, \ldots, \tilde{P}_{kl}$ over probabilities from group k is denoted $A_{\mathbf{P}_k}$, and the corresponding l-dimensional subspace is denoted $B_{\mathbf{P}_k}$.

Note that $A_{\mathbf{P}_k}$ includes the implicit constraint $\sum_i \tilde{P}_{ki} = 1$.

The constrained belief distributions are obtained by conditioning the original belief distributions on the total set of constraints:

---

[1] Unless explicitly stated otherwise, bold face symbols denote vectors throughout this paper.

[2] It is implicit that these constraints are coherent so that, for example, if $\tilde{U}_1 < \tilde{U}_2$ and $\tilde{U}_2 < \tilde{U}_3$, then it cannot hold that $\tilde{U}_3 \leq \tilde{U}_1$.

Definition 7. The constrained belief distributions are given by

$$(U_1, \ldots, U_n)' = (\tilde{U}_1, \ldots, \tilde{U}_n)' | A_{\mathbf{U}}$$

for the utilities, and by

$$(P_{k1}, \ldots, P_{kl})' = (\tilde{P}_{k1}, \ldots, \tilde{P}_{kl})' | A_{\mathbf{P}_k}$$

for probabilities of group k.

Our real interest lies in the constrained variables, since these take into account both the originally defined belief distributions and the total set of constraints for those distributions. Because the constraints introduce dependencies, one needs to operate on the joint belief distributions. If the decision maker has some explicit beliefs concerning interdependencies, not captured by the constraint sets $A_{\mathbf{U}}$ and $A_{\mathbf{P}_k}$, joint belief distributions should be specified already from the outset. Otherwise, which should be the more common scenario, the constraint sets contain all available information on dependencies, and the marginal unconstrained belief distributions are independent of each other. This independence then allows for easy calculation of the required joint density function for the unconstrained belief distributions. Exemplifying with the utilities, one gets

$$f_{\tilde{U}_1, \ldots, \tilde{U}_n}(u_1, \ldots, u_n) = f_{\tilde{U}_1}(u_1) \cdots f_{\tilde{U}_n}(u_n) \,. \tag{1}$$

The joint density function for the constrained belief distributions is obtained by reducing the support of the original belief distributions to $B_{\mathbf{U}}$ (in the case of utilities), and scaling up the density function for all points in $B_{\mathbf{U}}$ so that the function integrates to one in its support. This is a multivariate equivalent to truncating a univariate random variable. The joint density function for $\mathbf{U} = (U_1, \ldots, U_n)'$ is therefore

$$f_{\mathbf{U}}(\mathbf{u}) = \frac{f_{\tilde{\mathbf{U}}}(\mathbf{u})}{\Pr(\tilde{\mathbf{U}} \in B_{\mathbf{U}})} = \frac{f_{\tilde{\mathbf{U}}}(\mathbf{u})}{\int \cdots \int_{B_{\mathbf{U}}} f_{\tilde{\mathbf{U}}}(\mathbf{u}) \, d\mathbf{u}} \quad \text{for } \mathbf{u} \in B_{\mathbf{U}} \,. \tag{2}$$

The corresponding density for a probability group $\mathbf{P_k} = (P_{k1}, \ldots, P_{kl})'$ is

$$f_{\mathbf{P_k}}(\mathbf{p_k}) = \frac{f_{\tilde{\mathbf{P}}_k}(\mathbf{p_k})}{\Pr(\tilde{\mathbf{P}}_k \in B_{\mathbf{P}_k})} = \frac{f_{\tilde{\mathbf{P}}_k}(\mathbf{p_k})}{\int \cdots \int_{B_{\mathbf{P}_k}} f_{\tilde{\mathbf{P}}_k}(\mathbf{p_k}) \, d\mathbf{p_k}} \quad \text{for } \mathbf{p_k} \in B_{\mathbf{P}_k} \,. \tag{3}$$

If the original belief distributions were defined in terms of marginal distributions rather than as a joint one, it is possible to express (2) and (3) even more explicitly. For the utilities one obtains by combining (1) and (2)

$$f_{\mathbf{U}}(\mathbf{u}) = \frac{f_{\tilde{U}_1}(u_1) \cdots f_{\tilde{U}_n}(u_n)}{\int \cdots \int_{B_{\mathbf{U}}} f_{\tilde{U}_1}(u_1) \cdots f_{\tilde{U}_n}(u_n) \, d\mathbf{u}} \quad \text{for } \mathbf{u} \in B_{\mathbf{U}} \,. \tag{4}$$

Combining Second-Order Belief Distributions with Qualitative Statements

3.2 Marginal Constrained Belief Distributions

To see how the constraints altered the belief distribution over some variable, for example the j:th utility, it is necessary to compute the marginal constrained belief distribution U_j and compare this to the original marginal belief distribution $\tilde{U}_j$.

Any marginal distribution can be obtained by integrating the joint distribution over all variables except the one of interest. If $\mathbf{u_j^-} = (u_1, \ldots, u_{j-1}, u_{j+1}, \ldots, u_n)'$, then the marginal distribution of U_j is given by

$$f_{U_j}(u_j) = \int \cdots \iint \cdots \int_{B_U^{-j}} \left[\frac{f_{\tilde{U}}(\mathbf{u})}{\int \cdots \int_{B_U} f_{\tilde{U}}(\mathbf{u}) \, d\mathbf{u}} \right] d\mathbf{u_j^-} \quad \text{for } u_j \in B_U^j , \quad (5)$$

where B_U^{-j} is the $n-1$-dimensional subspace of B_U that arises by removing its j:th dimension.

Example 1. Consider a decision tree where the belief distributions over two utilities u_a and u_b are both the standard uniform distribution $U(0, 1)$:

$$f_{\tilde{U}_j}(u_j) = 1 \quad \text{for } 0 \le u_j \le 1 \text{ and } j \in \{a, b\} ,$$

and where there is only one constraint $\tilde{U}_a \le \tilde{U}_b$. Combining (1) and (5), the marginal belief distribution over u_b under the constraint can be computed:

$$f_{U_b}(u_b) = \int_0^{u_b} \left[\frac{1 \times 1}{\int_0^1 \int_0^{u_b} 1 \times 1 \, du_a du_b} \right] du_a = \int_0^{u_b} \frac{du_a}{1/2} = 2u_b \quad \text{for } u_b \ge u_a .$$

We recognize that $U_b \sim \text{Beta}(2, 1)$, which is a special case of a more general result (see Theorem 3). The shift in marginal belief distribution over u_b, imposed by the constraint, is depicted in Fig. 1. This marked alteration of the marginal belief imposed by the constraint is a demonstration that constraints carry a substantial amount of information about the decision situation at hand. $\qquad\square$

3.3 Belief Distribution Over Expected Utility

The quantity of main interest here is the expected utility for a given decision alternative, or the difference in expected utility between two alternatives. Within our proposed framework, the specified belief distributions over utilities and probabilities, in combination with the sets of comparative constraints, will all affect the resulting belief distribution over the expected utility.

The resulting distributions tend to become more and more warped around the mean as the depth and breadth of the tree increases. This phenomenon is due in part

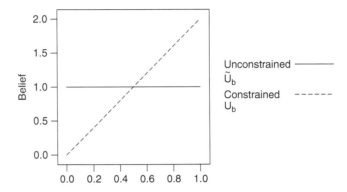

Fig. 1 Results from Example 1, where the considered decision tree contains two utilities whose unconstrained belief distributions were both standard uniform. Following the constraint $\tilde{U}_a \leq \tilde{U}_b$, the resulting (constrained) belief U_b over the second utility is now Beta(2, 1)

to the multiplication of distributions that takes place from root to leaf, and in part to the effects of convolution of the resulting leaf distributions Sundgren et al. (2009).

In view of Definition 4, if we impose belief distributions over the variables, the expected utility is really a transformation of random vectors. Therefore, to be able to analytically derive the resulting belief distribution over the expected utility of a given alternative, we need the following central result from probability theory:

Theorem 1 (The transformation theorem). *Let* $\mathbf{X} = (X_1, \ldots, X_n)'$ *be a continuous random vector with density function* $f_\mathbf{X}(\mathbf{x})$ *and domain* $V \subset \mathbb{R}^n$. *Let* $g = (g_1, \ldots, g_n)$ *be a bijection from* V *to a set* $W \subset \mathbb{R}^n$, *and define* $\mathbf{Y} = g(\mathbf{X})$. *Assume that g and its inverse h are both continuously differentiable. Then, the density function of* $\mathbf{Y}$ *is*

$$f_\mathbf{Y}(\mathbf{y}) = f_\mathbf{X}(h_1(\mathbf{y}), \ldots, h_n(\mathbf{y})) \times |\det(\mathbf{J})| \quad for\ \mathbf{y} \in W,$$

where

$$\mathbf{J} = \begin{pmatrix} \frac{\partial x_1}{\partial y_1} & \cdots & \frac{\partial x_1}{\partial y_n} \\ \vdots & \ddots & \vdots \\ \frac{\partial x_n}{\partial y_1} & \cdots & \frac{\partial x_n}{\partial y_n} \end{pmatrix}$$

is the Jacobian of the transformation. □

Theorem 1 can be applied to the expected utility of some alternative A_i:

Theorem 2. *Given a decision tree T, for the branch corresponding to* A_i, *label the (constrained) belief distributions over all utilities by* $\mathbf{U} = (U_1, \ldots, U_s)'$, *and label the belief distributions over all probabilities by* $\mathbf{P} = (P_1, \ldots, P_r)'$. *Further, let* $\mathbf{U}^- = (U_1, \ldots, U_{s-1})'$. *Finally, denote the index set for probabilities leading up*

Combining Second-Order Belief Distributions with Qualitative Statements   75

to u_j by C_j, and let $\Psi_j = \prod_{i \in C_j} P_i^3$. Then the belief distribution over $E(A_i)$ is given by

$$f_{E(A_i)}(z) = \int \cdots \int f_{\mathbf{P},\mathbf{U}}\left(\mathbf{p}, \mathbf{u}^-, \frac{z - \sum_{i=1}^{s-1} \psi_i u_i}{\psi_s}\right) \frac{1}{\psi_s} \, d\mathbf{p} \, d\mathbf{u}^- .$$

Proof. Let $Z = E(A_i)$ and consider the following transformation:

$$\begin{cases} (Y_1, \ldots, Y_r)' & = \mathbf{P} \\ (Y_{r+1}, \ldots, Y_{r+s-1})' = \mathbf{U}^- \\ Y_{r+s} & = Z = \Psi_s U_s + \sum_{i=1}^{s-1} \Psi_i U_i \end{cases}$$

which has the following inverse:

$$\begin{cases} \mathbf{P} = (Y_1, \ldots, Y_r)' \\ \mathbf{U} = \left(Y_{r+1}, \ldots, Y_{r+s-1}, \dfrac{Y_{r+s} - \sum_{i=1}^{s-1} \Psi_i Y_{r+i}}{\psi_s}\right)' \end{cases} .$$

The Jacobian of this transformation has the following determinant:

$$\det(\mathbf{J}) = \begin{vmatrix} \dfrac{\partial p_1}{\partial y_1} & \cdots & \dfrac{\partial p_1}{\partial y_{r+s}} \\ \vdots & \ddots & \vdots \\ \dfrac{\partial p_r}{\partial y_1} & \cdots & \dfrac{\partial p_r}{\partial y_{r+s}} \\ \dfrac{\partial u_1}{\partial y_1} & \cdots & \dfrac{\partial u_1}{\partial y_{r+s}} \\ \vdots & \ddots & \vdots \\ \dfrac{\partial u_s}{\partial y_1} & \cdots & \dfrac{\partial u_s}{\partial y_{r+s}} \end{vmatrix} = \begin{vmatrix} 1 & 0 & \cdots & 0 \\ 0 & 1 & \cdots & 0 \\ \vdots & \vdots & \ddots & \vdots \\ e_1 & e_2 & \cdots & 1/\psi_s \end{vmatrix} = 1/\psi_s$$

where $e_1, \ldots, e_{r+s-1}$ are partial derivatives that do not contribute to the determinant and therefore do not need to be calculated. According to Theorem 1, the joint density of $\mathbf{Y} = (Y_1, \ldots, Y_{r+s})'$ is

$$f_{\mathbf{Y}}(\mathbf{y}) = f_{\mathbf{P},\mathbf{U}}\left(\mathbf{p}, \mathbf{u}^-, \frac{z - \sum_{i=1}^{s-1} \psi_i u_i}{\psi_s}\right) \frac{1}{\psi_s}$$

---

[3] That is, Ψ_j is the aggregated belief distribution over the probability at the leaf of the j:th utility.

for some domain W. The marginal belief distribution of $E(A_i) = Z$ is derived by integrating out all other variables:

$$f_{E(A_i)}(z) = \int \cdots \int f_{\mathbf{P},\mathbf{U}}\left(\mathbf{p}, \mathbf{u}^-, \frac{z - \sum_{i=1}^{s-1} \psi_i u_i}{\psi_s}\right) \frac{1}{\psi_s} d\mathbf{p}\, d\mathbf{u}^-$$

where integration is over some subspace of $[0, 1]^{r+s-1}$ and the domain of $f_{E(A_i)}$ can be called W'. □

A few remarks are called upon. Firstly, $E(A_i)$ could have been expressed in terms of any U_j, however the choice of U_s is notationally convenient. Secondly, it should be noted that each ψ_j is not a constant, but rather a product of p_is. Finally, since $f_{\mathbf{P},\mathbf{U}}$ is the joint (constrained) density for the beliefs over all utilities and probabilities, it can be factorized into groups of independent variables. Specifically, all utilities will be independent of all probabilities, and probabilities from different groups will also be independent.

The following example shows how the transformation works in practice for an apparently simple situation.

Example 2. Consider the decision tree for some alternative A_i given in Fig. 2. The belief distribution over the expected utility is given by

$$E(A_i) = P_{11}P_{111}U_{111} + P_{11}P_{112}U_{112} + P_{12}U_{12}$$
$$= P_{11}P_{111}U_{111} + P_{11}(1 - P_{111})U_{112} + (1 - P_{11})U_{12} \ .$$

For simplicity, since it reduces the dimensionality of the problem, we assume that the reformulation of P_{112} and P_{12} is coherent with the constraints used. The transformation becomes

$$\begin{cases} (Y_1, Y_2, Y_3, Y_4)' = (P_{11}, P_{111}, U_{111}, U_{112})' \\ Y_5 \qquad\qquad = P_{11}P_{111}U_{111} + P_{11}(1 - P_{111})U_{112} + (1 - P_{11})U_{12} \end{cases}$$

with inverse

$$\begin{cases} (P_{11}, P_{111}, U_{111}, U_{112})' = (Y_1, Y_2, Y_3, Y_4)' \\ U_{12} \qquad\qquad\qquad = \dfrac{Y_5 - Y_1 Y_2 Y_3 - Y_1(1 - Y_2)Y_4}{1 - Y_1} \end{cases} \ .$$

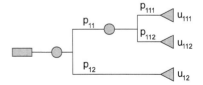

Fig. 2 Decision tree considered in Example 2

The determinant of the Jacobian is given by

$$
\det(\mathbf{J}) =
\begin{vmatrix}
\dfrac{\partial p_{11}}{\partial y_1} & \dfrac{\partial p_{11}}{\partial y_2} & \dfrac{\partial p_{11}}{\partial y_3} & \dfrac{\partial p_{11}}{\partial y_4} & \dfrac{\partial p_{11}}{\partial y_5} \\[2mm]
\dfrac{\partial p_{111}}{\partial y_1} & \dfrac{\partial p_{111}}{\partial y_2} & \dfrac{\partial p_{111}}{\partial y_3} & \dfrac{\partial p_{111}}{\partial y_4} & \dfrac{\partial p_{111}}{\partial y_5} \\[2mm]
\dfrac{\partial u_{111}}{\partial y_1} & \dfrac{\partial u_{111}}{\partial y_2} & \dfrac{\partial u_{111}}{\partial y_3} & \dfrac{\partial u_{111}}{\partial y_4} & \dfrac{\partial u_{111}}{\partial y_5} \\[2mm]
\dfrac{\partial u_{112}}{\partial y_1} & \dfrac{\partial u_{112}}{\partial y_2} & \dfrac{\partial u_{112}}{\partial y_3} & \dfrac{\partial u_{112}}{\partial y_4} & \dfrac{\partial u_{112}}{\partial y_5} \\[2mm]
\dfrac{\partial u_{12}}{\partial y_1} & \dfrac{\partial u_{12}}{\partial y_2} & \dfrac{\partial u_{12}}{\partial y_3} & \dfrac{\partial u_{12}}{\partial y_4} & \dfrac{\partial u_{12}}{\partial y_5}
\end{vmatrix}
=
\begin{vmatrix}
1 & 0 & 0 & 0 & 0 \\
0 & 1 & 0 & 0 & 0 \\
0 & 0 & 1 & 0 & 0 \\
0 & 0 & 0 & 1 & 0 \\
e_1 & e_2 & e_3 & e_4 & \dfrac{1}{(1-y_1)}
\end{vmatrix}.
$$

This evaluates to $\frac{1}{1-y_1} = \frac{1}{1-p_{11}}$, which coincides with the general description above, for $\psi_s = p_{12} = 1 - p_{11}$. The joint density of $\mathbf{Y}$ is given, for some domain W, by

$$
f_{\mathbf{Y}}(\mathbf{y}) = f_{P_{11},P_{111},U_{111},U_{112},U_{12}}\left(p_{11}, p_{111}, u_{111}, u_{112}, \frac{z-v}{1-p_{11}} \right) \frac{1}{1-p_{11}},
$$

with $v = p_{11} p_{111} u_{111} + p_{11}(1-p_{111})u_{112}$. This density can be factorized with respect to independent variables. Assuming that the utilities are not independent under the given constraints, one obtains

$$
f_{\mathbf{Y}}(\mathbf{y}) = f_{P_{11}}(p_{11}) f_{P_{111}}(p_{111}) f_{U_{111},U_{112},U_{12}}\left(u_{111}, u_{112}, \frac{z-v}{1-p_{11}} \right) \frac{1}{1-p_{11}}.
$$

Finally, the resulting belief distribution over $E(A_i)$ can be calculated from

$$
f_{E(A_i)}(z) = \iiiint f_{\mathbf{Y}}(\mathbf{y}) \, \mathrm{d}p_{11} \, \mathrm{d}p_{111} \, \mathrm{d}u_{111} \, \mathrm{d}u_{112}. \qquad \square
$$

However, the complexity of this operation is very high since both the integrand itself as well as the integration limits might be very difficult to derive. And even if these steps have been carried out, the sheer dimensionality of the integration might be prohibitive in practice. So in real cases more efficient methods of calculating this must be utilized. As we shall see, in certain cases this can be done at relative computational ease, even for moderately large trees.

4 Simulation from Expected Utilities

As shown in Theorem 2, the resulting distribution over the expected utility of an alternative can be expressed in terms of a multidimensional integral. However, in general such integrals should rarely be possible to compute analytically, and so to be able to benefit from the theoretical results presented thus far, approximate methods are called upon. One possibility would be to make use of numerical integration techniques. We have, however, opted for another solution, namely to use simulation.

To be able to simulate from the resulting belief distributions over the expected utilities of the alternatives, or from some function thereof, we need to sample from the respective constrained belief distributions over probabilities and utilities. Any utility is independent of any probability, and any two probabilities from separate groups are independent, because of the restricted set of constraints allowed. Therefore we can sample from the joint belief distribution over the utilities and from the belief distributions over the various probability groups separately.

The most straightforward approach would be to utilize (2) and (3) through a simple form of rejection sampling von Neumann (1963). Specifically, exemplifying with utilities, we could simply apply the following scheme:

Algorithm 1 (Rejection sampling).

Repeat until m samples are retained in total:
Sample $\tilde{\mathbf{u}} = (\tilde{u}_1, \ldots, \tilde{u}_n)'$ from $(\tilde{U}_1, \ldots, \tilde{U}_n)'$ and retain the sample if $\tilde{\mathbf{u}} \in B_{\mathbf{U}}$.

$\square$

Even if $\tilde{U}_1, \ldots, \tilde{U}_n$ are independent, so that sampling is straightforward, this approach has one major drawback: As n grows, the probability that a sample is accepted approaches 0, which in effect means that the real number of samples m' required to collect a nominal number of m samples approaches infinity. If the intended application is not interactive, however, this straightforward simulation scheme might suffice.

Another generic approach to sampling from a multivariate distribution is to factorize the joint density function into a series of univariate, conditional distributions. For our joint distribution of beliefs over the utilities this would correspond to the following:

$$f_{U_1,\ldots,U_n}(u_1,\ldots,u_n) = f_{U_n}(u_n) f_{U_{n-1}|U_n=u_n}(u_{n-1}) \cdots f_{U_1|U_2=u_2,\ldots,U_n=u_n}(u_1) . \quad (6)$$

For each of the univariate distributions, sampling can be performed according to the inverse transformation method Devroye (1986). This method requires draws from a standard uniform distribution to be inserted into the inverse of the distribution function. Therefore, in practice, this approach can only be really effective if the distribution functions that correspond to the density functions in (6) can be analytically inverted. The complete sampling scheme is summarized in the following:

Algorithm 2 (Multivariate inverse transform sampling).

---

1. Split up $f_{U_1,\ldots,U_n}$ according to (6).
2. Derive the distribution functions $F_{U_n}, F_{U_{n-1}|U_n=u_n}, \ldots, F_{U_1|U_2=u_2,\ldots,U_n=u_n}$.
3. Derive the inverse distribution functions $F_{U_n}^{-1}, F_{U_{n-1}|U_n=u_n}^{-1}, \ldots, F_{U_1|U_2=u_2,\ldots,U_n=u_n}^{-1}$.
4. Draw n vectors $\mathbf{x}_i = (x_i^1, \ldots, x_i^m)'$ of samples from $X \sim U(0, 1)$.

Combining Second-Order Belief Distributions with Qualitative Statements

5a. Sample from U_n:

$$(u_n^1, \ldots, u_n^m)' = (F_{U_n}^{-1}(x_n^1), \ldots, F_{U_n}^{-1}(x_n^m))'$$

5b. Sample from U_{n-1}:

$$(u_{n-1}^1, \ldots, u_{n-1}^m)' = (F_{U_{n-1}|U_n=u_n^1}^{-1}(x_{n-1}^1), \ldots, F_{U_{n-1}|U_n=u_n^m}^{-1}(x_{n-1}^m))'$$

$$\vdots$$

5n. Sample from U_1 :

$$(u_1^1, \ldots, u_1^m)' = (F_{U_1|U_2=u_2^1,\ldots,U_n=u_n^1}^{-1}(x_1^1), \ldots, F_{U_1|U_2=u_2^m,\ldots,U_n=u_n^m}^{-1}(x_1^m))' \qquad \square$$

The following example is supposed to delineate the fundamental principles of Algorithm 2:

Example 3. Consider again the situation in Example 1. According to (4), the joint constrained belief distribution is given by

$$f_{U_a,U_b}(u_a, u_b) = \frac{f_{\tilde{U}_a}(u_a) f_{\tilde{U}_b}(u_b)}{\int_0^1 \int_0^{u_b} f_{\tilde{U}_a}(u_a) f_{\tilde{U}_b}(u_b)\, du_a du_b} = \frac{1 \times 1}{\int_0^1 \int_0^{u_b} 1 \times 1\, du_a du_b}$$

$$= \frac{1}{1/2} = 2$$

for $u_b \geq u_a$. As will be demonstrated in Sect. 4.1, f_{U_a,U_b} can be split up as follows:

$$f_{U_a,U_b}(u_a, u_b) = f_{U_b}(u_b) f_{U_a|U_b=u_b}(u_a) = 2u_b \times (1/u_b) .$$

The corresponding distribution functions and their inverses are given by

$$\begin{cases} F_{U_b}(u_b) &= u_b^2 \text{ for } 0 \leq u_b \leq 1 \\ F_{U_a|U_b=u_b}(u_a) &= \frac{u_a}{u_b} \text{ for } 0 \leq u_a \leq u_b \end{cases} \text{ and } \begin{cases} F_{U_b}^{-1}(x) &= \sqrt{x} \text{ for } 0 \leq x \leq 1 \\ F_{U_a|U_b=u_b}^{-1}(x) &= xu_b \text{ for } 0 \leq x \leq 1 \end{cases}.$$

Assume we want $m = 5$ draws, and that we sample $\mathbf{x}_b = (0.94, 0.13, 0.83, 0.47, 0.55)'$ and $\mathbf{x}_a = (0.18, 0.70, 0.57, 0.17, 0.94)'$ from $X \sim U(0, 1)$. This then yields

$$\begin{bmatrix} u_b^1 \\ u_b^2 \\ u_b^3 \\ u_b^4 \\ u_b^5 \end{bmatrix} = \begin{bmatrix} \sqrt{0.94} \\ \sqrt{0.13} \\ \sqrt{0.83} \\ \sqrt{0.47} \\ \sqrt{0.55} \end{bmatrix} = \begin{bmatrix} 0.97 \\ 0.36 \\ 0.91 \\ 0.69 \\ 0.74 \end{bmatrix} \text{ and } \begin{bmatrix} u_a^1 \\ u_a^2 \\ u_a^3 \\ u_a^4 \\ u_a^5 \end{bmatrix} = \begin{bmatrix} 0.18 \times u_b^1 \\ 0.70 \times u_b^2 \\ 0.57 \times u_b^3 \\ 0.17 \times u_b^4 \\ 0.94 \times u_b^5 \end{bmatrix} = \begin{bmatrix} 0.17 \\ 0.25 \\ 0.52 \\ 0.12 \\ 0.70 \end{bmatrix} .$$

Clearly, to obtain samples that accurately represent the $(U_a, U_b)'$ distribution, one would need to set m considerably higher than in this illustrative example. □

An implicit assumption in Algorithm 2 is that each distribution function be strictly increasing, since otherwise its inverse would not exist. However, this criterion is fulfilled for the vast majority of distributions and should not present any issues in practice. There are other complicating matters of greater practical concern, such that for an arbitrary joint distribution, even step 1 in the scheme might be difficult to accomplish. Should step 1 be successful, step 3 might still prove difficult or impossible. Should this be the case, one could use numerical techniques for inverting the distribution functions, though this would slow down the procedure quite dramatically. We shall therefore make use of one particular distribution, the uniform distribution, where it is possible to carry out the complete scheme. The uniform distribution is appealing since it can be thought of as a direct probabilistic equivalent to using intervals. In other words, if the decision maker would choose an interval $[a, b]$, a natural choice in the context of the current approach would be the $U(a, b)$ distribution.

4.1 Special Case of Uniform Distributions

One important sub-case is when the distributions are uniform. We describe here an efficient way of sampling from the resulting belief distribution over the expected utility if one (a) uses arbitrary independent uniform belief distributions over all utilities; (b) uses an ordering constraint within each of an arbitrary number of disjoint subsets of the utilities[4]; and (c) uses no other constraints. Because there are no relations between utilities from different subsets, independence allows sampling from each subset separately. We can therefore, without loss of generality, describe the special case where there is just one subset $\{u_1, \ldots, u_n\}$ containing all utilities.

For the probabilities, the implicit sum constraint is a complicating factor. We will here use the Dirichlet distribution to sample from a probability group, and not impose any relational constraints. The Dirichlet distribution has been previously used in this context Ekenberg et al. (2007) and it does relate to the above: If we use standard uniform distributions over $p_{k1}, \ldots, p_{kl}$ and impose no further constraints than the implicit sum constraint, then $(P_{k1}, \ldots, P_{kl})'$ will be distributed according to the Dirichlet$(\alpha_1 = 1, \ldots, \alpha_l = 1)$ distribution. Sampling from a Dirichlet distribution is straightforward using standard statistical software.

Assume, for brevity, that we have named the utilities after their ordering. Thus, the constraint under consideration is $A_U : \tilde{U}_1 \leq \ldots \leq \tilde{U}_n$. The following result, given with a standard proof, is useful:

---

[4] The union of these subsets need not equate the set of all utilities.

Combining Second-Order Belief Distributions with Qualitative Statements

Theorem 3. *Let $X_1, \ldots, X_n$ be a sample from a distribution with density function f and distribution function F. Then the density function for the largest observation $X_{(n)}$ is $f_{\text{Beta}(n,1)}(F(x))f(x)$.*

Proof. The distribution function for $X_{(n)}$ is given by

$$F_{X_{(n)}}(x) = \Pr(X_1 \leq x, \ldots, X_n \leq x) = \prod_{k=1}^{n} \Pr(X_k \leq x) = (F(x))^n .$$

Differentiation then yields the density function:

$$f_{X_{(n)}}(x) = n(F(x))^{n-1} f(x) = f_{\text{Beta}(n,1)}(F(x))f(x) . \qquad \square$$

If all unconstrained belief distributions $\tilde{U}_j$ are $U(a, b)$, then the resulting belief distribution U_n is equivalent to $X_{(n)}$ in Theorem 3. It follows that

$$f_{U_n}(u_n) = n \left(\frac{u_n - a}{b - a} \right)^{n-1} \frac{1}{b - a} = \frac{n(u_n - a)^{n-1}}{(b - a)^n} .$$

Conditional on $U_n = u_n$, all remaining belief distributions are $U(a, b)$ truncated to the interval $[a, u_n]$, which means that $\tilde{U}_j | U_n = u_n \sim U(a, u_n)$ for $j < n$. Since $\tilde{U}_{n-1}$ is now the largest of the remaining variables, Theorem 3 gives

$$f_{U_{n-1}|U_n=u_n}(u_{n-1}) = \frac{(n-1)(u_{n-1} - a)^{n-2}}{(u_n - a)^{n-1}} .$$

By repeating the same argument for all belief distributions down to $\tilde{U}_1$, the joint density for the constrained belief distribution can be factorized, as required:

$$\begin{aligned}
f_{U_1, \ldots, U_n}(u_1, \ldots, u_n) &= \frac{n!}{(b - a)^n} \\
&= \frac{n(u_n - a)^{n-1}}{(b - a)^n} \frac{(n-1)(u_{n-1} - a)^{n-2}}{(u_n - a)^{n-1}} \cdots \frac{2(u_2 - a)}{(u_3 - a)^2} \frac{1}{u_2 - a} \\
&= f_{U_n}(u_n) f_{U_{n-1}|U_n=u_n}(u_{n-1}) \cdots f_{U_2|U_3=u_3}(u_2) f_{U_1|U_2=u_2}(u_1) .
\end{aligned}$$

All corresponding distribution functions can be derived, and it turns out that they are all analytically invertible:

$$\begin{cases} F_{U_n}^{-1}(x) &= x^{1/n}(b - a) + a \quad \text{for } k = n \\ F_{U_k|U_{k+1}=u_{k+1}}^{-1}(x) &= x^{1/k}(u_{k+1} - a) + a \quad \text{for } k = n-1, n-2, \ldots, 1 \end{cases} . \qquad (7)$$

Thus, all prerequisites for applying Algorithm 2 are fulfilled.

It would be a severe limitation to require that $\tilde{U}_1, \ldots, \tilde{U}_n$ follow the same uniform distribution. We can overcome this by once again making use of the fact that a truncated uniform variable still is uniform. Let each variable have its own uniform distribution, $\tilde{U}_j \sim U(a_j, b_j)$, and consider the following scheme:

Algorithm 3.

---

1. Order the set $\{0, a_1, \ldots, a_n, b_1, \ldots, b_n, 1\}$, and define all intervals $I_j = [c_j, d_j]$ from adjacent points in this ordered set. (Note that some points might be identical, in which case no interval results.) Denote the total number of intervals by N^I.
2. Construct all *configurations* C_i, i.e. all ways to distribute $\tilde{U}_1, \ldots, \tilde{U}_n$ between these intervals. Denote by N^C the total number of configurations where A_U can hold, and by N_{ij}^U the number of variables $\tilde{U}_k$ in interval I_j under configuration C_i.
3. FOR $i = 1$ TO $i = N^C$
       FOR $j = 1$ TO $j = N^I$
           FOR $k = 1$ TO $k = N_{ij}^U$
               Calculate $\bar{p}_{ijk} = \Pr(\tilde{U}_k \in I_j) = F_{\tilde{U}_k}(d_j) - F_{\tilde{U}_k}(c_j)$
           END FOR
           Calculate $\bar{p}_{ij} = \Pr(\tilde{U}_1, \ldots, \tilde{U}_{N_{ij}^U} \in I_j) = \prod_{k=1}^{N_{ij}^U} \bar{p}_{ijk}$
           Calculate $\widehat{p}_{ij} = \Pr(A_U \text{ holds in } I_j | C_i) = 1/N_{ij}^U!$
       END FOR
       Calculate $\bar{p}_i = \Pr(C_i) = \prod_{j=1}^{N^I} \bar{p}_{ij}$
       Calculate $\widehat{p}_i = \Pr(A_U | C_i) = \prod_{j=1}^{N^I} \widehat{p}_{ij}$
       Calculate $p_i = \widehat{p}_i \bar{p}_i$
      END FOR
      Calculate $p = \sum_{i=1}^{N^C} p_i$
4. FOR $r = 1$ TO $r = m$
       Sample $\mathbf{x^r} = (x_1^r, \ldots, x_{N^C}^r)'$ from $(X_1, \ldots, X_{N^C})' \sim \text{Multinom}(1; \; p_1/p, \ldots, p_{N^C}/p)$
       FOR the single i corresponding to $x_i^r = 1$
           FOR $j = 1$ TO $j = N^I$
               Draw a sample of the variables in I_j under configuration C_i, by setting $a = c_j$ and $b = d_j$ in (7), and using Algorithm 2
           END FOR
       END FOR
      END FOR   □

In other words, Algorithm 3 simulates from $f_U(\mathbf{u})$ by treating it as a mixture density over the configurations, with suggested mixture parameters p_i/p. The validity of this approach is asserted by the following theorem:

Combining Second-Order Belief Distributions with Qualitative Statements

Theorem 4. *Consider a decision tree T with a set of utility variables whose corresponding belief distributions $\tilde{U}_1, \ldots, \tilde{U}_n$ are independent and distributed as $\tilde{U}_j \sim U(a_j, b_j)$. Under the constraint $A_U : \tilde{U}_1 \leq \ldots \leq \tilde{U}_n$, Algorithm 3 can be used to sample from the resulting belief distribution $\mathbf{U} = (U_1, \ldots, U_n)'$.*

Proof. By the law of total probability and Bayes' theorem, and following the notation of Algorithm 3:

$$
\begin{aligned}
f_\mathbf{U}(\mathbf{u}) &= \sum_{i=1}^{N^C} \left[f_{\tilde{\mathbf{U}}|A_U, C_i}(\mathbf{u}) \times \Pr(C_i|A_U) \right] \\
&= \sum_{i=1}^{N^C} \left[f_{\tilde{\mathbf{U}}|A_U, C_i}(\mathbf{u}) \times \frac{\Pr(A_U|C_i)\Pr(C_i)}{\Pr(A_U)} \right] \\
&= \sum_{i=1}^{N^C} \left[f_{\tilde{\mathbf{U}}|A_U, C_i}(\mathbf{u}) \times \frac{\Pr(A_U|C_i)\Pr(C_i)}{\sum_{i=1}^{N^C} \Pr(A_U|C_i)\Pr(C_i)} \right] \\
&= \sum_{i=1}^{N^C} \left[f_{\tilde{\mathbf{U}}|A_U, C_i}(\mathbf{u}) \times \frac{p_i}{p} \right],
\end{aligned}
$$

which shows that p_i/p is the correct mixture parameter for configuration C_i.

Finally the claim follows by considering the distributions $f_{\tilde{\mathbf{U}}|A_U, C_i}(\mathbf{u})$. Under C_i, but not yet considering A_U, all variables in I_j are distributed as $U(c_j, d_j)$. Thus, since they are equidistributed, A_U can be introduced through Algorithm 2 and (7). Further, because the relative order of any two variables from different intervals is fixed given C_i, they are independent, and the complete distribution $(U_1, \ldots, U_n)'$ can be obtained by repeated interval-wise sampling from configurations drawn according to a Multinom$(1;\ p_1/p, \ldots, p_{N^C}/p)$ distribution. □

Note that the algorithm only includes configurations where A_U can hold. This is for practical reasons, and the validity follows immediately since $\Pr(A_U|C_i) = 0$ for any configuration not fulfilling this criterion.

If there are n variables and N^I resulting intervals, the total number of configurations where A_U can hold is $\binom{n+N^I-1}{N^I-1}$. Therefore, in practice, Algorithm 3 is only efficient if N^I is kept relatively low, which means that the variables need to share a few common endpoints. In our experience, it is very feasible to allow a resolution of 0.2, meaning that endpoints are chosen from $\{0, 0.2, 0.4, 0.6, 0.8, 1\}$, even for n relatively large (about 30). Obviously the computational complexity will also depend on how many samples that are desired in the simulation. Note that the tree as a whole can be much larger than n; it is really the size of the largest set of ordered utility variables that matters.

Example 4. Consider the decision tree with two alternatives given in Fig. 3. Assume that the decision maker has specified the following:

Fig. 3 Decision tree considered in Example 4

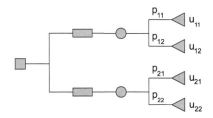

$$\begin{cases} u_{11}, u_{12}, u_{21} \in [0, 1] \text{ and } u_{22} \in [0, 0.5] \\ p_{11}, p_{12}, p_{21}, p_{22} \in [0, 1] \\ u_{11} \geq u_{21} \text{ and } u_{12} \geq u_{22} \end{cases}.$$

The intuitive interpretation of this situation is that alternative 1 is superior to alternative 2. However, an interval analysis would yield $E(A_1) - E(A_2) \in [-1, 1]$, indicating no discrimination whatsoever.

Now consider the approach described here, and assume that the decision maker instead specifies

$$\begin{cases} \tilde{U}_{11}, \tilde{U}_{12}, \tilde{U}_{21} \sim U(0, 1) \text{ and } \tilde{U}_{22} \sim U(0, 0.5) \\ (\tilde{P}_{11}, \tilde{P}_{12})' \sim \text{Dirichlet}(1, 1) \text{ and } (\tilde{P}_{21}, \tilde{P}_{22})' \sim \text{Dirichlet}(1, 1) \\ \tilde{U}_{11} \geq \tilde{U}_{21} \text{ and } \tilde{U}_{12} \geq \tilde{U}_{22} \end{cases}.$$

The resulting belief distribution over $(E(A_1), E(A_2))'$ was simulated using Algorithm 3 separately for $(\tilde{U}_{11}, \tilde{U}_{21})'$ and $(\tilde{U}_{12}, \tilde{U}_{22})'$.[5] $(\tilde{P}_{11}, \tilde{P}_{12})'$ and $(\tilde{P}_{21}, \tilde{P}_{22})'$ were also simulated separately. 1 million samples were drawn, and the results are displayed in Fig. 4. It is apparent from these figures that the belief in alternative 1 exceeds that of alternative 2 quite clearly, even though there is some overlap. Further, the simulation provides us with a straightforward quantification of the degree of discrimination. For example, we can extract a 90% probability interval for $E(A_1) - E(A_2)$ from the 5:th and 95:th percentiles of the simulated values, which happens to be $[0.03, 0.71]$. The interpretation of this interval is direct: Under the beliefs and constraints specified by the decision maker, there is a 90% probability that the difference in expected utility between alternatives 1 and 2 lies between 0.03 and 0.71. Based on this analysis we can confidently discriminate between the alternatives, and the result is coherent with intuition. □

A desirable extension of the proposed method would be to allow for other classes of distributions than the uniform. There are two inherent obstacles in the framework that precludes this. First, as can be realized from Theorem 3, the class of density functions for the distribution in question must obey rather strict form conventions, in

---

[5] For $(\tilde{U}_{11}, \tilde{U}_{21})'$, there is only one interval to consider, and so it really suffices with Algorithm 2. Note that $\tilde{U}_{11}$ and $\tilde{U}_{21}$ are precisely $\tilde{U}_b$ and $\tilde{U}_a$, respectively, from Examples 1 and 3.

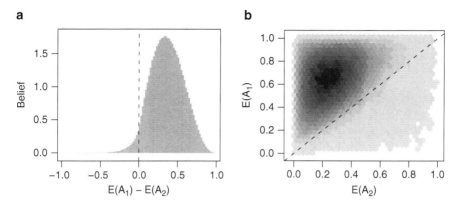

Fig. 4 Results from the simulation of Example 4. (**a**) Histogram over the simulated belief for the difference $E(A_1) - E(A_2)$. (**b**) Two-dimensional histogram over the simulated beliefs for $E(A_1)$ and $E(A_2)$. Darker color indicates a higher density of points

order to yield analytically invertible distribution functions for the ordered variables. Second, (7), as well as the fundamental idea behind Algorithm 3, rest on a subtle, yet very powerful property of the uniform distribution: A truncated uniform variable is still uniform, with parameters equal to the endpoints of the truncated interval. As far as we are aware, no other continuous distribution shares this property. Therefore, to be able to derive similar methods for other classes of distributions, fundamentally different approaches would be needed.

5 Summary and Conclusions

There is often a need in real-life decision analyses to allow for imprecise statements regarding probabilities and utilities. Various interval-based approaches have been suggested, but the resulting aggregations range within intervals as well, causing an, often unnecessary, information loss. There is therefore a need to extend the representation of the decision situation using more information, but keeping the requirement that the decision maker does not have to be more precise than what is possible.

To this end, we have proposed a solution where qualitative statements, i.e. relations between variables, are combined with second-order information in the form of belief distributions over probability and utility variables. Within the framework of belief distributions, we have demonstrated how qualitative statements translate to constraints, and how these constraints affect both marginal belief distributions as well as the resulting belief distributions over the expected utilities of the alternatives. Further, and of clear practical benefit, we have presented a computationally meaningful method where arbitrary uniform distributions can be used for the utility variables, in combination with ordering relations among these variables. This method rests on a series of fairly subtle, yet fundamental, theoretical arguments.

The results presented here have two distinct advantages. First, while preserving the ability to be deliberately imprecise and including qualitative relations, our approach allows a high degree of discrimination between alternatives compared to what interval-based approaches can accomplish. At the same time, we can make a semantically clear interpretation of the resulting beliefs within a probabilistic framework.

Acknowledgements The authors wish to thank Alina Kuznetsova for valuable suggestions that helped improve the contents of this paper.

References

Choquet, G. (1954). Theory of capacities. *Annales de l'institut Fourier, 5*, 131–295.

Danielson, M., & Ekenberg, L. (1998). A framework for analysing decisions under risk. *European Journal of Operational Research, 104*(3), 474–484.

Danielson, M., & Ekenberg, L. (2007). Computing upper and lower bounds in interval decision trees. *European Journal of Operational Research, 181*(3), 808–816.

Danielson, M., Ekenberg, L., & Larsson, A. (2007). Distribution of expected utility in decision trees. *International Journal of Approximate Reasoning, 46*(2), 387–407.

Danielson, M., Ekenberg, L., & Riabacke, A. (2009). A prescriptive approach to elicitation of decision data. *Journal of Statistical Theory and Practice, 3*(1), 157–168.

Dempster, A. P. (1967). Upper and lower probabilities induced by a multivalued mapping. *Annals of Mathematical Statistics, 38*(2), 325–399.

Devroye, L. (1986). Non-Uniform Random Variate Generation. New York, USA: Springer.

Ding, X., Danielson, M., & Ekenberg, L. (2010). Disjoint programming in computational decision analysis. *Journal of Uncertain Systems, 4*(1), 4–13.

Dubois, D., & Prade, H. (1988). *Possibility Theory.* New York: Plenum Press.

Ekenberg, L. (2000). The logic of conflicts between decision making agents. *Journal of Logic and Computation, 10*(4), 583–602.

Ekenberg, L., Andersson, M., Danielsson, M., & Larsson, A. (2007). Distributions over Expected Utilities in Decision Analysis. In 5th International Symposium on Imprecise Probability: Theories and Applications, Prague, Czech Republic.

Ekenberg, L., Danielson, M., & Thorbiörnson, J. (2006). Multiplicative properties in evaluation of decision trees. *International Journal of Uncertainty, Fuzziness and Knowledge-Based Systems, 14*(3), 293–316.

Ekenberg, L., & Thorbiörnson, J. (2001). Second-order decision analysis. *International Journal of Uncertainty, Fuzziness and Knowledge-Based Systems, 9*(1), 13–37.

Ekenberg, L., Thorbiörnson, J., & Baidya, T. (2005). Value differences using second-order distributions. *International Journal of Approximate Reasoning, 38*(1), 81–97.

Ellsberg, D. (1961). Risk, ambiguity, and the savage axioms. *Quarterly Journal of Economics, 75*(4), 643–669.

Good, I. J. (1962). Subjective probability as the measure of a non-measurable set. In *Logic, Methodology, and Philosophy of Science: Proceedings of the 1960 International Congress,* pp. 319–329. Stanford University Press.

Jaffray, J. Y. (1999). Rational Decision Making With Imprecise Probabilities. In *1st International Symposium on Imprecise Probabilities and Their Applications, Ghent, Belgium.*

von Neumann, J. (1963). Various techniques used in connection with random digits. In A. H. Taub (Ed.), *John von Neumann, Collected Works*, Vol. V. New York, USA: MacMillan.

Shafer, G. (1976). *A Mathematical Theory of Evidence*. Princeton: Princeton University Press.

Shapira, Z. (1995). *Risk Taking: A Managerial Perspective*. Russel Sage Foundation.

Smith, C. A. B. (1961). Consistency in statistical inference and decision. *Journal of the Royal Statistical Society. Series B (Methodological) 23*(1), 1–25.

Sundgren, D., Danielson, M., & Ekenberg, L. (2009). Warp effects on calculating interval probabilities. *International Journal of Approximate Reasoning, 50*(9), 1360–1368.

An Econometric Model Based on the Maxmin Expected Utility Model: An Application to Earthquake Insurance

Toshio Fujimi and Hirokazu Tatano

Abstract This study empirically investigates the influence of ambiguity on consumers' decision to buy a hypothetical earthquake insurance policy. Using survey data, it identifies effects of specific consumer characteristics on their decision based on the Maxmin Expected Utility (MEU) model. We develop an econometric model consistent with the MEU model derived from axioms. Our study provides three main results: First, respondents' preferences for the insurance when faced with 1%, 5%, and 10% appraisal risk are generally inconsistent with expected utility theory. Second, respondents demanded more than a 10% reduction in insurance premium as compensation for accepting each tier of appraisal risk. Third, the required discount is greatest among men who had previously purchased earthquake insurance and had experienced earthquake damage to their houses, and the required discount increases with age and education.

1 Introduction

Decisions are often made under uncertain circumstances. Broadly speaking, there are two types of uncertainties. The first is essentially random and is called "risk." The second, called "ambiguity," arises from imprecise, unreliable, or incomplete information or other factors that prohibit precise quantification of risk. This type of uncertainty leads to the nondeterministic nature of subjective risks. In this regard,

---

T. Fujimi (✉)
Graduate School of Science and Technology, Kumamoto University, Kumamoto, Japan
e-mail: fujimi@kumamoto-u.ac.jp

H. Tatano
Disaster Prevention Research Institute, Kyoto University, Kyoto, Japan
e-mail: tatano@imdr.mbox.media.kyoto-u.ac.jp

Y. Ermoliev et al. (eds.), *Managing Safety of Heterogeneous Systems*, Lecture Notes
in Economics and Mathematical Systems 658, DOI 10.1007/978-3-642-22884-1_5,
© Springer-Verlag Berlin Heidelberg 2012

Camerer and Weber (1992) thought experiment illustrates how ambiguity influences decisions. Suppose that in a game of betting on coins, participants need to choose one of the two coins. A participant wins if the coin lands head up. It is known to the participants that coin A was tossed 1,000 times and landed head up 500 times, while coin B was tossed twice and landed head up once. Even though participants believe both are *honest coins*, many of them choose coin A because they are more confident of its fairness, since they have greater evidence proving that coin A is an honest coin. Ambiguity about probability creates a kind of risk in betting on coin B: the risk of having the wrong belief. The tendency to avoid ambiguity is called "ambiguity aversion."

Ellsberg (1961) originally identified ambiguity aversion, and abundant empirical analyses have confirmed its robustness. However, most empirical studies have examined it only in experimental settings and have not studied the effect of ambiguity on people's evaluation of a real-world public policy under uncertainty. Only recently have econometric models using field data been developed to investigate whether and how individuals' support for a public policy involving risk is influenced by their perception regarding the ambiguity of its risk and their attitudes toward ambiguity in general.

Cameron (2005) has proposed an econometric model based on expected utility theory allowing for ambiguity about future global temperatures. He shows that perceived ambiguity affects willingness to pay (WTP) for programs that mitigate climate change. Riddel and Shaw (2006) have developed a model that incorporates preferences for mortality risk from nuclear waste transport when respondents face ambiguity about the risk and applied it to estimate the welfare cost. They found that negative externalities from the perceived risks and ambiguity of nuclear waste transport may be substantial.

This study presents a theoretically-consistent econometric model addressing the relationships among risk, ambiguity, and preference based on the maxmin expected utility (MEU) model developed by Gilboa and Schmeidler (1989). The MEU model assumes that people make decisions by considering the worst among the set of subjective probabilities that they consider plausible. The MEU model represents risk aversion by concavity of the utility function, and it represents perceived ambiguity and ambiguity aversion by the set of subjective probabilities. By corresponding to the MEU model in all these respects, our model allows clear interpretations of its estimation results.

The remainder of this paper is structured as follows: Sect. 2 briefly describes previous studies on ambiguity and MEU models that can deal with ambiguity. Section 3 explains the household survey data concerning hypothetical earthquake insurance involving appraisal risk. Section 4 explains the data, and Sect. 5 develops the econometric model consistent with the MEU model. Section 6 presents the estimated results. Section 7 presents implications from the results. Section 8 concludes.

2 Ambiguity and the Maxmin Expected Utility Model

Ellsberg (1961) originally identified ambiguity as a counterexample to subjective expected utility (SEU) developed by Savage (1954). SEU has a wide application. If one's preference satisfies several plausible axioms, one can construct a unique subjective probability distribution under Knightian uncertainty and follow expected utility theory by using it. Thus, Savage (1954) argued that the distinction between risk and uncertainty is not essential. However, SEU cannot explain observed decisions under some types of uncertainty, as Ellsberg (1961) hypothetical experiment showed: Suppose that you have to draw from an urn containing 30 red balls and 60 balls in an unspecified combination of black and yellow. There are four lotteries: X, Y, X', and Y'. In lottery X, you win \$100 if a red ball is drawn; in lottery Y, \$100 if a black ball is drawn; in lottery X', \$100 if a red or yellow ball is drawn; and in lottery Y', \$100 if a black or yellow ball is drawn. Otherwise, you win nothing. You must choose between X and Y and also between X' and Y'. You may choose X and Y'. Simple calculation shows that SEU cannot describe your decision through any subjective probability distributions. This tendency in these decisions is called "ambiguity aversion" because objective probability distributions are known in X and Y' and are unknown in X' and Y. Numerous empirical studies have examined its existence and extent and have recognized its robustness (Becker and Brownson (1961); Slovic and Tversky (1974); MacCrimmon and Larsson (1979); Einhorn and Hogarth (1985, 1986); Kahn and Sarin (1988); Curley and Yates (1986)).

Many researchers have defined ambiguity in different ways. Ellsberg (1961) defined it as "the quality depending on the amount, type, reliability, and unanimity of information." Einhorn and Hogarth (1985, 1986) and Hogarth and Kunreuther (1995) defined ambiguity as the intermediate state between complete lack of knowledge and risk in which a probability distribution is specified. Fellner (1961); Frisch and Baron (1988), and Camerer and Weber (1992) defined ambiguity as uncertainty about probability created by lack of relevant information that could be known. However, in all the definitions, ambiguity is considered as the situation in which a subjective probability distribution cannot be determined because of a lack of information.

A popular approach to formulate ambiguity is to define it as a distribution of multiple subjective probabilities perceived by a decision maker. This approach relaxes the "independence axiom," while SEU requires a unique subjective probability distribution. Gilboa and Schmeidler (1989) developed MEU in this framework.

MEU is constructed in the framework of Anscombe and Aumann (1963). Let S be a set of states, Σ be an algebra of S and X be a set of outcomes. Denote an act by $f : S \rightarrow X$, and denote by F the set of acts. MEU model is expressed as follows:

$$[\text{MEU}] \quad f \succeq g \Leftrightarrow \min_{P \in C} \int_S u(f(s))dP(s) \geq \min_{P \in C} \int_S u(f(s))dP(s) \quad (1)$$

where $P : \Sigma \rightarrow [0, 1]$ is a subjective probability distribution and C is the closed convex set of probability distributions.

3 Application: Earthquake Insurance in Japan

In this study, we focus on the ambiguity of appraisal risk in earthquake insurance to examine whether and how personal characteristics affect the perception of ambiguity. The purchase rate of earthquake insurance is low in Japan. Only 18% of Japanese households purchased earthquake insurance in 2005, primarily because they regard it as expensive. Insurers set premiums that customers find expensive because earthquakes are a low-frequency and potentially high-impact risk. Vulnerability to catastrophic claims means insurers must pay reinsurance fees or interest on catastrophe bonds to offset bearing enormous losses themselves, which in turn increases the insurance premiums. Also, lack of information due to the infrequent occurrence of earthquakes causes ambiguity in earthquake risk and provides less information with which insurers can quantify risk and the probability of damage. Kunreuther et al. (1995) surveyed actuaries, insurers, and reinsurers to show that insurance premiums are higher for ambiguous than for unambiguous risks.

It is generally assumed that households believe earthquake insurance is expensive because they underestimate the risk of earthquakes; however, many surveys have shown otherwise. Mazda et al. (2005) showed that only 18% of respondents purchase earthquake insurance even though 64% believed an earthquake will more than partially collapse their home within 25 years. Non-Life Insurance Rating Organization of Japan (2004) showed that 61% of respondents not intending to purchase earthquake insurance believe an earthquake will severely damage their house or town within 20 years. In our survey, 75% of respondents believe that an earthquake of seven-point intensity (equivalent to the Great Hanshin Earthquake) will occur with more than 10% probability within 25 years. Insurance premiums in the surveyed area were actuarially fair when probabilities of a more-than-partial home collapse exceeded 12% in 25 years. However, our survey showed that only 30% of households purchase earthquake insurance. In short, many households do not buy earthquake insurance even though they perceive earthquake risk to be sufficiently high enough that the insurance premium is actuarially inexpensive. These results imply that reasons other than cost and perceived risk explain reluctance to buy earthquake insurance. We hypothesize that the explanation is consumers' perception of ambiguity in how insurers assess earthquake damage and pay damage claims.

The amount of insurance payments is determined after earthquake damage occurs. However, the prospective purchaser of insurance may not understand the criteria for assessing damage as specified in the contract and must trust the insurer's appraisal of damage. Thus, the household faces the risk of receiving a lesser payment than it expected because the damage appraisal is unexpectedly strict. In this paper, we term this as "appraisal risk."

Kahneman and Tversky (1979) introduced the notion of "probabilistic insurance", namely an insurance policy which, in the event that the hazard occurs, pays off with some probability strictly less than one. Wakker et al. (1997) pointed out that most insurance policies are probabilistic because there is always a possibility that the insurer will not pay for reasons such as insolvency or fraud. Appraisal risk also

An Econometric Model Based on the Maxmin Expected Utility Model

93

makes an insurance policy probabilistic for a buyer because there is a possibility that he cannot get as much insurance payment as expected at the time of contraction because of unexpectedly strict appraisal of damage.

Ambiguity in appraisal risk may be more important for earthquake insurance than for other insurance types, such as car or accident insurance. Earthquake claims are less frequent than claims for other insurances such as car, fire, or accident, and with those incidents, the amount that insurers will pay is known beforehand to the purchaser. However, earthquake insurance payouts are too infrequent to provide information to estimate their amount even roughly. As a result of this ambiguity, we believe households overestimate appraisal risk and hesitate to buy earthquake insurance.

4 Survey Data

To explore our hypotheses, we mailed questionnaires to 3,000 households in Joyo City, Kyoto, in mid-January 2006. Recipients were selected randomly from the Nippon Telegraph & Telephone phone book, and 681 responses were collected (23.4% response). Table 1 compares arithmetic means of the sample and the population.

The questionnaires are structured as follows: First, the hypothetical situation is presented in the questionnaire. Then, the questionnaire asks about WTP for full-coverage insurance and for probabilistic insurance.

"Imagine you own a house worth 10 million yen and your other assets (e.g., cash, stocks, land) are worth 20 million yen. Suppose there is a 5% probability that an earthquake with a seismic intensity of 7 on the Japanese scale will occur in 25 years (0.205% probability per year). If such an earthquake occurs, there is a 50% probability that your house will be half-destroyed (5 million yen loss) and a 50% probability that it will be completely destroyed (10 million yen loss)."

(A) What is the most you would be willing to pay for an insurance policy that will cover all damages from an earthquake?

(B) Imagine you have been offered coverage identical to the previous policy, except there is **about** $\alpha\%$ appraisal risk. That is, there is a possibility of **about** $\alpha\%$ that your claim will not be paid if your house is half-destroyed and that only half of your claim will be paid if it is completely destroyed. This risk arises from the

Table 1 Comparison between sample and population means

	Sample mean	Population mean
Age of the head of the household	62	55
Income (thousand yen)	6,040	6,629
Number of household members	2.99	2.81
Penetration rate of earthquake insurance	0.17	0.15

adjustor's overly strict appraisal of earthquake damage. What is the most you would be willing to pay for probabilistic earthquake insurance?

We call the policy described in question (A) "full-coverage insurance" and the policy described in question (B) "probabilistic insurance". With reference to probabilistic insurance, $\alpha\%$ was presented to the respondents as a mean probability of the ambiguous appraisal risk, which we called "referenced probability of appraisal risk." Survey recipients were randomly divided into three groups. The stated value of $\alpha\%$ was set to 1% for the first group, 5% for the second, and 10% for the third. The word *about* appeared in a large, bold font to emphasize the ambiguity of appraisal risk.[1] The extent of the range of appraisal risk perceived by the respondents was left to them.

Respondents were asked about their WTP for full-coverage insurance described in (A) and probabilistic insurance described in (B) by using payment cards. They were presented 16 bids representing monthly insurance premiums from 0 to 100 thousand yen and were asked to select one as their WTP.[2]

5 Model

To examine the influence of ambiguity of appraisal risk on WTP, we analyze the data by using both the EU and MEU models. Although EU is widely used to model decision making under uncertainty, it cannot represent observed individual choices under ambiguity. MEU is the generalized expected utility model for dealing with ambiguity developed by Gilboa and Schmeidler (1989). First, we explain EU, then, we describe MEU.[3]

---

[1] Einhorn and Hogarth (1985, 1986) originally conveyed ambiguity in this manner. Mauro and Maffioletti (1996, 2004) examined whether responses differ with different ways of expressing ambiguity: the way mentioned above, the range (e.g., $a\%$–$b\%$) and the several probabilities ($a\%$, $b\%$, $c\%$). They found no statistical differences among them.

[2] Clearly, this survey is hypothetical. It is impossible to pay real incentives to the respondents. One could devise similar experiments involving real money. In the earthquake insurance setting, however, the probability of the event and its losses need to be considerably lower and larger, respectively, than the lottery choice in the experiment. Therefore, the stakes would have to be affordably low, which makes the experiment different from the earthquake insurance we wished to consider. Hence, in this domain, hypothetical experiments for large sums are more instructive than real experiments for pennies. Further, research indicates no significant difference in analytical results between respondents participating in an experiment with real money or playing a hypothetical game. Evidence is provided by Beattie and Loomes (1997); Grether and Plott (1979), and Binswanger (1981) and has been surveyed by Camerer (1995) and Camerer and Hogarth (1999).

[3] In this study, a decision is assumed to be made for one-year span. This assumption may be criticized because it ignores issues of underpayment and overpayment over potentially extended random time horizons. However, earthquake insurance is a one-year contract, so our approach is realistic.

5.1 Expected Utility Model

According to the hypothetical situation described in Sect. 4 that is presented to respondents, a decision maker with no insurance has the following prospect:

$$\Pi = (1 - \pi_1 - \pi_2, W + Y; \pi_1, W + Y/2; \pi_2, W) \tag{2}$$

where Y is the value of the house (10 million yen) and W is the value of other assets (20 million yen). π_1 is the probability of half destruction ($0.205\% \times 50\% = 0.1025\%$ per year) and π_2 is the probability of complete destruction ($0.205\% \times 50\% = 0.1025\%$ per year). The house's value becomes $Y/2$ in the former instance and zero in the latter.

Under EU, WTP for the full-coverage insurance w_f is determined by the equation

$$V_{EU}(w_f) \equiv u(W + Y - w_f)$$
$$V_{EU}(w_f) = \tilde{u} \tag{3}$$

where u is a utility function and $\tilde{u}$ is the expected utility without insurance, and is given as follows:

$$\tilde{u} = (1 - \pi_1 - \pi_2)u(W + Y) + \pi_1 u(W + Y/2) + \pi_2 u(W) \tag{4}$$

Further, we consider probabilistic insurance. If a household perceives no ambiguity in appraisal risk, it faces the prospect involving "referenced probability of appraisal risk" described in Sect. 4B. This is written as

$$Q = (q_0, W + Y; q_1, W + Y/2, q_2, W) \tag{5}$$

where $q_0 = 1 - \alpha(\pi_1 + \pi_2)$, $q_1 = \alpha(\pi_1 + \pi_2)$, $q_2 = 0$ and α is the referenced probability of appraisal risk. Under EU, the WTP for probabilistic insurance w_p is determined by

$$V_{EU}(w_p) \equiv q_0 u(W + Y - w_p) + q_1 u(W + Y/2 - w_p) + q_2 u(W - w_p)$$
$$V_{EU}(w_p) = \tilde{u} \tag{6}$$

5.2 Maxmin Expected Utility Model

Ambiguity of appraisal risk is considered using the MEU model. Ambiguity perceived by the decision maker is expressed as C: the set of subjective probability distributions he faces. Let us denote a subjective probability distribution by

$$P = (p_0, W + Y; p_1, W + Y/2, p_2, W), \tag{7}$$

where p_0, p_1, and p_2 are parameters endogenously determined to be the worst scenario for the decision maker under the constraint $P \in C$. In this setting, the MEU function in (1) can be specified as

$$V_{MEU}(w_p)$$

$$\equiv \min_{P \in C} \left[p_0 u(W + Y - w_p) + p_1 u(W + Y/2 - w_p) + p_2 u(W - w_p) \right] \quad (8)$$

WTP for the probabilistic insurance w_p is determined by the following equation

$$V_{MEU}(w_p) = \tilde{u} \quad (9)$$

As for the full-coverage insurance, MEU is reduced to EU; since it has no appraisal risk, no ambiguity exists.

5.3 Specification of Subjective Probability Distributions

To estimate the model, the forms of the set of subjective probability distributions C, and utility function u need to be specified. As for C, we apply the robust control theory of Hansen and Sargent (2001). The right side of (8) can be seen as "a constraint robust control problem" if C is specified as

$$C = \{P : R(P, Q) \le \eta\} \quad (10)$$

where $R(P, Q)$ is the relative entropy between P and Q. Parameter η represents the size of ambiguity.

$$R(P, Q) = \sum_{k=0}^{2} p_k \ln \frac{p_k}{q_k}. \quad (11)$$

Hansen and Sargent (2001) show that the constraint robust control problem has a same solution as "a multiplier robust control problem" as below

$$V_{MEU}(w_p)$$

$$\equiv \min_{P \in C} \left[p_0 u(W + Y - w_p) + p_1 u(W + Y/2 - w_p) + p_2 u(W - w_p) \right.$$

$$\left. + \theta R(P, Q) \right]. \quad (12)$$

The parameter θ in the previous problem can be interpreted as an implied Lagrange multiplier on the constraint. θ can be interpreted as ambiguity parameter

An Econometric Model Based on the Maxmin Expected Utility Model   97

because it has one-on-one correspondence to η that represents the size of ambiguity. The size of ambiguity becomes larger as the value of θ decreases. Since $R(P, Q)$ is convex in p_0, p_1 and p_2, the first-order condition gives the following solution:

$$P^* = (p_0^*, W + Y; p_1^*, W + Y/2, p_2^*, W),\tag{13}$$

where

$$p_0^* = \frac{q_0}{q_0 + q_1 e^{(u(W+Y-w_p)-u(W+Y/2-w_p))/\theta}}$$

$$p_1^* = \frac{q_1 e^{(u(W+Y-w_p)-u(W+Y/2-w_p))/\theta}}{q_0 + q_1 e^{(u(W+Y-w_p)-u(W+Y/2-w_p))/\theta}}$$

$$p_2^* = 0.\tag{14}$$

Thus, the probabilistic insurance purchase decision can be modeled by MEU, where w_p is determined by

$$V_{MEU}(w_p) \equiv p_0^* u(W + Y - w_p) + p_1^* u(W + Y/2 - w_p)$$

$$V_{MEU}(w_p) = \tilde{u}.\tag{15}$$

5.4 Specification of Utility Function

We used the constant relative risk attitude (CRRA) utility function. It is derived from assumptions that relative risk aversion remains constant for all levels of wealth. The CRRA utility function is widely used in econometric studies of risk; therefore, we can compare our estimation result with results in previous studies. Curvature of the utility function represents attitude toward risk. Therefore, using the CRRA utility function does not seriously bias conclusions because its curvature is sufficiently flexible to fit the plausible range of respondents' risk attitudes. The CRRA utility function is written as

$$u(x) = \frac{x^{1-\gamma}}{1 - \gamma},\tag{16}$$

where γ is the Pratt-Arrow coefficient of relative risk aversion. The effect of risk aversion on the decision to purchase the earthquake insurance is represented by this parameter.

Relative risk aversion may vary among demographic groups. Thus, we link it with respondents' social characteristics linearly

$$\gamma = \gamma_0 + \mathbf{x}'\gamma,\tag{17}$$

where γ_0 is an intercept, $\mathbf{x}$ is a column vector of respondents' characteristics variables, and γ is a parameter vector. Size of the ambiguity may vary across demographic groups. Hence, we link ambiguity parameter θ in (14) with respondents' characteristics linearly,

$$\theta = \theta_0 + \mathbf{x}'\theta, \tag{18}$$

where $\theta_{0.01}$, $\theta_{0.05}$ and $\theta_{0.10}$ are dummy variables (equal to one if the referenced probability of appraisal risk is 1%, 5%, and 10%, respectively), $\mathbf{x}$ is a vector of respondents' characteristics variables, and θ is a parameter vector.

5.5 Estimation Method

These models are estimated using Cameron (1987) grouped-data regression. This model is based on a random utility framework in which expected utility is assumed to consist of deterministic term V and error term ϵ representing observation errors. Here, V is V_{EU} in (3) or (6) if a respondent follows EU. And V is V_{MEU} in (15) if he follows MEU. Suppose that respondent i is willing to pay w_i for earthquake insurance. In our survey, he/she is shown bidding $B_1 < \cdots < B_J$ as the insurance premium. Note that $V_i(B_1) > \cdots > V_i(B_J)$. Respondent i choose $B_{j(i)}$ if

$$V_i(B_{j(i)+1}) + \epsilon_{j+1} < V_i(w_i) + \epsilon_i \quad \text{and} \quad V_i(w_i) + \epsilon_i \leq V_i(B_{j(i)}) + \epsilon_j, \tag{19}$$

where V_i is the deterministic term expected utility of respondent i, and ϵ_i, ϵ_j and ϵ_{j+1} are error terms. Assume that $\epsilon = \epsilon_i - \epsilon_j$ and that $\epsilon = \epsilon_i - \epsilon_{j+1}$ independently follows normal distribution with mean 0 and variance σ^2. The likelihood that respondent i choose $B_{j(i)}$ can be written using $V_i(w_i) = \tilde{u}$:

$$
\begin{aligned}
L_i &= \Pr\left[V_i(B_{j(i)+1}) - V_i(w_i) < \epsilon \leq V(B_{j(i)}) - V_i(w_i)\right] \\
&= \Pr\left[V_i(B_{j(i)+1}) - \tilde{u} < \epsilon \leq V(B_{j(i)}) - \tilde{u}\right] \\
&= \Phi\left(\frac{V_i(B_{j(i)}) - \tilde{u}}{\sigma}\right) - \Phi\left(\frac{V_i(B_{j(i)+1}) - \tilde{u}}{\sigma}\right)
\end{aligned}
\tag{20}
$$

where Φ is the standard normal distribution function. Thus, the log likelihood of all samples can be written as

$$\ln L = \sum_{i=1}^{N} \ln\left[\Phi\left(\frac{V_i(B_{j(i)}) - \tilde{u}}{\sigma}\right) - \Phi\left(\frac{V_i(B_{j(i)+1}) - \tilde{u}}{\sigma}\right)\right]. \tag{21}$$

This log likelihood is maximized to estimate parameters.

An Econometric Model Based on the Maxmin Expected Utility Model 99

Table 2 Estimation results of simple models

| Variables | EU | | | | MEU | |
| | Full-coverage insurance | | Probabilistic insurance | | Probabilistic insurance | |
	Coeff	P-value	Coeff	P-value	Coeff	P-value
γ_0	2.027	0.00	−2.483	0.00	1.437	0.00
σ	2.089E+7	0.20	8.756E-3	0.32	1.357E+8	0.00
$\theta_{0.01}$					1.247E+8	0.01
$\theta_{0.05}$					3.719E+8	0.07
$\theta_{0.10}$					3.041E+8	0.00
n	557		351		351	
$\ln L$	−1687		−1110		−1093	
AIC	6.050		6.313		6.199	

$\ln L$ is the maximized log-likelihood.

$\mathrm{AIC} = -\frac{2}{n}(\ln L - k)$, where n is sample size and k is the number of parameters.

6 Results

6.1 Estimation Results of Simple Models

Parameters in V_{EU} and V_{MEU} are estimated by maximizing the log likelihood in (21)[4]. First, we examine the estimation results of the models that only have constant terms (σ, γ_0, $\theta_{0.01}$, $\theta_{0.05}$, and $\theta_{0.10}$). This model is called the "simple model" and it ignores the effects of personal characteristics. The results are presented in Table 2. Estimated CRRA coefficients are 2.027 for full-coverage insurance and −2.483 for probabilistic insurance with 1% statistical significance. Ljungqvist and Sargent (2000) and Gollier (2001) established by a hypothetical experiment that the coefficient of relative risk aversion lies in the range of 1–4. The empirical literature supports this. Friend and Blume (1975) studied the demand for risky assets and conducted that γ generally exceeds unity and is probably greater than 2. Using expenditure data, Weber (1975) estimated γ to lie within a range of 1.3–1.8, and Szpiro (1986) obtained a similar range by using aggregate time-series data for property insurance. In a careful study of consumption, Hansen and Singleton (1982) found relative risk aversion parameters ranging from 0.68 to 0.97. Mankiw's study (1985) of consumption spending obtained relative risk aversion estimates from 2.44 to 5.26 for nondurable goods and from 1.79 to 3.21 for durable goods.

The estimate for the full-coverage insurance is a reasonable value. However, the estimate for probabilistic insurance is unreasonable because it implies risk-loving behavior. This suggests that the purchase decision for full-coverage insurance can be explained by EU, while decisions concerning probabilistic insurance cannot be explained by EU. The ambiguity presented by appraisal risk must be considered

---

[4]Estimation is carried out with GAUSS Maxlik.

as an element in purchasing probabilistic insurance. By applying the MEU in consideration of the ambiguity of appraisal risk, the estimated CRRA coefficient becomes a reasonable value: 1.486 for the probabilistic insurance. The CRRA coefficient γ, the standard deviation of error term σ, and ambiguity parameters $\theta_{0.01}$, $\theta_{0.10}$ are statistically significant at the 5% level, and $\theta_{0.05}$ is statistically significant at the 10% level. These imply that risk and ambiguity parameters are both necessary to express purchase decisions for the probabilistic insurance.

Then, we examine whether the EU model or the MEU model more consistently interprets the data concerning purchase of probabilistic insurance. The smaller value of the Akaike information criteria (AIC) indicates that the model is more consistent with the data. The AIC values of EU and MEU are 6.313 and 6.199, respectively, which shows that MEU is better for modeling probabilistic insurance. The log-likelihood ratio test can be applied because MEU nests EU. MEU is superior to EU at 1% statistical significance.

6.2 Estimated Effects of Personal Characteristics

Now we consider the effects of consumers' personal characteristics. Table 3 lists the independent variables **x** for (17) and (18), and Table 4 shows the estimation results. Under EU estimates of γ for full-coverage insurance and probabilistic insurance are 1.801 and -0.636, respectively. Negative value of γ for probabilistic insurance implies that EU is not suitable for explaining the data. Previous studies of risk

Table 3 Variables of personal characteristics

Variables	Description	Mean
Age	Age(in years)	62.0
Female	Dummy; 1 if a respondent is female, 0 otherwise.	0.088
Marriage	Dummy; 1 if a respondent is married, 0 otherwise.	0.945
Child	Dummy; 1 if a respondent has a child under 10 years old, 0 otherwise.	0.077
Education	Dummy; 1 if a respondent graduated from a university, 0 otherwise.	0.379
Unemployment	Dummy; 1 if a respondent is unemployed or retired, 0 otherwise.	0.279
Self-employment	Dummy; 1 if a respondent is self-employed, 0 otherwise.	0.103
Civil servant	Dummy; 1 if a respondent is civil servant, 0 otherwise.	0.073
Experience	Dummy; 1 if a respondent has experienced a economic loss from earthquake, 0 otherwise.	0.074
Purchase	Dummy; 1 if a respondent has purchased an earthquake insurance, 0 otherwise.	0.171
Never-Paid	Dummy; 1 if a respondent has never received any insurance payment, 0 otherwise.	0.337
Trust	Dummy; 1 if a respondent trusts insurance companies, 0 otherwise.	0.311

An Econometric Model Based on the Maxmin Expected Utility Model 101

attitude as related to personal characteristics revealed that female, older, married servant respondents are more risk-averse, while better-educated, self-employed, and unemployed respondents are less risk-averse than their co-respondents (Barsky et al. (1997); Binswanger (1980, 1981); Cramer et al. (2002); Donkers et al. (2001); Eisenhauer and Venturaz (2003); Halek and Eisenhauer (2001); Hartog et al. (2002); van Praag (1996); Riley and Chow (1992); Shubert et al. (1999); Siegal and Hoban (1982); Sunden and Surette (1998)). For full-coverage insurance, our results are mostly consistent with and supported by previous findings. For probabilistic insurance, inclusion of personal characteristics weakens the risk-loving attitude, which could be attributed to capturing the effects of ambiguity aversion. That is why signs of our coefficients differ from previous studies.

Table 4 Estimation results of EU and MEU including variables of personal characteristics

Variables	EU				MEU	
	Full-coverage insurance		Probabilistic insurance		Probabilistic insurance	
	Coeff	P-value	Coeff	P-value	Coeff	P-value
Estimation of γ						
γ_0	1.444	0.00	−1.229	0.00	1.844	0.00
Age	0.004	0.00	0.010	0.00		
Female	0.123	0.00	0.095	0.03	0.159	0.00
Education	−0.079	0.00	−0.120	0.00	−0.178	0.00
Experience	0.006	0.63	−0.098	0.06	−0.072	0.07
Marriage	0.021	0.16	−0.080	0.21	0.023	0.11
Unemployment	−0.114	0.00	−0.112	0.00	−0.032	0.17
Self-employment	0.018	0.38	0.013	0.59	0.015	0.54
Civil servant	0.078	0.04	−0.111	0.04		
Child	0.061	0.12	0.071	0.19	0.036	0.29
Estimation of θ						
$\theta_{0.01}$					0.840E+5	0.41
$\theta_{0.05}$					2.083E+5	0.00
$\theta_{0.10}$					4.719E+5	0.11
Age					0.024E+5	0.08
Female					−1.006E+5	0.00
Education					1.107E+5	0.10
Experience					0.576E+5	0.00
Purchase					1.926E+5	0.06
Never-Paid					0.051E+5	0.16
Trust					0.693E+5	0.19
σ	4.138E+7	0.19	0.1332	0.45	2.538E+7	0.11
$\hat{\gamma}$	1.801		−0.636		1.934	
n	557		308		306	
ln L	−1479		−908		−886	
AIC	5.350		5.966		5.915	

ln L is the maximized log-likelihood.

AIC$= -\frac{2}{n}(\ln L - k)$, where n is sample size and k is the number of parameter.

Under MEU, estimation results for probabilistic insurance are presented in the rightmost column in Table 4. The estimate of γ is 1.934, close to that for full-coverage insurance, and signs of coefficients except for *Experience* are congruent with it. Further, AIC indicates that MEU is better than EU in explaining their decisions regarding probabilistic insurance.

Next, we examine the relationship between the ambiguity parameter and personal characteristics. A positive sign for the coefficient shows that the correspondent variable reduces perceived ambiguity because a larger θ means lesser perceived ambiguity. At 5% statistical significance, the signs of *Female* and *Experience* are negative and positive, respectively, indicating that women perceive greater ambiguity and that respondents who previously suffered earthquake damage perceive less ambiguity. The signs for attributes *Age*, *Education*, and *Purchase* are positive at 10% significance, which shows that perceived ambiguity increases with age, education, and previous experience of purchasing earthquake insurance. This is because a higher-educated respondent can understand information about insurance and earthquakes. The characteristics *Never Paid* and *Trust* are not statistically significant at the 10% level.

7 Policy Implications

Estimation results above show that MEU surpasses EU in explaining respondents' choices of WTP for earthquake insurance under conditions of ambiguous appraisal risk. This implies that households tend to overestimate appraisal risk under ambiguity. To clarify the negative effects of ambiguity of appraisal risk on decisions to purchase earthquake insurance, we calculate the risk and ambiguity premium shown by Table 5. Risk and ambiguity premium are additional payments to buy earthquake insurance because of risk and ambiguity, respectively. Customers' WTP consists of their expected loss, risk premium, and ambiguity premium. The risk premium is calculated by WTP without ambiguity ($\theta = +\infty$) minus expected loss. The ambiguity premium is calculated by WTP with ambiguity ($\theta = \hat{\theta}$) minus WTP without ambiguity ($\theta = +\infty$). The ambiguity premium of appraisal risk with referenced probabilities 1%, 5%, and 10% are $-2,198$, $-2,098$, and $-2,988$ yen/year, respectively, and each reduces about 10% of the earthquake insurance value.

Estimation results above show that MEU is better than EU to explain respondents' choices of WTP for earthquake insurance with ambiguous appraisal risk. This

Table 5 Risk premium and ambiguity premium

Mean of appraisal risk	$\alpha = 1\%$	$\alpha = 5\%$	$\alpha = 10\%$
Expected loss	15,273	14,863	14,350
Risk premium	5,861	5,782	5,683
Ambiguity premium	$-2,198$	$-2,098$	$-2,988$
Willingness to pay	18,936	18,546	17,045
(yen/year)			

implies that a household tend to overvalue the appraisal risk in earthquake insurance under ambiguity. In order to clarify negative effect of ambiguity of appraisal risk on purchase decision of earthquake insurance, we calculate the risk and ambiguity premium shown by Table 5. Risk and ambiguity premium are additional payments to buy earthquake insurance because of risk and ambiguity, respectively. Here, a willingness to pay consists of expected loss, risk premium, and ambiguity premium. The risk premium is calculated by WTP without ambiguity ($\theta = +\infty$) minus expected loss. The ambiguity premium is calculated by WTP with ambiguity ($\theta = \hat{\theta}$) minus WTP without ambiguity ($\theta = +\infty$). The ambiguity premium of appraisal risk with reference probability 1%, 5%, and 10% are $-2,198$, $-2,098$, and $-2,988$ yen/year, respectively, and each reduces about 10% of the earthquake insurance value.

To increase sales of earthquake insurance, reducing the perceived ambiguity of appraisal risk is more effective than reducing appraisal risk itself because ambiguity premiums are not significantly different even though the referenced probability of appraisal risk (that represents appraisal risk itself) varies by 1%, 5%, and 10%. Estimation results shown in Table 4 are useful for reducing perceived ambiguity of appraisal risk. The characteristic *Purchase* takes the largest positive value among dummy variables, indicating that the purchase of earthquake insurance reduces the perceived ambiguity of appraisal risk. This is natural because the respondent who purchases it receives more knowledge about earthquake insurance than his co-respondents. *Experience* is positive, suggesting that respondents who have suffered earthquake damage better understand how insurance adjusters appraise earthquake loss. These findings suggest that information about earthquake insurance or about how earthquakes damage houses can reduce the ambiguity of appraisal risk. For example, photographs or videos of houses damaged by earthquakes may be useful in reducing customers' ambiguity of appraisal risk.

8 Conclusions

This study presents a new econometric model addressing the relationships among risk, ambiguity, and preference based on the MEU model developed by Gilboa and Schmeidler (1989). We applied this model to investigate the effect of ambiguity or appraisal risk on the decision to purchase earthquake insurance. Our model was estimated using data from a survey with a set of questions on a hypothetical earthquake insurance. The major results of this study may be summarized as follows: First, reluctance to buy earthquake insurance with ambiguous appraisal risk is better predicted by the MEU model than the EU model. Second, people dislike earthquake insurance with ambiguous appraisal risk: most respondents demanded more than a 10% reduction in premium to offset a 1% appraisal risk. Ambiguity premiums are not significantly different, even though the referenced probability of appraisal risk varies. Hence, reducing the ambiguity of appraisal risk is more effective than reducing appraisal risk itself for increasing sales of earthquake insurance. Third, the perceived ambiguity is less among men who previously had purchased earthquake insurance and had experienced earthquake damage to their houses. In addition, ambiguity decreases with age and education.

Acknowledgements We thank Professor Markku Kallio for reading the draft and making many helpful suggestions. Helpful comments from anonymous reviewers and participants at the December 2009 CwU Workshop are also acknowledged.

References

Anscombe, F. J., & Aumann, R. J. (1963). A definition of subjective probability. *The Annals of Mathematical Statistics 34*, 199–205.

Barsky, R. B., Juster, F. T., Kimball, M. S., & Shapiro, M. D. (1997). Preference parameters and behavioral heterogeneity: an experimental approach in the health and retirement study. *Quarterly Journal of Economics, 112*, 537–579.

Beattie, J., & Loomes, G. (1997). The impact of incentives upon risky choice experiments. *Journal Risk and Uncertainty, 14*, 155–168.

Becker, S. W., & Brownson, F. O. (1961). What price ambiguity? Or the role of ambiguity in decision-making. *Journal of Polit Economy, 72*, 62–73.

Binswanger, H. (1980). Attitudes toward risk: experimental measurement in rural India. *American Journal of Agricultural Economics, 62*, 395–407.

Binswanger, H. (1981). Attitudes toward risk: theoretical implications of an experiment in rural India. *Economic Journal, 91*, 869–890.

Camerer, C. F. (1995). Individual decision making. In J. H. Kagel, & A. E. Roth (Eds.), *The Handbook of Experimental Economics*. Princeton, NJ: Princeton University Press.

Camerer, C. F., & Hogarth, R. M. (1999). The Effects of financial incentives in experiments: a review and capital-labor-production framework. *Journal of Risk and Uncertainty, 19*, 7–42.

Camerer, C., & Weber, M. (1992). Recent develop-ments in modeling preferences: uncertainty and ambiguity. *Journal of Risk and Uncertainty, 5*, 325–370.

Cameron, T. A. (1987). The impact of grouped-data regression models. *Journal of Econometrics, 35*, 37–57.

Cameron, T. A. (2005). Individual option prices for climate change mitigation. *Journal of Public Economics, 89*, 283–301.

Cramer, J. S., Hartog, J., Jonker, N., & van Praag, C. M. (2002). Low risk aversion encourages the choice for entrepreneurship: an empirical test of a truism. *Journal of Economic Behavior and Organization, 48*, 29–36.

Curley, S. E., & Yates, F. J. (1986). An empirical evaluation of descriptive models of ambiguity reactions in choice situations. *Journal of Mathematical Psychology, 33*, 397–427.

Donkers, B., Melenberg, B., & van Soest, A. (2001). Estimating risk attitudes using lotteries: a large sample approach. *Journal of Risk and Uncertainty, 22*, 165–195.

Einhorn, H., & Hogarth, R. (1985). Ambiguity and uncertainty in probabilistic inference. *Psychological Review, 92*, 433–461.

Einhorn, H., & Hogarth, R. (1986). Decision making under ambiguity. *Journal of Business 59*: 225–250.

Ellsberg, D. (1961). Risk, ambiguity, and the savage axioms. *Quarterly Journal of Economics 75*, 643–669.

Eisenhauer, J. G., & Venturaz, L. (2003). Survey measures of risk aversion and prudence. *Applied Economics 35*, 1477–1484.

Fellner, W. (1961). Distortion of subjective probabilities as a reaction to uncertainty. *Quarterly Journal of Economics 75*, 670–694.

Friend, I., & Blume, M. E. (1975). The demand for risky assets. *The American Economic Review, 65*, 900–922.

Frisch, D., & Baron, J. (1988). Ambiguity and rationality. *Journal of Behavioral Decision Making, 1*, 149–157.

Gilboa, I., & Schmeidler, D. (1989). Maxmin expected utility with non-unique prior. *Journal of Mathematical Economics, 18*, 141–153.

Gollier, C. (2001). *The Economics of Risk and Time* (pp. 31–32). MIT Press

Grether, D. M., & Plott, C. R. (1979). Economic theory of choice and the preference reversal phenomenon. *The American Economic Review, 85*, 260–266.

Halek, M., & Eisenhauer, J. G. (2001). Demography of risk aversion. *Journal of Risk and Insurance 68*, 1–24.

Hansen, L. P., & Singleton, K. J. (1982). Generalized instrumental variables estimation of nonlinear rational expectations models. *Econometrica, 50*, 1269–1286.

Hansen, L. P., & Sargent, T. J. (2001). Robust control and model misspecification. *The American Economic Review 91*, 60–66.

Hartog, J., Ferrer-i-Carbonell, J., & Jonker, N. (2002). Linking measured risk aversion to individual characteristics. *Kyklos, 55*, 3–26.

Hogarth, R. M., & Kunreuther, H. (1995). Decision making under ignorance: arguing with yourself. *Journal of Risk and Uncertainty 10*, 15–36.

Kahn, B. E., & Sarin, R. K. (1988). Modelling ambiguity in decisions under uncertainty. *Journal of Consumer Research 15*, 265–272.

Kahneman, D., & Tversky, A. (1979). Prospect theory: an analysis of decision under risk. *Econometrica, 47*, 263–291.

Kunreuther, H. J., Meszaros, R. M., Hogarth, R., & Spranca, M. (1995). Ambiguity and underwriter decision processes. *Journal of Economic Behavior and Organization, 26*, 337–352.

Ljungqvist, L., & Sargent, T. J. (2000). *Recursive Macroeconomic Theory* (pp. 260–261). MIT Press

MacCrimmon, K. R., & Larsson, S. (1979). Utility theory: axioms versus paradoxes. In M. Allais & O. Hagen (Eds.), *Expected Utility and the Allais Paradox*. Dordrecht, Holland: Reidel

Mankiw, N. G. (1985). Consumer durables and the real interest rate. *Review of Economics and Statistics 67*, 353–362.

Mauro, C. D., & Maffioletti, A. (1996). An experimental investigation of the impact of ambiguity on the valuation of self-insurance and self-Protection. *Journal of Risk and Uncertainty 13*, 53–71.

Mauro, C. D., & Maffioletti, A. (2004). Attitudes to risk and attitudes to uncertainty: experimental evidence. *Applied Economics 36*, 357–372.

Mazda, Y., Tatano, H., & Okada, N. (2005). Estimation of risk premium for disaster risk by contingent valuation method. *Infrastructure Planning Review 22*, 325–334 (in Japanese).

Non-Life Insurance Rating Organization of Japan (2004). Survey on public awareness of catastrophic earthquake risks. *Earthquake Insurance Research 5* (in Japanese)

van Praag, C. M. (1996). *Determinants of Successful Entrepreneurship. Tinbergen Institute Research Series*, vol. 107. Amsterdam: Thesis Publishers.

Riddel, M., & Shaw, W. D. (2006). A theoretically-consistent empirical model of non-expected utility: an application to nuclear-waste transport. *Journal of Risk and Uncertainty 32*, 131–150.

Riley, W. B., & Chow, K. V. (1992). Asset allocation and individual risk aversion. *Financial Analysts Journal, 48*, 32–37.

Savage, L. J. (1954). *The Foundations of Statistics*. New York: Wiley.

Shubert, R., Brown, M., Gysler, M., & Brachinger, H. W. (1999). Financial decision-making: are women really more risk averse? *The American Economic Review 89*, 381–385.

Siegal, F. W., & Hoban, J. P. (1982). Relative risk aversion revisited. *The Review of Economics and Statistics, 64*, 481–487.

Slovic, P., & Tversky, A. (1974). Who accepts Savage's axiom? *Behavioral Science, 19*, 368 373.

Sunden, A. E., & Surette, B. J. (1998). Gender differences in the allocation of assets in retirement savings plans. *The American Economic Review, 88*, 207–211.

Szpiro, G. G. (1986). Measuring risk aversion: an alternative approach. *The Review of Economics and Statistics, 68*, 156–159.

Wakker, P. P., Thaler, R. H., & Tversky, A. (1997). Probabilistic Insurance. *Journal of Risk and Uncertainty, 15*, 7–28.

Weber, W. E. (1975). Interest rates, inflation, and consumer expenditures. *The American Economic Review, 65*, 843–858.

Part II
Modeling Uncertainties of Heterogeneous Systems

Modeling Technological Change Under Increasing Returns and Uncertainty

Andrei Gritsevskyi and Yuri Ermoliev

Abstract The aim of this paper is to analyze methodological challenges involved in modeling of endogenous technological changes with increasing returns and uncertainties by using stylized versions of models. Realistic versions of these models are analytically intractable making it difficult to comprehend the interplay of different assumptions on their outcomes. We demonstrate path-dependences of myopic evolutionary approaches, the infeasibility of straightforward "trial-and-error" processes, and the need for adequate long-term policy assistance. We also show why increasing returns and uncertainties radically offset the rationale for postponed investments in new technologies and how stochastic models cope with systemic risks implicitly induced by interdependencies among uncertainties, technologies, the structure of models, and decisions. The paper demonstrates possible misleading character of alternative models of uncertainties. It shows the need for proper modeling of long-term random horizons, corresponding discounting, security constraints and requirements of robustness by using systemic valuations and "distribution free" stochastic programming/optimization.

1 Introduction

The proper modeling of Technological Changes (TC) is decisive for the evaluation of the true socio-economic and environmental impacts of development policies. Traditional models assume that technological innovations are key factors of long-term economic growth and the prosperity of nations (Abramovitz 1993; Barnett

---

A. Gritsevskyi (✉)
International Atomic Energy Agency, Vienna, Austria
e-mail: A.Gritsevskyi@IAEA.org

Y. Ermoliev
International Institute for Applied Systems Analysis, Laxenburg, Austria
e-mail: ermoliev@iiasa.ac.at

Y. Ermoliev et al. (eds.), *Managing Safety of Heterogeneous Systems*, Lecture Notes in Economics and Mathematical Systems 658, DOI 10.1007/978-3-642-22884-1_6,
© Springer-Verlag Berlin Heidelberg 2012

and Morse 1967; Freeman 1994). However, on-going global changes, in particular, the pollution with potential catastrophic global climate changes, the increasing gap between the rich and poor, insecurity of food, water, energy, and countries, inspire great concerns about sustainable developments, equity and the welfare. Searching only for economic efficiency and growth produce adverse impacts of innovations which are impossible to evaluate by using traditional models.

The aim of this paper is to analyze methodological challenges involved in adequate modeling of TC by using simple versions of IIASA[1] path-breaking models (Arthur 1989; Arthur et al. 1987; Gritsevskyi and Ermoliev 1999; Gritsevskyi and Nakićenović 2002; Grübler and Gritsevskyi 2002). Realistic versions of these models are analytically intractable making it difficult to comprehend the interplay of different assumptions and their outcomes. The paper analyses the following basic issues.

In traditional economic models (see detailed analysis in (Cowan 1991; Gritsevskyi and Nakićenović 2002; Grübler and Gritsevskyi 2002; Nakićenović 1996; Ruttan 1997)) TC is represented by exogenous variables which improve performance of technologies through time independently of policies. As a consequence, such models strongly advocate to postpone investments in new technologies until they became cheap enough, i.e., "weight-and-see" policies. In reality, technological changes are (see discussion in (Grübler and Gritsevskyi 2002)) endogenous. They can be affected by deliberate policies related to urgent socio-economic, environmental and safety/security issues.

In other words, models with exogenous technological changes ignore the necessity to invest in new technology in order to make this technology cheaper and better with respect to desirable performance indicators. As technology becomes more widely adopted, the cheaper and better it becomes. This is so-called increasing returns phenomenon. Explicit modeling of increasing returns and uncertainties, as it was demonstrated in (Gritsevskyi and Ermoliev 1999; Gritsevskyi and Nakićenović 2002; Grübler and Gritsevskyi 2002), radically offsets the rational for postponed investments with crucial policy implications regarding timing of investments.

The modeling approaches with diminishing returns do not allow representation of these essential characteristics of technological developments. Despite this deficiency (see discussion e.g., in (Metcalfe 1987; Nakićenović 1996)), the diminishing returns dominate the standard models because such models are convex with simple concepts of global solutions and equilibriums. Contrary, increasing returns are associated with non-convexities, local solutions, disequilibriums, path dependencies and the concept of "lock-in" states of developments. This case requires significant remodeling of traditional approaches (Arrow 1962; Griliches 1996; Gritsevskyi and Ermoliev 1999; Grossman and Helpman 1991; Ruttan 1997). In particular, a new technology requires a vast variety of other technologies (Gritsevskyi and Nakićenović 2002; Grübler and Gritsevskyi 2002), including infrastructures. Nonetheless, these essential interdependencies (externalities) are represented in

---

[1] International Institute for Applied System Analysis.

the traditional models in an extremely simplified exogenous manner. The use of aggregate production and utility functions precludes representation of vital economic, social, environmental and technological heterogeneities.

Market-oriented, top-down models of TC attempt to analyze the economy-wide impacts of innovations assuming that the main driver is the price signal. Unfortunately, the price signals address only narrow market-related values of investments whereas the main purpose of them is often dictated by non-market pressures (Grübler and Gritsevskyi 2002; Nakićenović 1996). Besides, concepts of increasing returns (non-convexities), externalities (interdependencies), inherent uncertainties, including uncertain horizons of "break-even" points, destroy the basic assumptions about the existence of the general equilibrium prices allowing decentralization of interdependent economic processes into independent individual activities.

In this paper, we demonstrate that the current progress in modeling allows developments of so-called "bottom-up" social planner models, for integrated assessment and management of the most important interdependencies affecting TC.

Section 2 describes the representation of increasing returns by experience (learning) curves (Argote and Epple 1990) showing the need for adequate modeling of uncertainties and time horizons. Section 3 analyses path dependencies and lock-in phenomenon of purely myopic evolutionary TC processes based on behavioral and market principles. It uses new types of potential urn's schemes easily illustrating why uncertainties of markets without appropriate regulations preclude the emergence of robust technological structures. Section 3 also shows why increasing returns offset the rational for postponing investments in new technologies therefore technologies that appear initially unattractive may diffuse into the market under proper policy assistance. Section 4 analyzes shortcomings of widely used net present value and real option theory valuations ignoring a vast variety of interdependencies among technologies apart from price signals. In particular, standard discounting implicitly induces horizons of valuations which may dramatically mismatch random break-even horizons of technological changes. This section introduces new approaches to discounting focusing on random break-even horizons rather then horizons of capital markets. Section 5 outlines an integrated stochastic optimization dynamic model with increasing returns developed in (Gritsevskyi and Ermoliev 1999; Gritsevskyi and Nakićenović 2002; Grübler and Gritsevskyi 2002) for long-term valuations of investments in technological developments. Section 6 demonstrates impacts of critical systemic interdependencies between technologies, decisions, uncertainties and robust solutions. Uncertainty and induced systemic risks invoke changes within a set of interdependent technologies, some of which may be less economically efficient but have advantages regarding other indicators, such as security of supply. This section shows why STO models allow to deal with this type of endogenous technological developments. Section 7 analyzes alternative models of uncertainties. Section 8 demonstrates advantages of distribution free stochastic versus probabilistic models. Section 9 concludes.

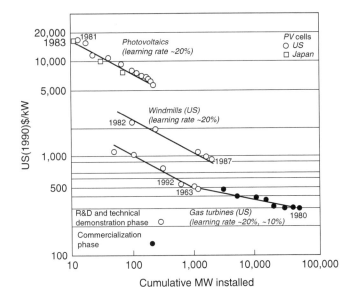

Fig. 1 Experience curves for gas turbines, windmills and photovoltaics. Cost improvements per unit installed capacity, in USD (1990) per kW, are shown against the cumulative installed capacity, in MWe, on logarithmic scale. Source: Nakićenović (1996)

2 Increasing Returns and Uncertainty

The increasing returns to scale effects have been discussed in the economic literature for long time. They were illustrated empirically by Wright (Argote and Epple 1990) in 1936 showing the decline of airframe production costs with experience measured by cumulative output. He showed that when plotted in log-log diagrams, unit costs for different inputs ("labor", "materials") appeared to be a linear function of cumulative production of airframes. Figure 1 illustrates these functions for different electricity generation technologies.

Figure 1 suggests that the unit cost $c(x)$ of the technology depends on the cumulative capacity x of this technology as exponentially declining function

$$c(x) = c_0 x^{-\alpha}, \qquad (1)$$

where c_0 is the cost of unit cumulative capacity $x = 1$ and α is a parameter that in general depends on x. The main conclusion is that unit costs are expected progressively to fall as producers and users gain experience with new technologies expressed by x, i.e., they exhibit the increasing returns. This relates to the path dependence (Arrow 1962; Arthur 1989; Arthur et al. 1987) of technological developments. The following sections demonstrate that as a consequence, uncoordinated

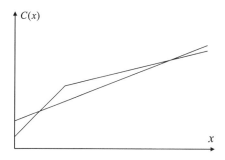

Fig. 2 Cumulative costs

purely evolutionary processes of technological developments may lead to different lock-in states with respect to cumulative (total) costs defined as the following.

The cost for incremental increase $x + \Delta$ of cumulative capacity x is calculated as $c(x)\Delta$ for small $\Delta > 0$. Therefore, the total cost function $C(x)$ for developing a capacity $x > 0$ is the integral

$$C(x) = \int_0^x c(z)dz. \qquad (2)$$

In the following we often use a piece-wise linear approximation of $C(x)$, e.g., shown in Fig. 2. Jumps at $x = 0$ usually reflect "set-up" costs incurred in research and developments (R&D) of technologies before their first deployments. Risk exposure to such extreme events as Chernobyl's disaster or BP's oil spills also lead to similar jumps. Breaks of linearity in $C(x)$ may correspond to significant breakthrough in technological developments.

It is widely recognized that increasing returns and uncertainties jointly play a decisive role in shaping future energy systems (Gritsevskyi and Ermoliev 1999; Gritsevskyi and Nakićenović 2002; Grübler and Gritsevskyi 2002; Grübler and Messner 1996; Messner et al. 1996; Romer 1986). Fundamental technological changes may be rather slow. Time horizons of a century or more are frequently adopted in energy systems studies. Modeling technological developments over such long-term horizons raises key issues regarding proper representation of inherent uncertainties (see Sect. 7). In particular, the transition to a new technology may take a rather uncertain amount of time as Fig. 3 illustrates.

Figure 3 shows the coexistence of cheap and expensive wind technologies. The wind technology slowly takes-off penetrating the market despite of higher overall costs. The cost-effective solutions of standard deterministic model would select the cheapest technology. The coexistence of both technologies can be justified only by explicit treatment of uncertainties and increasing returns (Sects. 6 and 7) within an appropriate time horizon: even if we know in 1980 the cost reduction trends for the wind technology for coming 20 years precisely, the additional 30 years may not be enough to justify its benefits. Uncertainties of trends dramatically affect conclusions. By extrapolating these trends we can find the amount of cumulative

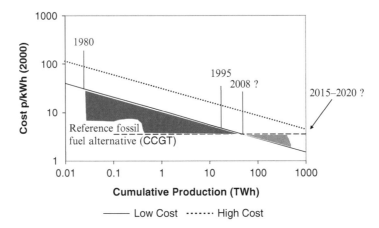

Fig. 3 Cost reduction for on-shore (low cost) vs. off-shore (high cost) wind energy technologies in the EU. Source (Study of Learning Curves 2000)

capacity needed for the two technologies to break-even with the conventional fossil fuel technology. If there is an uncertainty in the trends, than the break-even point may raise significantly the time when the new technology becomes commercial. In other words, the cost-effectiveness of the new technology can be justified only within an appropriate time horizon.

Uncertainties of trends depend on costs of technologies. IIASA's inventory (Strubegger and Reitgruber 1995) shows that probability distributions of costs are not symmetric with both rather "optimistic" and "pessimistic" views on future costs per unit capacity requiring adequate representations (Sect. 7). Interdependencies among technologies and their uncertainties are essential for achieving robust solutions. The market uncertainty and changes in costs of deployed technologies may produce savings in the cases when the cost of conventional technology unexpectedly rise, e.g., due to external shocks. Therefore, the installation of new expensive technology may have a considerable insurance value besides its potential long-term cost-efficiency (Sect. 6). Another uncertainty is associated with magnitudes and costs of energy reserves, resources and their extraction and production costs (Rogner 1997). All these issues restrict robustness of purely evolutionary technological developments.

3 Myopic Evolutionary Processes

The evolutionary approach for modeling the dynamics of technological changes was inspired by Schumpeter (1934) insights (see discussion in (Grübler and Gritsevskyi 2002; Silverberg et al. 1988)) that technological changes occur due to local search of firms for improvements and imitations of practices of other firms. As models

Modeling Technological Change Under Increasing Returns and Uncertainty 115

in (Arthur 1989; Arthur et al. 1983, 1985a, 1987) demonstrated very natural myopic rules of purely evolutionary (without policy support) technological change processes with increasing returns may lead to a vast variety of lock-in states of technological developments. This implies that products with increasing returns to scale are subjects to positive feedbacks and path-dependencies, i.e., once a particular technological path is led down, it is difficult to move to an alternative path without strong policy support. Let us consider this with some details.

3.1 Behavioral Models, Urn's Scheme

Every form of evolutionary economic behavior is shaped by trial and error. Adoption of new technologies (Arthur 1989; Arthur et al. 1983, 1985a, 1987) in this case can be viewed in terms of additions of units (technologies) channeled according to certain behavioral rules into a pool of existing units. Urn's schemes provide a family of such processes to model adaptive evolutionary dynamics of rather different at first glance discrete processes. We can think of an "urn" (a pool or portfolio of technologies) containing various types of products (technologies) used by consumers. A consumer makes his choice of product to be adopted by asking other consumers using the products. His decision can be based on a simple rule: ask (sample) a number of consumers and adopt the product which is used by the majority of them. Of course, behind such a rule we can see an attempt to make a good decision: if a technology is used by the majority, then it should be good.

Formally, a rather general cumulative process can be modeled as the following urn's scheme. Let $x^1 = (x_1^t, x_2^t, \ldots, x_N^t)$ be proportions of balls of different colors (different technologies). At each time interval $t = 1, 2, \ldots$, a new ball is added at random with probability dependent on t and x^t. This probability is usually not given explicitly, but rather it is defined implicitly, e.g., by a sampling rule channeling additions of balls. Let i_t be the ball added to the urn at time t. Then changes of balls in the urn follow the dynamics

$$x^{t+1} = x^t + \frac{1}{b+t}[\beta^t - x^t], \beta^t = (0, \ldots, 0, i_t, 0, \ldots, 0), t = 1, 2, \ldots \quad (3)$$

where b is the initial total number of balls.

It is striking (see discussion in (Arthur et al. 1987)) that even simplest cases of these processes demonstrate path-dependencies. In 1923 Polya and Eggenberger introduced the following urn scheme. Starting with one red and one white ball in the urn, add a ball each time according to the rule: Choose a ball in the urn at random and replace it; if it is red, add red; if it is white, add white. Polya proved that proportions of balls tend to a limit with probability 1. But the limiting proportions of red (or white) balls are random variables uniformly distributed between 0 and 1. In other words, if this Polya process were run once, the proportion of red balls may settle down say to 22.3927... percent, and never change; if run again, it might settle to 81.4039...percent. A third time it might settle to 42.0641...percent. And so on. This dynamics of urn processes demonstrates a very important feature. There exist significant fluctuations in the proportion of balls at initial steps caused by random sampling rule. But in time total number of balls growth and proportions of balls fluctuate less and less, and, since it does not drift, it settles down (limiting state or structure emerge). Where it settles, of course, depends completely on its early even insignificant random movements. Unfortunately for general urn's schemes the emergence of the technological structure (limiting states) can be analyzed only for systems with two or three variables. As we show further, directions of movements according to (3) can be associated with a stochastic gradient of a function (potential) to be optimized (minimized or maximized). This provides a powerful approach for analyzing long-term implications of such evolutionary processes.

3.2   Market Processes

In the case of urn processes, the increasing return phenomenon occurs due to uncertainty involved in myopic decision rule relying on random sampling. The same type evolutionary dynamics is typical for market driven myopic rules. Various observations from cognitive science indicate that economic agents drive allocation of resources towards increasing their utilities. Accordingly, important myopic rule of evolutionary technological changes can be the cost effectiveness of purchased products: adopt the product which has cheapest unit cost. This again induces of type (3) evolutionary dynamics.

Let us assume that there are technologies (balls), $i = 1, 2, \ldots, n$ with the unit cost $c_1(x_1), c_2(x_2), \ldots, c_n(x_n)$ dependent on the composition of technologies at time t defined by a vector $x^t = (x_1^t, x_2^t, \ldots, x_n^t)$. A new technology (ball) i_t at time t is added to the pool of existing technologies according to the rule

$$i_t = Arg \min_i c_i(x_i^t), \tag{4}$$

i.e., the technology for which the unit cost is currently minimal. It is clear that changes in proportions of technologies again follow (3). The long-term outcomes of such processes depend now on the cost structure $c_i(x_i), i = \overline{1, n}$. Figure 4 illustrates the fundamental difference between diminishing and the increasing returns to scale. In Fig. 4 (left) both technologies exhibit diminishing returns.

Comparative advantage of technology 1 at the initial state, $c_1(0) < c_2(0)$, leads to incremental adoptions according to the market rule (4) up to the level b_1. Beyond this level, the cost effectiveness of technology 1 diminishes, technology 2 becomes dominant with respect to its cost effectiveness up to the level b_2; after this the market again switches to technology 1, and so on. Thus the myopic search for improvements by using "natural" market rule (4) (choice of cheapest technologies) in the case of diminishing returns leads to coexistence of rather different technologies in the overall cumulative portfolio. The composition of them depends on the demand that

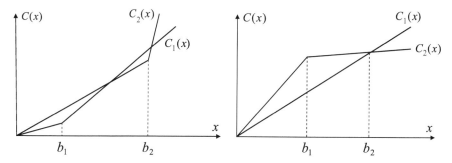

Fig. 4 Diminishing (*left*) and increasing (*right*) returns

is also connected with the time horizons. If the time horizon is such that demand does not exceed b_1, only technology 1 would dominate the market.

The situation significantly changes in the case of increasing returns. The right part of Fig. 4 shows cost of two technologies: mature technology 1 having constant marginal cost and new, therefore more expensive technology 2 at the first stages of developments.

Mature technology 1 is dominant at the initial stage, $c_1(0) < c_2(0)$. It is always advantages to choose according to market rule (4) this technology. Competing technology 2 becomes advantageous only after the support (despite market forces) of its development till the level b_1. Yet, this depends on the planning time horizon and the demand. Only with proper time horizon depending on potential demand exceeding level b_1, the development of technology 2 is advantageous. Yet, this depends also on various uncertainties. In other words, robust solutions require proper treatment of uncertainties and time horizons which may dramatically affect the convergence of evolutionary processes defined by (3), (4), as the next section demonstrates.

3.3 Potential Urn's Schemes, "Trial-and-Error" Experiments

Process defined by (3), (4) is a new type of urn's schemes which can be viewed as potential urn's scheme. In contrast to general schemes in (Arthur 1989; Arthur et al. 1983, 1985a, 1987), the potential urn's schemes are connected with optimization (minimization or maximization) of a function (a potential) providing a powerful approach for studying long-term outcomes of TC and emerging technological structures.

Let us show, that in deterministic case without uncertainties the process (3)–(4) converges to local solutions minimizing total social cost $f(x) = \sum_i C_i(x_i)$. In particular, in the case of diminishing returns this process converges to cost-effective global (equilibrium) solutions. Uncertainties break down this convergence even for convex functions $C_i(x_i)$ unless proper additional market regulations are introduced.

Assume that x_i is the proportion of total cumulative capacity used by technology i. Variables $x = (x_1, \ldots, x_n), x \in X$, where X is defined as

$$\Sigma_i x_i = 1, x_i \geq 0, i = \overline{1, n}.$$

We can always assume that the total capacity of technological pool is bounded by 1 in an appropriate scale. Then vector $\beta^t = (0, \ldots, 0, i_t, 0, \ldots, 0)$ defined by (4) minimizes linear cost-function

$$\sum_i f_{x_i}(x_i^t)x_i = \sum_i c_i(x_i^t)x_i \qquad (5)$$

for $x \in X$, where $f_x = (f_{x_1}, f_{x_2}, \ldots, f_{x_n})$ is the gradient of $f(x)$. In other words, the evolutionary process defined by (3), (4) is a gradient type process

$$x^{t+1} = x^t + \rho_t(\beta^t - x^t), t = 1, 2, \ldots \qquad (6)$$

with the step size $\rho_t = r/t$ for a constant $0 < r < 1$. Its convergence follows from general results on the convergence of stochastic linearization methods (Proposition 1) illustrating shortcomings of market-based evolutionary approaches.

Uncertainties characterized by a random vector ω^t affect unite costs $c_i(x_i^t, \omega_t)$, therefore the myopic market rule (4) has the form

$$i_t = \text{Arg} \min_i c_i(x_i^t, \omega^t) \qquad (7)$$

that selects technology i_t minimizing random linear cost function

$$\sum_i c_i(x_i^t, \omega^t)x_i. \qquad (8)$$

Unfortunately, straightforward stochastic version (6), (7), (8) of the processes (3), (4) does not converge in general. Its convergence can be achieved by essential modification of rule (4). Let us define more stable smooth tendency of random costs by the following averaging operation

$$\alpha_i^{t+1} = \delta_t c_i(x_i^t, \omega^t) + (1 - \delta_t)\alpha_i^t, t = 1, 2, \ldots. \qquad (9)$$

where $0 \leq \delta_t \leq 1$. Accordingly, the vector β^t is defined by the rule

$$i_t = \text{Arg} \min_i \alpha_i^t, \qquad (10)$$

i.e., β^t minimizes $\sum_i \alpha_i^t x_i, x \in X$.

Proposition 1. *Consider the sequence $\{x^t\}$ defined by (3), (9)–(11). Assume that expected costs $C_i(x) = EC_i(x, \omega)$ are continuous functions of x; $F(x) = \sum_i C_i(x)$;*

$$0 \leq \delta_t \leq 1, t\delta_t \to \infty, \sum_{t=1}^{\infty} \delta_t^2 < \infty. \qquad (11)$$

Then the sequence $\{F(x^t)\}$ converges with probability 1 and the sequence $\{x^t\}$ converges with probability 1 to the set of local solutions minimizing the total social cost-function $F(x)$. If $c_i(x_i, \omega)$ is not affected by uncertainties, then the convergence takes place with $\delta_t = 1, t = 1, 2, \dots$.

The proof of this proposition follows from general results discussed in (Barnett and Morse 1967; Ermoliev and Wets 1988). Condition (11) can be fulfilled by choosing $\delta_t = 1/k$ in some time intervals $[\tau_k, \tau_{k+1}]$, $k = 1, 2, \dots$, where $\tau_{k+1} - \tau_k \to \infty$, $k = 1, 2, \dots$.

This Proposition shows that even in the case of diminishing returns the promise of uncertainty require rather deliberate price stabilization mechanisms defined by variables α_i^t. A fundamental issue restricting the cost-effectiveness of myopic evolutionary processes is also the irreversibility of decisions. All these call for coordinated assistance of TC by using decision support tools based on proper long-term valuations and integrated STO models (Sect. 5).

Purely evolutionary passive approaches are often modified by more active "trial-and-error" experiments. Figure 3 illustrates that the assistance of new technologies to break-even points may take long time, i.e. results of the trial-and-error experiments are not observable immediately. In addition, results of experiments at different locations and within different time intervals have significant spatio-temporal heterogeneities. Besides, the feasibility of these experiments is critically restricted by their dimensions. To "hit" purely at random even the set of non-negative solutions in N-dimensional space of decision variables is N^N. Therefore, even for $N = 10$ independent variables, the straightforward trial-and-error approach is able to discover that a non-negative solution provides the desirable outcome would have an extremely large number of failures: on average only once in 10^{10} trials outcomes belong to the set.

4 Valuation of Technological Changes

New technologies are usually more expensive than traditional mature technology. Yet in the future new technologies may become more advanced than traditional technologies with respect not only to purely economic efficiency but also various other indicators reflecting broad social, economic and environmental impacts of new technologies. How can we valuate investments in the development of technologies which may become beneficial within a long time horizon. Short-term horizons of purely market-based valuations may lead to lock-in states of technological developments and misleading policy implications as Sect. 3 shows. This and the next section provide a short comparative analysis of commonly used valuation frameworks. Market-based approaches include Net Present Value (NPV) analysis and the Real Option Theory. The so-called top-down macro models attempt to valuate economy-wide impacts of innovations in a highly aggregate manner. The bottom-up micro or social planner models place emphasis on detailed description of TC processes.

4.1 Net Present Value Analysis, Discounting

This is the most commonly used investment valuation framework. In order to determine the Net Present Value (NPV) of new technologies, we need to estimate technology-generated future expected cash flows V_t, $t = 0, 1, 2, \ldots$ on the basis of an appropriate market model and assumptions about the rate of new technologies deployment. Assume that r is a constant prevailing market interest rate, then alternative investments are compared by $V = V_0 + d_1 V_1 + \ldots$, where $d_t = d^t, d = (1 + r)^{-1}, t = 0, 1, \ldots$, is the discount factor, r is the discount rate, and V denotes NPV. Disadvantages of this criterion are well known. In particular, the NPV critically depends on the prevailing interest rate which may not be easily defined in practice. In addition, the NPV does not reveal the temporal variability of cash flow streams. Two alternative streams may have the same NPV despite the fact that in one of them all the cash is clustered within a few periods, but in another it is spread out evenly over time. Positive cash flows will arise when new technologies have lower market prices or costs comparing with traditional technologies. As Fig. 3 illustrates, there are uncertainties in assessing the deployment time, or break-even time when the new technology becomes economic. This creates a vital difficulty in assessing a traditional constant discount rate.

It was shown (see e.g., (Ermoliev et al. 2010)) that a discounted valuation $\sum_{t=0}^{\infty} d^t V_t$ with discount factor $d < 1$ is equivalent to the undiscounted valuation $E \sum_{t=0}^{\tau} V_t$ with a random τ dependent on d. Namely, let $q = d$, $p = 1 - q$, and τ be a random variable with the geometric probability distribution $P[\tau = t] = pq^t$. Then $d_t = P[\tau \geq t]$, and for $E v_t = V_t$ we have

$$\sum_{t=0}^{\infty} d_t V_t = E \sum_{t=0}^{\tau} v_t.$$

The same is true for general discount factor d_t. We can think of τ as a random "stopping time" moment. The expected duration of τ for standard discount rates d obtained from capital markets does not exceed a few decades and, as such, these rates may easily mismatch uncertain break-even points. The expected duration of τ, $E\tau = 1/p = 1 + 1/r \approx 1/r$. Therefore, for r related to the market interest rate of 3.3%, $r = 0.033$, the expected duration of the stopping time horizon is $E\tau \approx 30$ years, i.e., this rate orients the policy analysis on an expected 30-year time horizon. Certainly, this rate has no relation with break-even point in Fig. 3.

Since NPV valuations use only mean values V_t therefore this approach is also unable to quantify potential insurance value of new technologies which arise when prices of conventional technologies jump up due to some shocks. The use of stopping time criterion (Ermoliev et al. 2010) $E \sum_{t=0}^{\tau} v_t$, with random v_t, and in characterizing the break-even time moments in combination with stochastic decision support models (Sect. 5) allows to overcome these shortcomings.

4.2 Real Option, Top-Down and Bottom-Up Models

In contrast to simple "now-or-never" NPV decision framework, the Real Option approach allows to introduce the timing of decisions. The decision maker may be advised to postpone the development of new technology until the investments become economic. This framework also allows to add insurance value of new technologies in the event of severe conventional technology price increase. The central limitation of this framework, as well as NPV approach, is that new technologies are evaluated from a narrowly defined market-based perspectives. Only combinations of increases and decreases of conventional technologies prices create a chance for new technologies to be adopted and become winners. Definitely other potential socio-economic and environmental benefits of new technologies which are not presented in price signals may be significantly greater than the market-based benefits. This approach deals only with a given option, that is not evaluated together with a set of other interdependent technologies as in social planner models of Sect. 5.

Computable general economic equilibrium (Manne and Richels 1994; Metcalfe 1987; Nordhaus 1973; Rosenberg 1982) macro models have become the standard tool for the analysis of the economy-wide impact of technological changes. It is assumed that the equilibrium prices provide the unique sources of information on which decisions of all economic units are coordinated in a fully decentralized consistent manner. All non-price driven improvements in technology are represented by an exogenous autonomous efficiency improvement parameter or by exogenous assumption about future so-called "backstop" technologies. Therefore, the technological change in these models is described as gradual replacement of existing technologies as relative prices of alternative technologies are changed. Unfortunately, only under very strong assumptions it is possible to show that such equilibrium prices exist. In general, there is no unique equilibrium. It is not possible (without breaking the equilibrium) to include properly in the model increasing returns (non-convexities), inherent uncertainties and risks especially if they are endogenous.

Top-down macro models are usually based on rather aggregate notions such as capital, labor, materials, energy, knowledge (see e.g. discussion in (Arrow 1962; Cowan 1991; Freeman 1994; Metcalfe 1987)) which simplifies the analysis by ignoring critical spatio-temporal, social, economic, environmental and technological heterogeneities. Ideally, these models must be combined with the bottom-up models. The advantages of computers and mathematical modeling tools allow developing new models in which the aggregation is achieved by taking into account directly detailed observable technological, human, economic and environmental aspects. These bottom-up models place emphasis on a detailed description of TC processes, increasing returns, interdependencies among different technologies and feasibility constraints, e.g., pollution reduction targets. They allow to present inherent uncertainties and different agents having geographically heterogeneous risk exposures. In other words, new models are able to provide truly integrated decision support for valuation of policies guiding technological changes towards desirable targets.

5 Systemic Valuations

In the following we outline main features of the bottom-up modeling framework developed at IIASA on the basis of the energy systems-engineering model MESSAGE (see (Gritsevskyi and Ermoliev 1999; Gritsevskyi and Nakićenović 2002; Grübler and Gritsevskyi 2002; Messner et al. 1996)). In general, it is a multi-region, multi-agent model involving uncertain increasing returns on technological developments and other uncertainties in which technology choice takes place (e.g. demand, resource availability, environmental targets). The starting global (Messner et al. 1996) version of the MESSAGE model includes more than 100 different energy extraction, conversion, transport, distribution and end-use technologies. The future costs of all technologies were assumed to be uncertain with cost distributions based on the IIASA energy technology inventory (Strubegger and Reitgruber 1995).

Overall approach is based on the idea of representing energy systems developments as a dynamic network where flows from one energy form to another correspond to energy technologies such as electricity generation from coal or gas power plants. Figure 5 illustrates this network. Five different stages of energy flows are shown – energy extractions from energy resources, primary energy conversion into secondary energy forms, transport and distribution of energy to the point of end use that results in the delivery of final energy, and finally the conversion at the point of end use into useful energy forms that fulfill the specific demands. All possible connections between the individual energy technologies are also specified in Fig. 5. Various demands for useful energy are shown for different sectors of the economy.

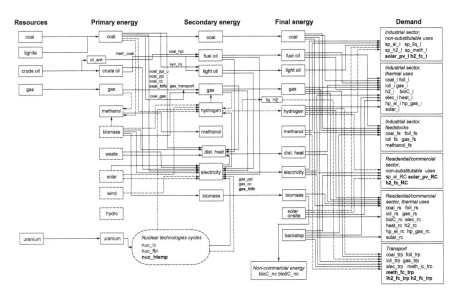

Fig. 5 Schematic diagram illustrating network structure of the energy model. Source: Gritsevskyi and Nakićenović (2002)

Each technology in the system is characterized by unite costs, efficiency, lifetime, emissions, etc. In addition to various balance constraints, there are limitations imposed by the resource availability as a function of (uncertain) costs. Multi-agent modeling environment allows to analyze spillover effects. The overall objective is to fulfill various demands by the utilization of technologies and resources with the minimal total discounted system costs.

The modeling approach relies on two-stage dynamic stochastic optimization model. It is important that this model incorporates both anticipative long-term decisions and their adaptive short-term adjustments once new information is revealed. Sections 6 and 7 show that these two main mechanisms (anticipation and adaptation) for coping with uncertainties ensure robust flexible policies characterized by endogenously defined risk measures. These implicitly induced risk measures depend on the structure of the model including decision variables and distributions of uncertainties. Conventional approaches of the control theory are not applicable in general for this type of models.

Formally the basic model is formulated as the minimization of cost function

$$F(x) = \sum_{t=0}^{T} d^t \left[E \left\langle c^t(x|_0^t, \omega), x^t \right\rangle + \left\langle a^t(x|_0^t, \omega), y^t(\omega) \right\rangle \right] \tag{12}$$

subject to constraints

$$\sum_{k=0}^{t} A_k(\omega) x^k + B_t(\omega) y^t(\omega) = b^t(\omega), t = 0, 1, \ldots, T, \tag{13}$$

$$x^t \in X_t \subseteq R^n, y^t \in Y_t \subseteq R^m, \tag{14}$$

where x^t is a vector of decision variables at time $t = 0, 1, \ldots, T$, $x|_0^t = (x^0, x^1, \ldots, x^t)$, $x = x|_0^T$; ω defines random (uncertain) variables and d^t is a discount factor at time t that can be substituted also by random stopping time (horizon) τ (Sect. 4.1) associated with break-even points; $C^t(x|_0^t, \omega)$ are stochastic unit costs a given technology path $x = x|_0^T$; matrices $A_t(\omega)$, $B_t(\omega)$ and vectors $b_t(\omega)$ reflect uncertain relations for resources, links between technology activities, energy demands, and environmental constraints. This model is a specific case of the dynamic two-stage STO models (see (Ermoliev and Wets 1988)) with the first stage decision vector x, and the second stage decision vector $y(\omega) = (y^0(\omega), y^1(\omega), \ldots, y^\tau(\omega))$. To model the increasing returns, marginal cost $c^t(\cdot, \omega)$ is represented by experience curves (Sect. 2), therefore expected value $E \left[\langle c^t(x|_0^t, \omega), x^t \rangle \right]$ is in general a non-convex non-smooth function. Because probability distributions of costs have multimodal character (Strubegger and Reitgruber 1995), the use of expected costs may be misleading. Instead, it is natural to use such robust characteristics as median and other percentiles. The next sections illustrate this by using simple stylized models.

6 Interdependencies and Endogenous Risks

This section shows that proper representation of interdependencies among uncertainties, and ex-ante and ex-post decisions of the model (12)–(14) induces endogenous risk aversion and collective robust risk sharing solutions.

6.1 Multiagent Framework Under Uncertainty

The classical assertion about gains from production by most efficient technology, specialization and intensification is true only if uncertainties are not taken into account. In reality, production technologies may be exposed to different contingencies. In this case, of particular importance is a properly organized network of producers (technologies) allowing to diversify risks and provide mutual insurance. In this case as this section shows, the less efficient technology (producer) may provide the supply of production and enhance stability of the system. Let us illustrate this with stylized versions of social planner model outlined in Sect. 5.

For the sake of clarity suppose that there are only two technologies $i = 1, 2$, producing the same good. Let x_i denote the production level of the technology i, c_i is production unit cost. The production can also be adjusted by a back-stop technology with cost b for unit produce y. Assume $c_1 < c_2 < b$, i.e., the cheapest is the first technology. The energy security constraint is to satisfy the exogenous inelastic demand d.

In the absence of uncertainty we assume there is no distortion of energy production and no additional regulations on the size of the production capacities x_1, x_2. The model is formulated as the minimization of the total cost function:

$$c_1 x_1 + c_2 x_2 + by \tag{15}$$

subject to feasibility constraints

$$x_1 + x_2 + y = d, x_1 \geq 0, x_2 \geq 0, y \geq 0, \tag{16}$$

i.e., the model assumes interdependence and possible cooperation between producers using different technologies. The optimal solution to the problem is $x_1^* = d$, $x_2^* = 0$, $y^* = 0$, i.e., the production is undertaken by the most efficient technology that accords with classical views.

Consider more realistic problem of planning energy production under uncertainty of outputs which may reduce the production x_1, x_2. In this case, (16) is transformed to constraint

$$a_1 x_1 + a_2 x_2 + y = d, \tag{17}$$

where a_1, a_2 are contingencies or shocks to x_1, x_2 due to hazardous events. In this section, we assume that a_1, a_2 are random variables $0 \leq a_i \leq 1, i = 1, 2$, i.e., $\omega = (a_1, a_2)$. Other representations of uncertainty are analyzed in Sect. 7.

Modeling Technological Change Under Increasing Returns and Uncertainty 125

Uncertainty ω is revealed between periods 1 and 2. The ex-ante decisions x_1, x_2 are made in period 1, whereas the ex-post back-stop decision y is made in period 2 using available ω, i.e., y is a function of ω, $y(\omega)$. The social planner model (12)–(14) is formulated now as the minimization of total expected cost

$$c_1 x_1 + c_2 x_2 + bE y(\omega) \tag{18}$$

subject to energy security constraint

$$a_1 x_1 + a_2 x_2 + y(\omega) = d$$

for all ω. If endogenous supply $a_1 x_1 + a_2 x_2$ falls short of demand d, the residual amount $d - a_1 x_1 - a_2 x_2$ must come from back-stop technology at unit cost b. Clearly, the optimal period 2 decisions is $y(x, \omega) = \max\{0, d - a_1 x_1 - a_2 x_2\}$, that is, it depends non-smoothly on period 1 decisions (path-dependencies) and ω, providing strong cross-period random interactions among decisions. Therefore, minimization of cost-function (18) of the social planner model is equivalent to the following stochastic minimax problem: minimize

$$F(x) = c_1 x_1 + c_2 x_2 + bE \max\{0, d - a_1 x_1 - a_2 x_2\},$$

where $bE \max\{0, d - a_1 x_1 - a_2 x_2\}$ is the expected back-up cost when the demand d exceeds random supply $a_1 x_1 + a_2 x_2$.

6.2 Induced Systemic Risks

The following shows why the less-efficient producer (technology) is able to stabilize the overall production of the system. Conversely, inadequate behavior of this producer generates insecurity of the energy supply system. Assume that only the efficient technology is at risk, i.e., $a_2 = 1$. Let function $F(x)$ have continuous derivatives, e.g., the cumulative distribution function of a_1 has a continuous density function. It is easy to see that the optimal positive decision $x_1^* > 0$, $x_2^* > 0$ exists in the case when $F_{x_1}(0,0) < 0$, $F_{x_2}(0,0) < 0$. We have $F_{x_1}(0,0) = c_1 - bEa_1$, $F_{x_2}(0,0) = c_2 - b$. Therefore, somewhat surprisingly, the less efficient technology must be active unconditionally (since $c_2 - b < 0$). The cost efficient technology is inactive in the case $c_1 - bEa_1 \geq 0$, leaving production entirely to the higher-costs technology 2 ($c_2 > c_1$). Only in the case $c_1 - bEa_1 < 0$ both technologies are active. The "less cost-efficient" technology 2 is able to stabilize the aggregate production in the presence of uncertainties affecting the "more cost-efficient" technology 1. It is important to derive the production share of the technology 2. The derivative $F_{x_2}(x_1, x_2) = c_2 - bP[d > a_1 x_1 + x_2]$ can be found by using formulas for optimality conditions of stochastic minimax problems (see, e.g., (Ermoliev and Wets 1988), and references therein). This means that the optimal production level $x_2^* > 0$ of technology 2 is a quantile defined by the equation

$$P[d > a_1 x_1^* + x_2^*] = c_2/b, \tag{19}$$

assuming $x_1^* > 0$ (otherwise $x_2^* = d$). This is an endogenously induced risk aversion due to interdependencies among ex-ante and ex-post decisions subject to the security constraints. Thus, the market share of the risk-free higher-cost technology 2 is determined by the quantile of the distribution function describing uncertainties a_1 of the risk-exposed technology 1 and by the ratio of c_2/b, i.e., of production cost c_2 and back-up cost b. It also depends on x_1^*. Although not at risk ($a_2 = 1$), the optimal production level of technology 2 is defined by (19) through interdependencies among technologies participating in the same energy supply system. We can call it as a systemic risk. Interdependencies induce endogenous systemic risks and quantile type energy security constraints (19). Therefore, apart from exogenous risks, the production and the market are subject to endogenous risks dependent on the level of x_1, x_2. In financial applications (Rockafellar and Uryasev 2000) these constraints characterize Value-at-Risk (VaR). Optimal value of stochastic minimax model $F(x*)$ characterize conditional Value-at-Risk (CVaR).

In the case when both technologies (producers) are exposed to risks, the existence of optimal positive production follows from similar equations

$$F_{x_1}(0,0) = c_1 - bEa_1 < 0,$$

$$F_{x_2}(0,0) = c_2 - bEa_2 < 0.$$

The structure of optimal solution conceptually is similar as in case Sect. 6. In particular, there may be a situation with $c_2 - bEa_2 \geq 0$, when producer 2 is inactive, but the cost effective producer 1 is active now with the insurance provided by the back-stop technology.

These examples emphasize that market shares of technologies are to a larger extent determined by the contingencies and interdependencies of technologies, in which case the less efficient but with lower risk technology will likely have a higher share than a more efficient, but with higher risk exposure technology.

6.3 Uncertainty and Increasing Returns

Negative impacts of standard technologies may incur high implicit costs induced by regulations. Although this creates favorable conditions for new technologies, uncertainties of the demand may affect them. Let us consider the right chart of Fig. 4. If energy demand d does not exceed the break-even point b, then technology 1 is the cost-efficient solution. In this case, new expensive technology 2 may become cost-efficient only under additional feasibility constraints, say, on permitted secure pollution level l of the type

$$l_1 x_1 + l_2 x_2 \le l$$

where l_i is the pollution from unit production by technology i, $l_1 a > \gamma_1, l_1 > l_2$. If the planning time horizon is such that the demand exceeds b in Fig. 4, then the optimal solution is to develop new technology 2 with increasing returns starting from $t = 0$.

Uncertainties and risks may significantly affect this conclusion. The break-even point b characterizes only an expected value, say, $b = pb_1 + (1-p)b_2$, where b_1, b_2 are break-even points associated with two probable slopes of cost function $C_2(x)$ characterized by probabilities p and $1 - p$. Therefore, the value b is represented by an optimistic scenario $b_1 < b$ and a pessimistic scenario $b_2 > b$. Assume p is large enough, i.e., $b_1 < b$ has a large probability and the risk $1 - p$ of extreme event b_2 can be ignored. Then mature technology 1 provides a cost-efficient solution with the safety level p and the value at risk b_2 with probability $1 - p$. Conversely, if $1 - p > p$, and we can ignore extreme event b_1, then the cost-efficient solution requires immediate development of new technology 2 that will have the safety level $1 - p$ and the value at risk b_1 with probability p. The value at risk is understood in the sense of potential higher costs associated with the use of new technology. The safety constraints in this example are similar to risk indicator, see Sect. 8.1.

7 Decisions Under Uncertainty

This section analyzes alternative representations of uncertainty in the simplest case of two-period bottom up model (12)–(14) with uncertain demand.

7.1 Scenario Analysis, Pareto Optimality

Assume again that there is only two time intervals: the current and the future, $t = 1, 2$. Let us denote by x a total energy production level which is feasible to achieve in period $t = 1$ in order to meet uncertain demand $d(\omega)$ in period 2. Let us denote $d(\omega)$ for the simplicity by a random variable θ.

The main question then is the following: in which sense x satisfied uncertain demand θ, i.e.,

$$\text{``}x = \theta\text{''}. \tag{20}$$

The standard deterministic models assume that θ is known, therefore, in this case equation (20) raises no questions about its solution, $x^* = \theta$.

As long as there is uncertainty about the demand θ, then for any fixed x, there will be risks associated with the underestimation of the demand, when $x < \theta$ and with its overestimation, when $x \ge \theta$. The situation $x \ge \theta$ is associated with sunk costs, in other cases, with holding costs. Let us assume that these costs can be characterized

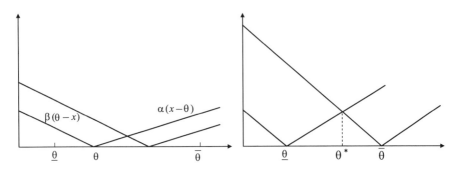

Fig. 6 Uncertain costs (*left*) and worst-case solution (*right*)

by a linear function $\alpha(x - \theta)$, where α is the unit surplus-cost. The situation $x < \theta$ is characterized by a linear function $\beta(\theta - x)$, where β is the unit shortage cost.

The cost function $C(x, \theta)$ associated with a decision x can be written as $C(x, \theta) = \gamma x + \alpha(x - \theta)$, when $x \geq \theta$ and $C(x, \theta) = \gamma x + \beta(\theta - x)$ when $x < \theta$, or in short

$$C(x, \theta) = \gamma x + \max\{\alpha(x - \theta), \beta(\theta - x)\}, \tag{21}$$

where γ is the unite production costs. Figure 6 (left) shows this function for $\gamma = 0$. It is possible now to reformulate the symbolic (20) as the minimization of $C(x, \theta)$ for all uncertain θ. This is the simplest decision problem under uncertainty that can be used as a test problem for evaluating perspectives of different approaches for designing robust solutions. Let us consider some of them. For the sake of simplicity we often assume that $\gamma = 0$.

According to the traditional scenario analysis, the model, in our case the minimization of cost function $C(x, \theta)$ "for all ω" is solved for every scenario $\theta_1, \theta_2, \dots$. Figure 6 (left) shows that for any given θ the optional solution is $x(\theta) = \theta$. Therefore, scenarios $\theta_1, \theta_2, \dots, \theta_N$ produce trivial solutions

$$x(\theta_1) = \theta_1, x(\theta_2) = \theta_2, \dots, x(\theta_N) = \theta_N,$$

which do not tell us how to choose some decision x that will be a reasonably good (robust) whatever be the uncertain demand θ.

We can also view the minimization of $C(x, \theta)$ for all θ as a multi-objective optimization problem with the set of criteria $C(x, \theta), \theta \in \Theta$. Figure 6 (left) shows that any solution $x(\theta) = \theta$ is a Pareto optimal solution, i.e., any point θ from the set of uncertain demand Θ is a Pareto-optimal solution. Again the main question is about a Pareto-optimal solution that will be in a sense robust against all potential scenarios of demand from Θ. Let us also realize that any scenario θ may have the likelihood 0.

7.2 Interval Uncertainty, Attainable Sets, Worst-Case Analysis

One clear and easy way to characterize uncertainty is by an interval or even sets of possible values – set-valued estimates. In other words, we can think of uncertainty θ as a variable assuming values from a set Θ. For example, we may assume that uncertain demand θ is between a lowest pessimistic value $\underline{\theta}$ and highest optimistic value $\overline{\theta}$. The interval $[\underline{\theta}, \overline{\theta}]$ characterizes also the set of attainable solutions, i.e., any point in $[\underline{\theta}, \overline{\theta}]$ cannot be excluded as a candidate for robust solution. Further research proceeds often by assigning a certain reference point μ for real demand θ, usually the middle point or the centroid of Θ; the optimal robust solution is calculated by $\min_x C(x, \mu)$. In our case, the optimal solution $x(\mu)$ minimizing $C(x, \mu)$ is $x(\mu) = \mu$ with optimal value $C(x(\mu), \mu) = 0$. The sensitivity analyses is usually used to examine values $C(x(\mu), \theta)$ for $\theta \neq \mu$. For our problem, the worst case situation occurs when $\theta = \underline{\theta}$ if $\alpha > \beta$ or $\theta = \overline{\theta}$ if $\alpha < \beta$, i.e., $x(\mu)$ leads to large variations of $C(x(\mu), \theta)$, $\theta \neq \mu$.

Our simplest decision problem under uncertainty shows that the straightforward calculation of optimal solutions for every scenario provides no clue about a robust solution although each of them can be a Pareto optimal solution. Set-valued estimates of uncertainty by intervals show that an arbitrary solution such as the middle point μ may have large variations for $\theta \in \Theta$. The worst-case approach chooses a solution x minimizing impacts of the worst-case scenario, i.e., the function

$$F(x) = \max_{\theta} C(x, \theta).$$

Figure 6 (right) shows that if the uncertainty of θ is characterized by an interval $[\underline{\theta}, \overline{\theta}]$, the worst-case solution is $\theta^* = (\beta\overline{\theta} + \alpha\underline{\theta})/(\alpha + \beta)$, which can be viewed as the mean value of two extreme situations defined by the lowest $\underline{\theta}$ and the highest $\overline{\theta}$ demands with weights $\alpha/(\alpha + \beta)$, $\beta/(\alpha + \beta)$. Thus, the worst case analysis is in a sense equivalent to assigning positive probability weights $\alpha/(\alpha + \beta)$, $\beta/(\alpha + \beta)$ only to the extreme situations $\underline{\theta}, \overline{\theta}$ and calculating then the mean value of $\underline{\theta}$ and $\overline{\theta}$ with respect to these weights. This approach may be considered as a step back from standard deterministic models when the,, solution of (20) is chosen as the mean value $E\theta$ of θ calculated on the basis of historical observations or/and questionnaires about all possible θ.

Unfortunately the characterization of uncertainty only by sets does not provide any additional information about more or less reasonable values of uncertain variables. In this situation the guaranteed solutions $x = \overline{\theta}$ is often recommended which satisfies any demand θ from the interval $[\underline{\theta}, \overline{\theta}]$ although $\overline{\theta}$ practically has probability 0.

7.3 Weights

The simple and easy characterization of uncertainties by sets of scenarios with straightforward scenario analysis and worst-case calculations arose serious skepticism about robust solutions. The following example illustrates more deep concerns regarding these approaches.

Consider a two type of situations with uncertainty facing by a group of 10 producers. In the first situation the disruption of supply may occur at once from all producers due to some systemic failure in one out of ten years. In the second situation the disruption of supply may occur independently from each of the ten producers also in one out of ten years. The set-values estimates of the uncertainty for the first type of shock is the set 0,10, whereas the set-values estimate for the second type of shock is the set $0, 1, \ldots, 10$. The scenario analysis and the worst case calculations would not emphasize the first type of disruption. But actually this is the most destructive situation, although the set of scenarios $0, 10$ is a subset of much richer set $0, 1, \ldots, 10$. In fact, we have more information then just a set of scenarios. The chance of disruption from all 10 producers is 10^{-1} in the first type of shock and only 10^{-10} in the second type of shock. Definitely, from this information the main attention in the policy recommendation must be given to the security of the supply under the first type of shock.

A more general idea to characterize the uncertainty is not only by identification a set Θ of admissible scenarios $\theta \in \Theta$, but also by ranking them according to the frequency of θ or a degree of our belief. For example, if there is only a finite number of scenarios $\theta_s, s = 1, \ldots, N$, then the main idea is to derive a number p_s, between 0 and 1, to indicate the degree of support $p_s, p_s > 0, \sum_{s=1}^{N} p_s = 1$, from existing evidences (historical observations, questionnaires, experts opinions) for a scenario θ_s to be the true scenario of the uncertainty. In particular, if there is no evidences strongly supporting some of scenarios then they may receive equal weights $p_s = 1/N$.

Weights can be also derived not only for a particular scenario, but also for sets of scenarios. For example, if two experts characterize uncertainty by two different partially overlapping interval $[\theta_1^1, \theta_2^1], [\theta_1^2, \theta_2^2], \theta_1^1 < \theta_1^2 < \theta_2^1 < \theta_2^2$, then we can distinguish three sets: $[\theta_1^1, \theta_1^2], [\theta_1^2, \theta_2^1]$, and $[\theta_2^1, \theta_2^2]$. Intervals $[\theta_1^1, \theta_2^1], [\theta_1^2, \theta_2^2]$ received support by both experts. Therefore, we can support these intervals with weights (1/4,2/4,1/4), where 4 is the total number of "votes" given by experts. These weights can be interpreted as probabilistic degrees of believes providing a powerful approach for coping with uncertainties as Sects. 6 through 8 demonstrate. Weights may also be characterized inequalities, e.g., $p_1 + p_3 \leq p_2 + p_5 + p_8, p_8 \leq p_9$, and so on.

7.4 Probabilistic Degree of Belief, Fuzzy Sets

For several centuries the idea of numerical degree of belief has been identified with the idea of probability. The use of probability or probabilistic calculus (the rule for chances) in the theory of beliefs usually called the Bayesian theory was formalized by the English clergyman T. Bayes (1702-1761). The notion of probability as the numerical degree of support of evidences is viewed either as indeed objective determined by given evidences, or subjective (personalized) view, which can be

Modeling Technological Change Under Increasing Returns and Uncertainty

discovered by observing individuals opinions. The subjective view has become predominant in the Bayesian theory after F. Ramsey, B. deFinetti, and especially since L.J. Savage published his foundations of statistics in 1954 (see discussion in (Shafer 1976)).

The ranking of alternative scenarios of uncertainty by probability weights has essential advantages for modeling of complex systems. First of all, this allows us in a consistent manner to represent within the same model both statistical uncertainty from "hard" statistical evidences (observations) and from "soft" public and expert opinions. Secondly it allows the use of powerful stochastic Monte Carlo simulation methods to model the propagation of uncertainties through complex systems and the valuations of robust solutions by using distribution free stochastic optimization methods.

Thirdly, the probabilistic approach allows us to formulate the concept of learning processes with sequential resolution of uncertainties and their interdependencies in time and space. This has decisive consequences for designing flexible robust policies incorporating within the same model ex-ante (anticipative) decisions x jointly with their ex-post (adaptive) adjustments y activated when new information is arrived.

The model of Sect. 5 and its simple versions in Sect. 6 incorporates such decisions. From Sect. 6 follows that the coexistence of such decisions within the same two-stage model induces robust solutions characterized by quantiles defined by the structure of the model. This co-existence creates a key feature of robust policies – their flexibility of adapting to new information. Unfortunately, other approaches e.g. the fuzzy-set theory have no such possibilities. Besides, there is no well established empirical method to quantify fuzziness (or vagueness) similar to frequency analysis of real observations, experiments, results of questionnaires or expert judgments as it exists in the probabilistic approaches.

8 Probabilistic and Stochastic Models

The power of probabilistic approaches stems also from the ability to represent uncertainty by either probabilistic or stochastic models. In other words, scenarios of uncertainties can be characterized by probability distributions or/and by random variables often in the form of scenario generators. For example, by using Monte Carlo simulations, a random set of scenarios can be easy sampled (generated) for any decisions x from analytically intractable multivariate dependent on x probability distributions. Stochastic optimization methods allow then to find robust solutions of complex analytically intractable models by using only these observations. In contrast to stochastic, probabilistic models attempt to characterize the uncertainty completely and explicitly in terms of analytically tractable characteristics of probability distribution functions. For example, expected values that lead often to solving rather complex partial and integro-differential equations analytically and even numerically tractable only for small number of variables and specific simple structures of models.

Even for two random variables θ_1 and θ_2 with known probability distributions, the evaluation of probability distribution for the sum $\theta_1 + \theta_2$ is already an analytically intractable (in general) task. In the decision analysis this sum depends on some decision variables e.g., we have $x_1\theta_1 + x_2\theta_2$, where $x_1 \geq 0$, $x_2 \geq 0$ are decision variables. The distribution of $x_1\theta_1 + x_2\theta_2$ is significantly affected by decision variables requiring the evaluation of a family of distribution functions: compare the situations $x_1 = 0, x_2 = 1$, and $x_1 = 1, x_2 = 0$. Stochastic optimization models deal directly with observations of random variables of type $x_1\theta_1 + x_2\theta_2$ for different x_1, x_2 without exact evaluation of their distributions (Ermoliev and Wets 1988; Marti 2005).

8.1 Risk Measures

It is important to see how the analyses of robust solutions changes with the introduction of probabilistic demand θ. Consider again function $C(x, \theta)$ defined by (20). For the simplicity of calculations let us assume that uncertainty θ is characterized by a continuous probability density function, therefore the expected cost function

$$F(x) = \gamma x + E \max\{\alpha(x - \theta), \beta(\theta - x)\} \tag{22}$$

has continuous derivatives. The minimization of $F(x)$ defines a solution that is optimal in a sense with respect to all θ. The function $F(x)$ defined by (22) is the simplest case of so-called stochastic minimax models (see (Ermoliev and Leonardi 1982; Ermoliev and Wets 1988)); If α, β, γ are deterministic unit cost, and $\beta > \gamma$, then the unique optimal solution x^* minimizing $F(x)$ exists and it satisfies the following optimality condition (see (Ermoliev and Leonardi 1982; Ermoliev and Yastremskii 1979; Rockafellar and Uryasev 2000)).

$$P\{\theta \geq x\} = \frac{\alpha + \gamma}{\alpha + \beta}. \tag{23}$$

In other words, the robust solution x^* is a quantile of the probability distribution of θ which guarantees that only with probability $(\alpha + \gamma)/(\alpha + \beta)$ actual demand θ exceeds the production level x^*. In financial applications solution x^* satisfying (23) is characterized by important Value-at-Risk (VaR) risk measure (see discussion in (Rockafellar and Uryasev 2000)). In other words, simplest stochastic optimization version (22) of minimization function (21) induces risk aversion in the form of quantile characterized by (23). It is important that x^* characterized by (23) utilizes the whole distribution of θ, and not only two worst case scenarios $\underline{\theta}, \overline{\theta}$ as in the worst-case approach of Sect. 7.2. Thus probabilistic model of uncertainty induces risk aversion dependent on the whole structure of the decision model including the distribution of uncertainty, and the structure of costs.

It is interesting to compare concepts of solution defined by (23) for uniform distribution of θ in $[\underline{\theta}, \overline{\theta}]$ and the worst-case approach defined in Sect. 7.2. From (23)

Modeling Technological Change Under Increasing Returns and Uncertainty

for $\gamma = 0$ it follows that the optimal solution x^* of (23) for the uniform distribution in $[\underline{\theta}, \overline{\theta}]$ satisfies the equation

$$\frac{\overline{\theta} - x^*}{\underline{\theta} + \overline{\theta}} = \frac{\alpha}{\alpha + \beta}$$

or

$$x^* = \overline{\theta} \frac{\beta}{\alpha + \beta} + \underline{\theta} \frac{\alpha}{\alpha + \beta},$$

i.e., it is identical with the worst-case solution.

8.2 Two-Stage Model

It is important now to see connections between induced risk aversion of type (23) and adaptive ex-post decisions $y(\theta)$ of STO model defined by (12)–(14). The minimization of function (22) can be reformulated in the form of model (12)–(14) with ex-ante anticipative decision x and ex-post adaptive decision $y(\theta) = (y_1(\theta), y_2(\theta))$ defined by minimizing

$$\gamma x + \alpha E y_1(\theta) + \beta E y_2(\theta)$$

subject to security of supply constraints

$$x_2 + y_1(\theta) - y_2(\theta) = \theta.$$

This is a simplest version of stochastic model defined by (12)–(14), where $y_1(\theta)$, $y_2(\theta)$ are ex-post decisions acting after observation of real demand θ. Since $\alpha > 0$, $\beta > 0$ then the optimal $y_1^*(\theta) = \max\{0, \theta - x\}$ and $y_2^*(\theta) = \max\{0, x - \theta\}$, what leads to the choice of ex-ante optimal x^* minimizing function (22).

In other words, the co-existence of ex-ante x and ex-post decisions $y(\theta)$ in model (12)–(14) induces systemic risk aversion endogenously defined by the structure and decisions of the whole related energy system that in the simplest case reduces to (23).

9 Concluding Remarks

Analyzed simple models of the TC under uncertainty and increasing returns easily demonstrate that new technologies require dedicated efforts. Such technologies are initially unattractive, but they offer uncertain potential for future improvements. In this sense, technological change arises out of the rationality (from social, economic, environmental, energy security perspectives) pursuing investments in anticipation of future returns within an appropriate inter-temporal optimization framework. Simplicity of the model easily shows the need for proper treatment of uncertainty and robust solutions calling for systemic rather than standard NPV or real option theory valuations. New approaches to endogenous discounting allow to focus on uncertain long-term horizons of break-even points dependent on policies which affect break-even points, feedback discounting, and so on. Analyzed simplified models provide insights for designing realistic large scale and long-term models.

For example, they demonstrate why simple scenario-by-scenario and/or decision-by-decision evaluations may be misleading, and why stochastic models allow to address jointly anticipative and adaptive stages of robust decision processes.

Acknowledgements We appreciate the collaboration with IIASA colleagues, especially with Brian Arthur, Gordon J. MacDonald, Alexandr Golodnikov, Arnulf Grübler, Nebojsa Nakićenović, Sabine Messner, and Manfred Strubegger, on various issues discussed in this chapter. Initial tests of the methodology were performed jointly with Alexandr Golodnikov, Arnulf Grübler, Sabine Messner. Nebojsa Nakićenović, Sabine Messner and Gordon J. MacDonald worked on one of the authors on the grant from the National Energy Research Scientific Computing Center at Lawrence Berkeley National Laboratory funded by the US Department of Energy; we are grateful for this financial support which was necessary for a realistic model implementation on the Gray T3E-900 supercomputer. We also thank the participants of the workshop on Coping with Uncertainty for important and constructive discussions, and the two anonymous referees for comments which led us to improvements of this chapter.

References

Abramovitz, M. (1993). The search for the sources of growth: areas of ignorance, old and new. *Journal of Economic History, 52*(2), 217–243.

Argote, L., & Epple, D. (1990). Learning curves in manufacturing. *Science, 247*, 920–924.

Arrow, K. (1962). The economic implications of learning by doing. *Review of Economic Studies, 29*, 155–173.

Arthur, W. B. (1989). Competing technologies, increasing returns, and lock-in by historical events. *The Economic Journal, 99*, 116–131.

Arthur, W. Brian, Ermoliev, Y. M., & Kaniovski, Y. M. (1983). A generalized urn problem and its applications (in Russian), *Kibernetika, 19*, 49-56. English translation in *Cybernetics, 19*, 61–71.

Arthur, W. Brian, Ermoliev, Y. M., & Kaniovski, Y. M. (1985a). Strong laws for a class of path-dependent urn processes. In Arkin, Shiryaev, & Wets (Eds.), *Proceedings of the International Conference on Stochastic Optimization, Kiev 1984.* Springer: Lecture Notes in Control and Information Sciences.

Arthur, W. Brian, Ermoliev, Y. M., & Kaniovski, Y. M. (1987). Path-dependent processes and the emergence of macro-structure. *European Journal of Operational Research, 30*, 294–303.

Barnett, H. J., & Morse, C. (1967). *Scarcity and Growth. The Economics of Natural Resource Availability.* Baltimore, USA: John Hopkins University Press.

Cowan, R. (1991). Tortoises and hares: choice among technologies of unknown merit. *The Economic Journal, 101*, 801–814.

Ermoliev, Y., Ermolieva, T., Fischer, G., Makowski, M. (2010). Extreme events, discounting and stochastic optimization. *Annals of Operations Research, 177*(1), 9–19.

Modeling Technological Change Under Increasing Returns and Uncertainty

Ermoliev, Y. M., & Leonardi, G. (1982). Some proposals for stochastic facility location models. *Mathematical Modelling, 3*, 407–420.

Ermoliev, Y. M., & Yastremskii, A. I. (1979). *Stochastic Models and Methods in Economic Planning*. Moscow: Nauka [in Russian].

Ermoliev, Y., & Wets, R.J.-B. (1988). *Numerical Techniques for Stochastic Optimization*. Berlin, Germany: Springer.

Freeman, C. (1994). The economics of technical change. *Cambridge Journal of Economics, 18*, 463–514.

Griliches, Z. (1996). The discovery of the residual: a historical note. *Journal of Economic Literature, XXXIV*, 1324–1330.

Gritsevskyi, A., & Ermoliev, Y., (1999). An Energy Model Incorporating Technological Uncertainty, Increasing Returns and Economic and Environmental Risks. Proceedings of International Association for Energy Economics 1999 European energy Conference "Technological progress and the energy challenges", Paris, France.

Gritsevskyi, A., & Nakićenović, N. (2002). modeling uncertainty of induced technological change. In Grübler, A., Nakićenović, N., & Nordhaus, W. D. (eds.), *Technological Change and the Environment, Resources for the Future and International Institute for Applied Systems Analysis*, pp. 251–279, Published by Resources for the Future, Washington, USA.

Grossman, G. M., & Helpman, E. (1991). *Innovation and Growth in the Global Economy*. Cambridge, StateMA, USA: MIT Press.

Grübler, A., & Gritsevskyi, A. (2002). A model of endogenized technological change through uncertain returns on innovation, pp. 280–319. In A. Grübler, N. Nakićenović, & W. D. Nordhaus, (Eds.), *Technological Change and the Environment, Resources for the Future and International Institute for Applied Systems Analysis*, Published by Resources for the Future, Washington, United States of America.

Grübler, A., & Messner, S. (1996). Technological uncertainty. In N. Nakićenović, W. D. Nordhaus, R. Richels & F. L. Toth (Eds.), *Climate Change: Integrating Science, Economics, and Policy, CP-96-001, International Institute for Applied Systems Analysis*, Laxenburg, Austria.

Manne, A. S., & Richels, R. G. (1994). The cost of stabilizing global co2 emissions: a probabilistic analysis based on expert judgment. *The Energy Journal 15*(1), 31–56.

Marti, K. (2005). *Stochastic Optimization Methods*. Berlin, Germany: Springer.

Messner, S., Golodnikov, A., & Gritsevskyi, A. (1996). A stochastic version of the dynamic linear programming model MESSAGE III. *Energy, 21*(9), 775–784.

Metcalfe, S. (1987). Technical change. In J. Eatwell, M., Milgate & P. Newman (Eds.), *The New Palgrave, A Dictionary of Economics*, Vol. 4, pp. 617–620. London, UK: Macmillan.

Nakićenović, N. (1996). Technological change and learning. In N. Nakićenović, W. D. Nordhaus, R. Richels & F. L. Toth (Eds.), *Climate Change: Integrating Science, Economics, and Policy, CP-96-001, International Institute for Applied Systems Analysis*, Laxenburg, Austria.

Nordhaus, W. D. (1973). The allocation of energy resources. *Brookings Papers on Economics Activity, 3*, 529–576.

Rogner, H.-H. (1997). An assessment of world hydrocarbon resources. *Annual Review of Energy and Environment, 22*, 217–262.

Romer, P. M. (1986). Increasing returns and long-run growth. *Journal of Political Economy, 94*, 1002–1137.

Rockafellar, T., & Uryasev, S. (2000). Optimization of conditional-value-at-risk. *The Journal of Risk, 2*, 21–41.

Rosenberg, N. (1982). *Inside the Black Box: Technology and Economics*. Cambridge, UK: Cambridge University Press.

Ruttan, V. W. (1997). Induced innovation, evolutionary theory and path dependence: sources of technical change. *The Economic Journal, 107*, 1520–1529.

Shafer, G. A. (1976). *Mathematical Theory of Evidence*. Princeton: Princeton University Press.

Schumpeter, J. A. (1934). *The Theory of Economic Development: And Inquiry into Profits, Capital, Credit, Interest and the Business Cycle*. Cambridge, MA, USA: Harvard University Press.

Silverberg, G., Dosi, G., & Orsenigo, L. (1988). Innovation, diversity and diffusion: a self-organization model. *The Economic Journal, 98*, 1032–1054.

Strubegger, M., & Reitgruber, I. (1995). *Statistical Analysis of Investment Costs for Power Generation Technologies, WP-95-109*. Laxenburg, Austria: International Institute for Applied Systems Analysis.

Study of Learning Curves (2000). IEA 264.

Stochastic Programming Perspective on the Agency Problems Under Uncertainty

Alexei A. Gaivoronski and Adrian Werner

Abstract We study the application of the stochastic programming framework to the analysis of complex agency problems under exogenous and endogenous uncertainty, presenting several models that deal with different types of such uncertainty. We demonstrate that the utilization of this framework extends the possibilities for the definition of parameters of incentive schedules. In this paper, we often refer to a model in a telecommunication environment consisting of a regulator (principal) and a service provider (agent) and study different aspects of regulation and licensing. However, the results can easily be generalized to other principal agent relationships.

1 Introduction

Agency or principal-agent problems refer to the multitude of economic and organizational situations where one or more agents perform actions involving management financial and other resources on behalf of other agent called *principal*. The principal has to employ incentive mechanisms in order to align the objectives of managing agents and make the results conforming to his objectives. The classical example of such situation is the relationship between shareholders of a firm and its management. Similar and important example is the relationship between the regulatory bodies striving to maximize the public good and industrial actors belonging to some industrial branch like telecommunications. Important topic of the current debate on

---

A.A. Gaivoronski (✉)
Department of Industrial Economics and Technology Management,
Norwegian University of Science and Technology, Trondheim, Norway
e-mail: Alexei.Gaivoronski@iot.ntnu.no

A. Werner
Department of Applied Economics and Operations Research, SINTEF Technology and Society
Trondheim, Norway
e-mail: Adrian.Tobias.Werner@sintef.no

Y. Ermoliev et al. (eds.), *Managing Safety of Heterogeneous Systems*, Lecture Notes
in Economics and Mathematical Systems 658, DOI 10.1007/978-3-642-22884-1_7,
© Springer-Verlag Berlin Heidelberg 2012

financial regulation is how to modify the current incentives in investment banking which favored the excessive short term risk taking without paying due regard to the long term performance and which are regarded by many analysts as important contributing factors to the recent recession.

The purpose of this paper is to provide a stochastic programming perspective on principal-agent problems. Beginning from the 70ties, such problems have emerged in the economic theory as some of the most important tools to analyze rational economic behavior of economic actors in the presence of asymmetric information. For the exposition of these problems and their role in the modern economic theory one can consult Laffont and Marimort (2002), see also Baron and Myerson (1982); Carlier (2001); Grossman and Hart (1983); Holmstrom and Milgrom (1987); Laffont (1994); Rochet and Choné (1998); Shavell (1979); for more recent developments see Arifovic and Karaivanov (2010); Bond and Gomes (2009); Karni (2008); Strausz (2006) and many others.

Two types of principal-agent models with asymmetric information have been studied: moral hazard and adverse selection.

In the case of *moral hazard* the principal does not observe the actions of the agent, but can observe their consequences. The agent and the principal possess the common knowledge of the probability distribution $H(x; a)$ of the outcomes x of the agent's action $a \in A$. The principal has to offer the incentive contract $\phi(x)$ to the agent and he finds such contract by maximizing his expected utility:

$$\max_{\phi(x)\in\Phi} \mathbb{E}_{H(x;a(\phi))} u^P(\phi(x), x) \tag{1}$$

where the expectation is taken with respect to the distribution of outcomes $H(x; a(\phi))$ that depends on the agent's response $a(\phi(x))$ to the incentive $\phi(x)$ that she obtains (and the principal predicts) by maximization of the agent's utility:

$$\max_{a\in A} \mathbb{E}_{H(x;a)} u^A(\phi(x)) + v^A(a) \tag{2}$$

where $u^A(w)$ is the agent's utility of money and $v^A(a)$ is the agent's utility of action. In addition, the principal solves (1) with the following participation constraint:

$$\max_{a\in A} \mathbb{E}_{H(x;a)} u^A(\phi(x)) + v^A(a) \geq \bar{u} \tag{3}$$

where $\bar{u}$ is the expected utility that the agent can obtain elsewhere in the economy. It is important that in this formulation the principal possess a substantial amount of information about the agent: the description of her set of actions, the agent's beliefs about the outcomes of her actions, the agent's utility function with respect to actions and money and, finally the agent's substitution utility level.

The *adverse selection* deals with the principal's uncertain knowledge about important characteristics of the agent. It is assumed that the agent(s) is characterized by the vector of parameters θ known to the agent, but unknown to the principal (like internal costs). The principal's knowledge about these parameters is described by some apriori distribution $H(\theta)$. The principal wants to induce the agent to obtain desirable outcome (perform desirable action) that depends on θ and for this purpose he designs a contract $\phi = (x(\theta), t(\theta))$ that specifies the monetary transfer $t(\theta)$ to the agent with parameters θ provided she obtains the outcome $x(\theta)$. He finds this contract by maximizing his expected utility

$$\max_{x(\theta), t(\theta)} \mathbb{E}_\theta u^P \left(P(x(\theta)) - t(\theta) \right) \tag{4}$$

where $P(x)$ is the monetary effect for the principal of the agent's outcome x. The principal knows the utility function of the agent and selects the contract in such a way that the agent will select the action and the reward that corresponds to its actual parameters θ, i.e.,

$$h(\theta, x(\theta)) + t(\theta) = \max_a \left(h(\theta, x(a)) + t(a) \right) \tag{5}$$

where $h(\theta, x)$ is the monetary effect for agent with parameters θ of the outcome x. In addition, the principal solves (4)–(5) taking into account the following participation constraint:

$$h(\theta, x(\theta)) + t(\theta) \geq 0, \ \forall \theta \tag{6}$$

The economic research cited above placed the emphasis on the study of the properties of the solutions of problems (1)–(3), (4)–(6), their specifications, extensions and combinations. Considerable attention was dedicated to the study of the characteristics of the optimal contracts with the emphasis on analytical results and their economic interpretation with consequences for specific application fields. Indeed, the principal-agent problems with asymmetric information are the natural paradigm for modeling of many economic phenomena. Insurance theory assumes that the insurant's level of caution can not be observed by the insurer Arnott and Stiglitz (1991); Spence and Zeckhauser (1971). In innovation or employment processes firm owners may not be able to observe the effort researchers or employees exert, see Aghion and Tirole (1994); Holmström (1999). Investors may have limited or no investment information and hire therefore an adviser in order to optimize their investments. Then this adviser's effort on the investment return must be distinguished from general market effects as noted in Baron and Holmström (1980). Another large area of application examples can be found in liberalized industrial sectors such as telecommunications Armstrong (1998); Audestad et al. (2006) or power management Pettersen (2004). Regulation represents a particularly important agency problem in such industrial sectors, see Laffont (1994); Verikoukis et al. (2004).

However, the emphasis on the study of the analytical properties of the principal-agent problems in economic literature has imposed certain limits and simplifications on the nature of the problems under study, and in particular:

1. *Limited possibilities for the treatment of uncertainty.* Both moral hazard and adverse selection principal-agent models aim at the modeling of uncertain knowledge of the principal about agent's actions and characteristics. However, in order to preserve the analytical tractability they introduce additional assumptions

about the information in the possession of the principal. For example, in the moral hazard models it is assumed that the principal knows the agent's beliefs about the consequences of her actions. In adverse selection models the emphasis is on the creation of the incentives for the agent to reveal her private information, suggesting implicitly that the agent is in possession of precise knowledge about her economic environment. Often these and other informational assumptions of such models are not satisfied in practice due to uncertain nature of many economic phenomena like users demand, price development, technological progress, external factors like weather, catastrophic shocks and disruptions, etc.

2. *Limits on the structure and complexity of decision models.* Again, the emphasis on the analytical results and their economic interpretation creates an incentive to study simpler optimization models with specific types of the objective functions and constraints. In particular, quite often the problem formulations (1)–(3), (4)–(6) are too general and simpler problems are studied in order to preserve the analytical tractability. For example, in Carlier and Dana (2005) it is taken $v^A(a) = -a$ in (2), in Carlier (2001) it is chosen $u^P(P(x(\theta)) - t(\theta)) = P^\top x(\theta) - t(\theta)$ in (4), in Rochet and Choné (1998) it is taken $h(\theta, x) = \theta^\top x$ in (5). The decision spaces of the principal and agent is also very simple: for the principal it is the monetary incentive to the agent, and for the agent it is her action, often without further structure. Consequently, the constraints in the principal-agent problems are of two types: incentive compatibility constraints (2),(5) and participation constraints (3),(6). However, the relationship between principals and agents very often have more structured and complex decision space. For example, the agents may select portfolios of industrial projects, plan production of goods and provision of services in order to achieve the aims set by the principal, the principals may decide about the resources available to agents. This requires inclusion of other types of constraints like resource constraints, production and other constraints that may depend on decision variables of both principal and agent.

Some economists recognize these limitations of the traditional principal-agent theory, observing that the adequate modeling of uncertainty can result in more realistic contracts. Suggesting the directions for future research in incomplete contracts, Tirole (1999) writes that the robustness of the theoretical contracts relative to often much simpler observable contracts is an important issue. He adds: "By robustness, I mean that these simple contracting forms are likely not to be very suboptimal when the parties make mistakes in their view of the world (this of course requires a theory of bounded rationality) or in the execution of the contract".

In this paper we aim at the relaxation of the limits described above by taking the approach of stochastic programming, which is the optimization methodology developed specifically for modeling of optimal decisions under uncertainty, see Birge and Louveaux (1997); Ermoliev and Wets (1988); Kall and Wallace (1994). The stochastic programming is less concerned with the study of the structure of the optimal solutions and obtaining of analytical results, but places the emphasis on the development of the models and numerical methods for their solution that are adequate for specific problem classes, incorporating if necessary the rich structure of decision space, uncertainty and constraints. This is the complementary approach to the principal-agent problems compared to one studied in the economic literature, but it utilizes some of the concepts developed there.

The contribution of this paper consists in the development of a series of stochastic programming models of different instances of principal-agent problems, that take as the starting point some of the themes found in the economic literature and supplement them with the treatment of uncertainty customary to stochastic programming. Besides, we show how additional constraints can be incorporated in the principal-agent models. After this we consider briefly one solution approach for such models. In order to do this, it was necessary to enhance the stochastic programming framework with the concepts of bilevel programming. There exists a substantial literature on bilevel programming, see Colson et al. (2005); Dempe (2002) and references there. This literature is concerned mainly with the deterministic case, Wolf and Smeers (1997) and Patriksson and Wynter (1999) being among the few exceptions. In order to be more specific, we often refer to the telecommunication environment, where an important example of the principal-agent problem is the relationship between the regulation authority and industrial actors. We do not aim here to provide a detailed exposure of the regulation in telecommunications, an interested reader is referred to the recent papers by Cambini and Jiang (2009) and Noam (2010). Exposition of the traditional economic-theoretical view on the principal-agent problems in regulation in general one can find in Baron and Myerson (1982) and Laffont (1994).

The rest of the paper is organized as follows. We start by formulating the general deterministic principal-agent model in Sect. 2 where we illustrate the importance of adequate treatment of uncertainty using a popular example from economic literature. Section 3 discusses the utilization of concepts of stochastic programming with bilevel structure for modeling and analyzing of different features of the agency relationship. In Sect. 4 one solution approach is outlined and Sect. 5 rounds up with conclusions.

2 Deterministic Case and Importance of Uncertainty

A general formulation of the principal-agent relationship can be given as follows. Assume that the principal determines an incentive schedule $\phi \in \Phi \subseteq \mathbb{R}^n$ inducing a response $a \in A \subseteq \mathbb{R}^m$ of the agent. The agent maximizes her utility $U_A(a, \phi)$ whereas the principal maximizes his utility $U_P(a, \phi)$, both depending on the decisions a and ϕ. Furthermore, assume that the principal's and the agent's decisions are subject to constraints $g_A(a, \phi)$ and $g_P(a, \phi)$, respectively. Consequently, the principal solves the following problem

$$\max_{\phi \in \Phi} U_P(a, \phi) \tag{7}$$

$$g_P(a, \phi) \geq 0$$

where the agent's decision a is found as the optimal solution of her problem

$$\max_{a \in A} U_A(a, \phi) \tag{8}$$

$$g_A(a, \phi) \geq 0$$

In economic terms, this formulation covers both the moral hazard problem (1)–(3) and the adverse selection problem (4)–(6).

The regulation represents a particularly important application field for the agency problem (7)–(8), see Laffont (1994); Verikoukis et al. (2004). The interference of a governmental authority with the behavior of industrial agents can have a catalytic but also constraining effects on the economic development. Therefore, regulatory policies should be designed carefully. Likewise the effects on the considered industry sector and possible interrelations with other fields of public economics have to be analyzed. Generally, the policy of a regulator consists of an incentive schedule and of obligations inducing the regulated firm(s) to follow certain policy guidelines. In a liberalized sector, such guidelines may comprise issues of customer protection, control of the entry of new competitors and of the market participants' behavior toward each other or ensuring efficiency and fast implementation of new technology. Different aspects of regulation are a subject of considerable research activity. Important issue that remained outside of the scope of this activity is inherent uncertainty of industrial environment, like the customer's acceptance of new services and the pace of the technological progress.

The interaction between both decision makers can be described in terms of a bilevel programming problem (BLP) or Stackelberg game or, more generally, as a mathematical program with equilibrium constraints (MPEC). For the analysis of realistic problems it is frequently necessary to control the feasibility of the actors' decisions by more complex constraints than only their domains. So far, the theoretical analysis was often simplified considerably by ignoring most of the constraints, in particular those on the agent's decisions. Also the principal's decisions are usually studied subject only to a participation constraint. Such a constraint reflects that the agent would not take part if a given decision of the principal would leave her with a too low utility. Recent economic approaches often utilize first-order optimality conditions on the agent's response. This is essentially a reformulation to a one-level nonlinear programming problem. If the agent's decision problem has no constraints such a one-level problem can be analyzed easily. But in practice an incentive schedule may influence not only the optimality of the agent's decisions via her utility function but also their feasibility by means of constraints. In our exposition we will therefore explicitly assume the existence of constraints on both actors' decisions.

Both objective functions and constraints in the problems (7)–(8) may depend on random parameters. One conceivable approach present in the economic literature is to substitute the random parameters by their expectations and solve resulting deterministic optimization problem. The following example sheds some light on the inadequacy of such approach.

Stochastic Programming Perspective on the Agency Problems Under Uncertainty

Example 1. Moral hazard model of output sharing. This model is present prominently in the economic literature, see Arifovic and Karaivanov (2010); Dutta and Zhang (2002); Holmstrom and Milgrom (1991); Stiglitz (1974) and references there for this problem, its extensions and applications. It has many applications to sharecropping in agriculture, franchising, licensing, publishing contracts, leasing of equipment, etc. In order to be specific, let us assume that the principal here is a telecommunication network provider who is in possession of the network infrastructure. The agent is a service provider that needs to lease this infrastructure for provision of her services. She employs the effort z that results in profit $y = z + \varepsilon$ where ε is distributed normally with zero mean and variance σ^2. The agent's effort is unobservable to the principal, but the profit is observable. The principal is risk neutral, and we shall assume the risk neutrality for the agent too. The case of risk averse agent considered in Arifovic and Karaivanov (2010) and others exhibit the same phenomena as the risk neutral case, so we consider the risk neutrality for transparency purposes.

The principal has to design a contract for leasing of the infrastructure to the agent. This contract consists of two components: fixed upfront payment f and share $1 - s$ of the agent's profits, $s \in [0, 1]$. The principal wants to select the contract that maximizes his expected profit, i.e. he solves the problem

$$\max_{s,f} \mathbb{E} (1 - s) y + f = \max_{s,f} (1 - s) z + f \tag{9}$$

The effort $z = z(s, f)$ is the decision variable of the agent. The agent selects it in order to maximize her profit net of payments to the principal, this is done by solving the problem

$$\max_{z} \mathbb{E} \left(sy - f - \frac{1}{2} cz^2 \right) = \max_{z} \left(sz - f - \frac{1}{2} cz^2 \right) \tag{10}$$

That is, the agent experiences the cost of effort that is proportional to the square of effort, as it is assumed often in the economic literature cited above. Following this literature we shall take $c = 1$. In addition, both the agent and the principal know the possible profit u that the agent can get by refusing the contract and engaging in some other activity, for example leasing the infrastructure from some other source or building her own infrastructure. Therefore the principal solves the problem (9)–(10) with additional participation constraint:

$$\max_{z} \left(sz - f - \frac{1}{2} cz^2 \right) \geq u \tag{11}$$

Observe that the problem (9)–(11) is a specific case of the general problem (7)–(8). The solution (s', f') of this problem is well known in the economic literature and consists of franchising: the principal demands the fixed upfront payment and renounces to obtain any share of profit resulting from the agent's activity:

$$s' = 1, \quad f' = \frac{1}{2} - u$$

where we have used our convention $c = 1$. So far we have followed Arifovic and Karaivanov (2010) in the setup of this example. Let us now introduce some additional uncertainty in this example and investigate its effects. Suppose that neither principal nor agent know exactly the alternative level of the profit u from (11). Instead, the knowledge of the principal about u is described by some probability distribution. In order to be specific, let us assume that u is distributed uniformly on $[0, 2\bar{u}]$, $\bar{u} \le 0.5$.

The contracting process in this case unfolds as follows. At the first stage the principal draws the contract (s, f) and offers it to the agent. Having this contract, the agent performs additional study of her opportunities and gets to know the exact value of her alternative profit u. On the basis of this knowledge the agent decides whether to accept the contract or not: if she can get superior profit from the principal (constraint (11) is satisfied) then the agent accepts the contract and the expected profit of the principal is defined by (9). Otherwise the agent declines the contract and the expected profit of the principal becomes zero. Let us derive the optimal contract for this contracting process, taking $c = 1$.

First of all, the optimal effort $z(s, f)$ of the agent can be obtained from (10) explicitly:

$$z(s, f) = s, \quad \max_z \left(sz - f - \frac{1}{2}z^2 \right) = \frac{1}{2}s^2 - f$$

Therefore the participation constraint (11) takes the form:

$$\frac{1}{2}s^2 - f \ge u$$

that yields the following expression for the expected profit of the principal $\pi(s, f)$, with expectation being taken with respect to both ε and u:

$$\pi(s, f) = \begin{cases} (1 - s)s + f & \text{if } \frac{1}{2}s^2 - f \ge 2\bar{u} \\ \frac{1}{2\bar{u}}((1 - s)s + f)\left(\frac{1}{2}s^2 - f\right) & \text{if } 0 \le \frac{1}{2}s^2 - f \le 2\bar{u} \end{cases}. \quad (12)$$

It remains to maximize this function with respect to f and $s \in [0, 1]$. After some elementary but tedious algebraic transformations we obtain the following expression for the optimal contract (s^*, f^*), that is again a franchising scheme:

$$s^* = 1, \quad f^* = \begin{cases} \frac{1}{2} - 2\bar{u} \text{ if } \bar{u} \le \frac{1}{8} \\ \frac{1}{4} \quad \text{if } \bar{u} \ge \frac{1}{8} \end{cases}, \quad \pi(s^*, f^*) = \begin{cases} \frac{1}{2} - 2\bar{u} \text{ if } \bar{u} \le \frac{1}{8} \\ 1/(32\bar{u}) \text{ if } \bar{u} \ge \frac{1}{8} \end{cases} \quad (13)$$

Let us compare this optimal contract with the contract that the principal would offer by substituting the expected value $\bar{u}$ of alternative profit u into the participation constraint (11) and solving the problem (9)–(11) with this constraint. Substituting this solution into expression (12) we obtain the expected value of the principal's profit (s', f') for the contract, designed with the averaged participation constraint:

$$s' = 1, \quad f' = \frac{1}{2} - \bar{u}, \quad \pi\left(s', f'\right) = \frac{1}{4}(1 - 2\bar{u}) \tag{14}$$

We see that the difference between the optimal stochastic contract (s^*, f^*) and the optimal average contract (s', f') is considerable. Both contracts renounce the profits from the agent's activity (this will be different in the case of the risk averse agent), but they adopt very different policies with respect to the fixed upfront payment. The difference between the two contracts is shown in Figs. 1 and 2.

We see that the optimal contract exhibits two distinct modes of behavior. If the possible range of the alternative profits for the agent is relatively small ($\leq 1/4$) then the optimal contract tries to win the agent in all cases by satisfying all possible participation constraints and demanding considerably less upfront payment than the contract for the average value of the alternative profit. Even though the upfront payment is smaller, the average value of the principal's profit is considerably larger because the agent will always accept the contract. If the range of the possible alternative profits exceeds the threshold of 1/4 then the optimal strategy becomes different: the principal demands a constant upfront payment irrespective of the

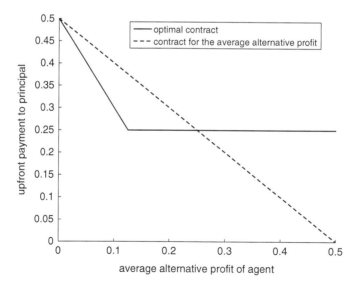

Fig. 1 Dependence of the fixed upfront payment f on the average alternative profit of the agent for the optimal contract (s^*, f^*) and the optimal contract (s', f') with averaged participation constraint

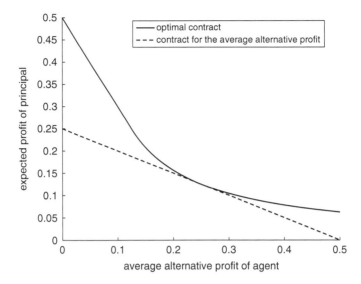

Fig. 2 Dependence of the principal's profit $\pi(s, f)$ on the average alternative profit of the agent for the optimal contract (s^*, f^*) and the optimal contract (s', f') with averaged participation constraint

upper bound on the possible alternative profits, discarding the agent in the case of her too high expectations. By contrast, the contract for the average value of uncertain alternative profit can not capture the structure of uncertainty and yields considerably lower profit for the principal. □

Example 1 highlights the importance of the adequate modeling of uncertainty in the principal-agent models. By interpreting the agency problems as stochastic programming problems equipped with bilevel structure, the incomplete information about several aspects of the principal agent relationship can be taken into account more adequately. The next sections are dedicated to the development of this viewpoint having as a reference the telecommunications environment.

3 Stochastic Programming Formulations of Agency Problems

This section demonstrates the utilization of stochastic programming techniques for the modeling and the analysis of agency problems. We give a closer characterization of different types of uncertainty and discuss specific aspects of the relationship in more detail.

Consider the deterministic agency model (7)–(8). This formulation indicates that the incentive schedule ϕ may assume a number of different shapes while modeling concrete and less concrete goals. Concrete requirements are, for example, the bounds on market share or minimum coverage rates. They influence the feasibility of the agent's decisions and can therefore be modeled as constraints on the agent's decision problem. Other goals, such as the best possible performance, affect the optimality of the agent's decisions. They can be modeled by penalties or rewards, and in this way they become a part of the agent's objective function. Also tax breaks or subsidies, possibly with selective application in order to encourage investments into unpopular regions, represent such kind of rewards. In our discussion we shall focus on these so-called explicit incentive schedules. In multi-period formulations of agency models additional implicit incentives, such as reputation or ratchet effects discouraging the agent's effort, may become important, see Meyer and Vickers (1997).

When the principal is perfectly informed about the agent's decision behavior and about all model parameters he can predict her response a on a given schedule ϕ and find an optimal incentive schedule. Often such a perfect knowledge can not be assumed and the agency problem contains uncertain parameters. This issue will be addressed in the following subsection. We specify the general agency problem (7)–(8) in more detail, taking into account uncertain parameters and constraints on both actors' decisions and discuss methods to treat both aspects. In Sect. 3.2 we outline an alternative approach to incorporate the uncertainty about environment parameters into a two-stage agency model. Section 3.3 is concerned with the principal's uncertainty about the agent's decision behavior. This type of imperfect information can be treated similarly, but the analysis is more elaborate. We consider the ways to perform the estimation of the agent's decision model and the ways to incorporate benchmarks and monitoring processes. Finally, Sect. 3.4 indicates an approach for modeling some aspects of licensing. Here, we focus on the possibility of a license withdrawal if regulatory obligations are not met.

3.1 Formulation of Agency Problems with Uncertain Parameters

In the agency problems found in a modern telecommunications environment, sources of uncertainty about the model parameters may be, for example, technological innovation, uncertain demand or Quality of Service issues such as failures or network congestion. Also the behavior of other actors which are not considered in the agency relationship may represent a source of uncertainty. Due to its situation, the regulated firm is typically better informed than the regulator, such that both actors may have different perceptions of the uncertain parameters. In what follows we shall often refer to the principal as regulator, bit the presented concepts are valid also for other types of principal.

The uncertainty about the environment parameters can be expressed by the help of random variables, say $\omega \in \Omega \subseteq \mathbb{R}^p$, with known or estimated probability distribution defined on appropriate probability space. Then the regulator's decision problem (7) can be formulated as

$$\max_{\phi \in \Phi(a)} \mathbb{E}_{\omega} U_P(a, \phi, \omega) \tag{15}$$

where the agent's decision a is obtained as the optimal solution of the regulator's perception of the agent's problem

$$\max_{a \in A(\phi)} \mathbb{E}_\omega U_A(a, \phi, \omega) \tag{16}$$

The sets $\Phi(a)$ and $A(\phi)$ of feasible decisions of the principal and the agent, respectively, are described in more detail below. The agent's decision problem (16) may contain constraints which are part of the incentive schedule and thus depend on both the agent's decision a and the incentive ϕ. Additionally, the problem often comprises constraints which are independent of the incentive, for example concerning the agent's technology or environment conditions. Also the regulator faces constraints on components of the incentive schedule. One such constraint is the participation constraint, which was considered in Example 1. Generally, the actors' sets of feasible decisions can be described by

$$\Phi(a) = \{\phi \mid g_P(a, \phi) \geq 0\} \tag{17}$$

$$A(\phi) = \{a \mid g_A(a, \phi) \geq 0\} \tag{18}$$

The constraints g_P and g_A may comprise deterministic constraints $g_{PD}(a, \phi)$ and $g_{AD}(a, \phi)$ as well as stochastic constraints $g_{PS}(a, \phi, \omega)$ and $g_{AS}(a, \phi, \omega)$. The satisfaction of the stochastic constraints involving random parameters ω is contingent on the exact realization of these parameters. Dependent on their meaning several deterministic equivalent formulations are conceivable. Such constraints may be satisfied on average, such as coverage rates or certain quality of service requirements. This results in a deterministic equivalent formulation

$$\mathbb{E}_\omega g_{PS}(a, \phi, \omega) \geq 0$$

Often the reliability or coverage requirements require a satisfaction with given minimal probability α. This yields a probabilistic constraint of the form:

$$\mathbb{P}\{\omega | g_{PS}(a, \phi, \omega) \geq 0\} \geq \alpha$$

The satisfaction of still other constraints may be required for any realization of the random variables. Examples of such constraints are the nonnegative customer numbers or capacity constraints. Another example of such constraints is given by the participation constraint. The general form of the participation constraint depends on the timing of the agent's decision. If the agent takes decision *before* uncertainty is revealed then the appropriate participation constraint is

$$\max_{a \in A(\phi)} \mathbb{E}_\omega U_A(a, \phi, \omega) \geq \mathbb{E}_\omega \bar{U}_A(\omega),$$

where $\bar{U}_A(\omega)$ is the utility that the agent can obtain by engaging in alternative activities. If the agent takes decision *after* uncertainty is revealed then the participation constraint becomes

$$\max_{a \in A(\phi)} U_A(a, \phi, \omega) \geq \bar{U}_A(\omega) \text{ almost sure,}$$

example of the latest case is the participation constraint (11) from Example 1.

Observe that in addition to the explicit constraints (17) the principal's problem may have implicit constraints that require nonemptyness of the agent's feasible set (18). These implicit constraints should be taken into account during design of numerical methods.

One can notice a certain similarity between the bilevel problem (15)–(16) and stochastic problems with recourse as in Kall and Wallace (1994) or Birge and Louveaux (1997). Unfortunately, this similarity extends only to a certain point because there is also a fundamental difference: stochastic problems with recourse put the averaged optimal value of the recourse function in the integrated objective, while the problem (15)–(16) depend on the optimal values of the agent's variables. This makes the problem (15)–(16) considerably more difficult numerically.

Another possibility to treat uncertainty in (15)–(16) is to consider minimax approach of robust optimization as in Ben-Tal et al. (2009). This leads to quite different problem formulations, which development is beyond the scope of this paper.

3.2 Two-Stage Stochastic Programming Problem with Bilevel Structure

There are many situations when the principal-agent interaction is distributed over several time periods and uncertainty gradually reveals itself during this decision time horizon. In the case of two periods such principal-agent decision process proceeds as follows.

1. At the beginning of the first period the principal optimizes his utility on the basis of information about the distribution of exogenous random parameters and apriori knowledge about the agent's preferences and utility function, that allow the principal to predict the agent's response. The agent optimizes her utility having her own apriori information about the distribution of exogenous random parameters. The decisions are implemented by both actors.
2. Additional information about exogenous uncertainty arrives to the actors that may differ between the actors. Besides knowing more about the exogenous uncertainty, the principal may learn more about the agent both from her actions and from external sources.
3. In the light of this new information the initial decisions ϕ and a may no longer be optimal for both actors, and they may even violate some constraints. Therefore both actors take corrective (or recourse) decisions that take into

account newly arrived information. The agent optimizes her second period utility utilizing updated distribution of remaining random parameters, while the principal optimizes his second period utility utilizing in addition the updated knowledge about the agent and predicting her recourse action.

The initial decisions in this framework should be taken by optimization of respective global utilities that integrate the immediate utility of the initial decision with the utility of recourse action.

Example 2. Spectrum licensing and development of mobile network. The regulator (principal) awards the license for a new portion of spectrum to the network operator (agent) that will establish a high speed data network of new generation. The license involves a substantial fee to be paid by the network operator. The decision about the fee amount is taken well before the demand for new services is known and even well before the whole menu of the data services is clear to both actors. After that the development of the network and services starts and the uncertainty in demand gradually reveals itself. One possible scenario is that the revenues and profits from the new network are smaller than expected and the firm has problems with servicing its debts incurred due to the necessity to pay the license fees. This may result in the recourse actions of both regulator and firm that may involve tax breaks, renegotiation of license conditions or even abandonment of the license. □

Example 3. Relationship between Virtual Network Operator (VNO) and Network Operator (NO), see Audestad et al. (2002) and Curwen and Whalley (2007) for definitions and discussion. The VNO (agent) will provide mobile telephony to a population of customers, but she does not possess the mobile network. Therefore she leases the network infrastructure from the NO (principal) for a fee that is determined by a leasing contract defined by the NO. The future evolution of the demand patterns is unclear at the time of the contract establishment. The VNO accepts the contract and starts the service provision. Gradually the demand and other uncertain factors become clear that may result both in overestimation and underestimation of the revenue potential. In the case of underestimation the VNO will face a lack of capacity and she may want to lease the additional capacity (recourse decision) from the NO on new terms. In the case of overestimation the recourse decision may consist in the renegotiation of the license fee or even the abandonment of the operation altogether. □

These situations can be described by the general stochastic principal-agent problem with recourse that is a stochastic program with recourse equipped by bilevel structure. This is an extension of the traditional stochastic problems with recourse as in Kall and Wallace (1994) or Birge and Louveaux (1997). The initial decision of the principal is obtained by solving the problem

$$\max_{\phi} \mathbb{E}_{\omega}\{U_P(\phi, a, Q_P(\phi, a, \omega)\} \tag{19}$$

$$g_P(\phi, a) \geq 0$$

with $a = a(\phi)$ is the optimal solution of

$$\max_{a} \mathbb{E}_\omega \{U_A (\phi, a, Q_A (\phi, a, \omega))\} \tag{20}$$

$$g_A(\phi, a) \geq 0$$

where $Q_P(\phi, a, \omega)$ is the optimal value of the principal's recourse problem (21)–(22) with the solution $\phi_R = \phi_R (\phi, a, \omega)$ and $Q_A(\phi, a, \omega) = Q_A(\phi, \phi_R (\phi, a, \omega), a, \omega)$ is the optimal value of the agent's recourse problem (23)–(24).

$$Q_P(\phi, a, \omega) = \min_{\phi_R} F_P(\phi, \phi_R, a, a_R, \omega) \tag{21}$$

$$f_P(\phi, \phi_R, a, a_R, \omega) \geq 0 \tag{22}$$

with $a_R = a_R (\phi, \phi_R, a, \omega)$ being the solution of the agent's recourse problem

$$Q_A(\phi, \phi_R, a, \omega) = \min_{a_R} F_A(\phi, \phi_R, a, a_R, \omega) \tag{23}$$

$$f_A(\phi, \phi_R, a, a_R, \omega) \geq 0 \tag{24}$$

The first-stage decisions can be interpreted as long-term anticipative decisions while the second-stage recourse decisions represent short-term adaptive decisions adjusting the strategy to the observed environment state.

Recourse problems may also represent a penalty for a violation of the stochastic constraints where the penalty vector p may be chosen according to the importance of satisfaction of the single constraints. Especially important in the agency theoretic context is the interpretation of recourse problems as a penalty incurred by the principal for unwelcome behavior of the agent, see for example the models in Sects. 3.3.1 and 3.4. Additionally, the parameters of the agent's recourse function may be a part of the principal's first-stage decision variables. For example, the principal's incentive schedule may comprise also the magnitude of the penalty for violating certain regulatory obligations.

Problem (19)–(22) represents a two-stage stochastic programming problem with recourse and bilevel structure. Several other formulations of such decision problems are analyzed in Gaivoronski and Werner (2007). A specific feature of two-stage agency problems is that the agent's recourse problem is part of the principal's problem not only indirectly, through the agent's objective function, but also directly due to the participation constraint. Consequently, the principal's decision is also affected by the agent's uncertainty about the model parameters. Let us illustrate this dependence by further extending Example 1 with the moral hazard model of output sharing.

Example 4. Two stage contract negotiation of output sharing with moral hazard. Let us consider on a somewhat more general level the principal-agent problem of Example 1, extending it to two periods with the special emphasis on the participation constraint. Namely, the principal takes the initial decision knowing only the distribution of the agent's alternative utility and modifies his offer after uncertainty about this utility reveals itself. In this case it will consist of the following steps:

- At the beginning the principal offers to the agent the *initial* contract (s_1, f_1) knowing the distribution of her alternative utility u.
- After receiving the initial offer (s_1, f_1) the agent observes her alternative utility u and communicates it to the principal (or, the principal learns this value by his own means).
- The principal extends to the agent the final offer (s, f) that he defines by computing the solution to the problem

$$\max_{s, f, s_1, f_1} \left(-U'_P (s - s_1, f - f_1) + \mathbb{E}_u \mathbb{E}_y \left\{ U_P (y, s, f) \, \mathbb{I}_{A(s, f, e)} \right\} \right) \quad (25)$$

where the output y is distributed according to the distribution $H(y; e)$ that depends on the effort e of the agent. Here $U_P (y, s, f)$ is the principal's utility of the output y and the contract (s, f), the function $U'_P (s - s_1, f - f_1)$ is the principal's disutility of the modification of the contract from (s_1, f_1) to (s, f), $\mathbb{I}_{A(s, f, e)}$ is the indicator function of the set

$$A (s, f, e) = \left\{ u \mid \max_{e \in E} \mathbb{E}_y U_A (s, f, y, e) \geq u \right\}$$

that represents the random participation constraint and $e = e (s, f)$ is obtained from the solution of the agent's problem

$$\max_{e \in E} \mathbb{E}_y U_A (s, f, y, e) \quad (26)$$

where $U_A (s, f, y, e)$ is the agent's utility of the contract (s, f), output y and the effort e. This is the specific case of the two stage principal-agent problem (19)–(24) where the agent does not has any recourse decisions and the principal has the initial decision (s_1, f_1) and the recourse decision (s, f).

$\square$

3.3 Imperfect Knowledge of the Principal About the Agent

We turn now to consideration of imperfect information of the principal about the agent's decision behavior. The incentive schedule shall control the decisions which the agent actually implements. However, it must be designed based on the predictions of the agent's decisions, which the principal obtains by solving his perception of the agent's decision problem. In the case of perfect information this prediction coincides with the actual decision of the agent, but often an information asymmetry between principal and agent exists that may result in incorrect incentive schedule. Typically, the principal has limited insight into the agent's decision process and can observe only part of the implemented decisions or results of the agent's decision problem, for example her achieved welfare. The agent is often better informed about the environment and faces a disutility in terms of effort, money or competitive disadvantage for the provision of additional information to the regulator. Therefore the principal can not properly evaluate the agent's response on a given incentive schedule. In addition, the principal may face exogenous uncertainty, which will further complicate the identification of the agent's decision behavior. All this will complicate the determination of an efficient and precise incentive schedule. Consider for example an incentive schedule inducing the agent to efforts toward low network congestion. The principal can not determine clearly if an observed low congestion is due to actual efforts of the agent, such as capacity extension or implementation of more efficient transfer technology, or if it is due to decreased user demand. Such problems of moral hazard have been the subject of economic research, see Holmström (1999); Mirrlees (1999).

Some aspects of the principal's asymmetric information about agent's parameters are considered in the economic literature on adverse selection, for example in Baron and Myerson (1982); Carlier (2001); Rochet and Choné (1998). The approach taken in this literature is to derive a contract that will induce the agent to reveal its private information to the principal. Such contracts require, however, additional knowledge about the agent's decision process and assume that the agent has the perfect knowledge about her parameters, like costs.

One possible stochastic programming model that takes into account the uncertainty about the agent's decision is considered below. For the sake of transparency we assumed here the perfect information about all other model parameters. Suppose that the principal has a conjecture about the agent's decision problem and obtains a guess of her response on a given incentive schedule ϕ by solving

$$\max_{a} U_A(\phi, a) \tag{27}$$

$$g_A(\phi, a) \geq 0$$

However, he assumes that this solution is an imperfect estimation of the response actually implemented by the agent. Therefore, the principal corrects his guess a by a random variable $\eta \in E \subseteq \mathbb{R}^m$ representing the noise or estimation error. Utilizing the corrected response $a + \eta$ in his decision problem he determines an optimal schedule as a solution of

$$\max_{\phi} \mathbb{E}_\eta U_P(\phi, a + \eta) \tag{28}$$

$$\mathbb{E}_\eta g_P(\phi, a + \eta) \geq 0$$

Additional information about the agent's decision process may be obtained by utilizing benchmarks or by a monitoring process. In the following we will study these issues closer.

3.3.1 Benchmarks

Benchmarks, such as performance of comparable firms or characteristics deduced by observations or by theoretical analysis may help to evaluate decisions implemented by the agent. An example is a licensing process where the license taker's actual performance is measured after some time. In order to assess the agent's behavior at this second stage, the regulator compares the agent's performance $X(a, \omega)$ against suitable benchmarks $X_P(\omega)$ taking into account the realized state ω of the environment. Then, a penalty for deviations from the target $X_P(\omega)$ is imposed, which affects the utility functions of both decision makers. A possible model formulation is the following.

The principal finds an incentive schedule ϕ_1 and a penalty parameter ϕ_2 such that his expected utility is maximized taking into account the penalty utility $Q_P(a, \phi_2, \omega)$ incurred to the agent

$$\max_{\phi_1, \phi_2} \mathbb{E}_\omega \{W_P(a, \phi_1, \omega) + Q_P(a, \phi_2, \omega)\} \tag{29}$$

$$g_P(\phi_1, \phi_2, a) \geq 0$$

with a being the optimal solution of the agent's decision problem

$$\max_a \mathbb{E}_\omega \{W_A(a, \phi_1, \omega) - Q_A(a, \phi_2, \omega)\} \tag{30}$$

$$g_A(\phi_1, \phi_2, a) \geq 0$$

Here $Q_A(a, \phi_2, \omega)$ is the agent's disutility resulting from the penalty incurred by the principal. It is determined at the second stage when the random variable ω and the agent's characteristics $X(a, \omega)$ become known

$$Q_A(a, \phi_2, \omega) = \min_y \{F_2(\phi_2, y) \mid y = X_P(\omega) - X(a, \omega)\} \tag{31}$$

where $F_2(\phi_2, y)$ is some measure of distance between y and zero. With this formulation the agent's decision problem (30) is a two-stage stochastic programming problem with the recourse problem (31). However, this recourse problem affects also the principal's decision problem (29), and the principal's decisions ϕ_1 and ϕ_2 represent parameters of the agent's first- and second-stage problems, respectively.

The second-stage problem may take on various shapes. If only down-side deviations from the target are penalized, the recourse problem is

$$Q_A(a, \phi_2, \omega) = \min_y F_2(\phi_2, y) \tag{32}$$

$$y = \max\{0, X_P(\omega) - X(a, \omega)\} \tag{33}$$

Constraint (33) represents a nonsmooth constraint which may complicate an analysis of the agency problem (29)–(30). Under certain assumptions it can, however, be substituted by the pair of smooth inequality constraints

$$y \geq X_P(\omega) - X(a, \omega)$$
$$y \geq 0$$

Often, the function F_2 is quadratic or linear in the deviation y, but other structures are conceivable as well. It should be chosen carefully in order to avoid dominance of the penalty Q_A over the principal's decision problem. In such a case the maximization of the welfare W_P would play a minor role, and the principal would induce under-performance of the agent. However, often one of the principal's goals is the maximization of social welfare. Bad performance of the agent decreases therefore the principal's utility while increasing the penalty Q_A. An optimal regulatory strategy should therefore represent a trade-off between both features. Furthermore, by setting the performance target ϕ_2 artificially high, the agent is forced to under-perform for any decision a. This issue must be taken into account when modeling the principal's decision problem, for example by specific constraints. The target ϕ_2 should comply with the agent's actual possibilities and with the general regulatory requirements of the considered sector.

Example 5. Two stage moral hazard with performance target. Assume that the agent's effort a yields a Quality of Service (QoS) level that can be described by a process $f(a)$. The principal does not know the exact description of this process, and it contains therefore a random parameter η. However, at the end of the first stage the actual value $f(a, \eta)$ of the quality achieved by the agent's decision a under the environment state η can be observed. The principal raises a penalty if a specified minimum quality ϕ_2 is not achieved. Assuming that this penalty is proportional to the deviation from the target and the second-stage decision is not subject to further constraints, the following stochastic programming problem with bilevel structure can be formulated.

$$\max_{\phi_1, \phi_2} \mathbb{E}_\eta \{U_P (\phi_1, f(a, \eta)) + p \max\{0, \phi_2 - f(a, \eta)\}\} \tag{34}$$

$$W_A(\phi_1, a) - p\mathbb{E}_\eta \max\{0, \phi_2 - f(a, \eta)\} \geq u \tag{35}$$

$$g_P(\phi_1, \phi_2, a) \geq 0$$

$$\phi_2 \geq 0$$

where $U_P (\phi_1, f)$ is the principal's utility of contract ϕ_1 and the QoS f; u is the agent's alternative utility. The agent's decision a is the optimal solution of the problem

$$\max_a W_A(\phi_1, a) - \mathbb{E}_\eta p \max\{0, \phi_2 - f(a, \eta)\} \tag{36}$$

$$g_A(\phi_1, \phi_2, a) \geq 0$$

where $W_A(\phi_1, a)$ denotes the agent's utility, and $p > 0$ is a fixed penalty parameter. In this model we assumed perfect information of the principal about the agent's decision process. If this is not the case, the formulation (36) should reflect this uncertainty since the penalty is based on the actually implemented decision and not on the prediction obtained as solution of (36). This may be modeled as suggested in problem (28).

If the function $f(a, \eta)$ is concave in a for all $\eta \in E$ then the penalty function

$$Q(\phi_2, a, \eta) = p \max\{0, \phi_2 - f(a, \eta)\}$$

is convex in a and therefore the agent's objective function (36) is concave. Furthermore, $Q(\phi_2, a, \eta)$ is continuously differentiable almost everywhere and, if the random variable η is absolutely continuously distributed, the expectation $\mathbb{E}_\eta Q(\phi_2, a, \eta)$ is continuously differentiable. □

3.3.2 Monitoring

A monitoring process can provide the principal with additional and/or more accurate information. It helps therefore to reduce the information asymmetry between the principal and the agent. By designing the monitoring as a part of the incentive schedule the agent may be induced to reveal more information about her decision process or about the characteristics of the considered sector. Utilizing the additional information, a more precise incentive schedule can be determined. This yields a reduction of agency costs and the agent may be induced to a better performance. However, typically the costs of such a process are increasing with its intensity and hence with the quality of the additionally obtained information. The principal must therefore find an optimal monitoring intensity in relation to the optimal incentive schedule, for example by minimizing the sum of the costs of monitoring and of the necessary incentives. Guangzhou Hu (2003) studies a one-period model for determining the optimal monitoring intensity in an employment process. Monitoring can be interpreted as a learning process and becomes therefore especially important in a dynamic setting of Holmström (1999). The following example indicates a stochastic programming formulation of an agency model with monitoring.

Example 6. Monitoring that reduces uncertainty about the agent's decision. Suppose that the regulator is aware of his imperfect information about the agent's decision process and takes this uncertainty into account as outlined in (27),(28). For the purpose of transparency assume further that there is a perfect information about the model parameter ω and that the agent's welfare is defined as follows

$$W_A(\phi, a) = f_A(\phi_1, a, \omega) = -(\omega a)^2 - (\phi_1 + \omega)a + \phi_2 \tag{37}$$

where $\phi = (\phi_1, \phi_2)$ is the contract offered by the principal. That is, following the economic literature cited above the agent's cost is proportional to the square of effort

Stochastic Programming Perspective on the Agency Problems Under Uncertainty   157

and it has also the linear part, to which the component ϕ_1 of the contract contributes (for example, environmental or QoS requirements of the principal that increase the agent's costs). The component ϕ_2 of the contract describes the monetary transfer (or tax) to the agent. Then the optimal schedule is a solution of the principal's problem

$$\max_{\phi} \mathbb{E}_\eta U_P \left(\phi, a + \eta \right) \tag{38}$$

$$\mathbb{E}_\eta W_A(\phi, a + \eta) = -\omega^2 (\bar{\delta}^2 + (\mu + a)^2) - (\phi_1 + \omega)(\mu + a) + \phi_2 \ge u \tag{39}$$

$$\mathbb{E}_\eta g_P(\phi, a + \eta) \ge 0$$

where u is the alternative utility of the agent, and the agent's response a solves the problem

$$\max_a W_A(\phi, a) \tag{40}$$

$$g_A(\phi, a) \ge 0$$

and

$$\bar{\delta}^2 = \mathbb{D}_\eta^2 = \mathbb{E}_\eta\{(\eta - \mathbb{E}_\eta)^T (\eta - \mathbb{E}_\eta)\}, \ \mu = \mathbb{E}_\eta.$$

Thus, $\bar{\delta}^2$ is the inherent or original variance and mean of the noise η, respectively. Let us assume, further that the principal's utility $U_P (\phi, a + \eta)$ is such that

$$\mathbb{E}_\eta U_P \left(\phi, a + \eta \right) = U \left(\phi, a, \mu, \bar{\delta} \right)$$

A monitoring process with the intensity $\theta \in [0, 1]$ reduces the variance

$$\delta^2 = (1 - \theta)\bar{\delta}^2$$

while there is no influence on the mean μ of the noise. The intensity is defined such that $\theta = 0$ when no monitoring takes place and increasing values of θ indicate higher intensity. Intensity $\theta = 1$ corresponds to the case of perfect information about the agent's decision process. Typically, the monitoring process causes costs $c(\theta)$ that increase with the intensity and diminish the principal's utility.

Taking into account this monitoring process the principal finds the optimal incentive schedule ϕ and the optimal monitoring intensity θ by solving the problem

$$\max_{\phi, \theta} \mathbb{E}_\eta\{U_P \left(\phi, a + \eta \right) - c(\theta)\}$$

$$- \omega^2((1 - \theta)\bar{\delta}^2 + (\mu + a)^2) - (\phi_1 + \omega)(\mu + a) + \phi_2 \ge u$$

$$\mathbb{E}_\eta g_P(\phi, a + \eta) \ge 0$$

$$\theta \in [0, 1]$$

with the response $a = a(\phi)$ being a solution of problem (40).   □

3.4 Licensing

Licensing represents an important tool for the regulation of liberalized environments such as the modern telecom sector and it is the subject of widespread studies, see Gruber (2002); Louta et al. (2003). The licensing can be an important regulatory tool. The provision of a right to use telecom infrastructure can help to impose regulatory goals such as the schedule of network roll out or coverage and development obligations in order to satisfy political, social or economic objectives. The allocation of licenses may therefore follow several guidelines, for example pricing constraints, efficient infrastructure utilization, access to bottleneck facilities or contributions toward universal service, but also the generation of additional wealth for the regulator by cream skimming. The fee paid for a license may have a quite serious and restrictive impact on the agents' decision behavior. For example, in order to recover this fee they will be motivated rather by short-term considerations. Therefore the regulation should direct the license takers' attention also to long term goals such as growth of the respective industry sector.

The allocation of the license and the size of the license fee add further dimensions to the principal's decision problem. The fee is an entry fee paid by the agent usually once and prior to her decisions. Licensing can then be included into the agency relationship by means of obligations the agent has to meet when holding a license. The satisfaction of these obligations may be controlled by an incentive schedule comprising constraints, penalties, rewards or even the withdrawal of the license. The following example provide a stochastic programming model that takes a part of these considerations into account.

Example 7. Licensing with penalties. Assume that, in order to provide service, the agent acquires a license for a license fee L_0. After a certain period of time her performance is evaluated by measuring the provided Quality of Service (QoS). Bad performance is penalized if a certain level ϕ_2 set by the principal is not reached. If the QoS level is only slightly lower than the target, a penalty is raised as described in Example 5, and the agent can proceed with service provision in the subsequent period. If, however, the quality is too poor and even an absolute minimum level ϕ_3 is not reached, the license is withdrawn and no further service provision can take place. Additionally, a fine $P \geq 0$ is raised. For the sake of simplicity we assume no reevaluation at the end of the second stage. The agent can alter her decision a_2 in the second stage but, in order to simplify the exposition, we assume that her strategy $a = (a_1, a_2)$ of first- and second-stage decisions is found at the beginning of the first stage. The principal finds an optimal decision $\phi = (\phi_1, \phi_2, \phi_3)$ consisting of the values for his first- and second-stage decisions $\phi_1 = (\phi_{11}, \phi_{12})$ and QoS levels ϕ_2 and ϕ_3 by solving the problem

$$\max_{\phi, P, L_0} \left(L_0 + U_P^1(\phi_{11}, a_1) + \alpha \mathbb{E}_\eta \{ l Q(\phi_2, a_1, \eta) + U_P^2(\phi_{12}, l a_1) + (1-l)P \} \right) \tag{41}$$

$$- L_0 + W_A(\phi_{11}, a_1) + \alpha \mathbb{E}_\eta \{ l(-Q(\phi_2, a_1, \eta) + W_A(\phi_{12}, a_2)) - (1-l)P \} \geq u \tag{42}$$

$$P - p(\phi_2 - \phi_3) \geq 0 \tag{43}$$

$$\phi_2 \geq \phi_3 \geq 0$$

$$g_P(\phi, a) \geq 0$$

with $\alpha \in [0, 1]$ being a discounting factor and $U_P^i(\phi_{1i}, a_i), i = 1, 2$ being the principal's utility resulting from the agent's effort a_i and the monetary transfer or tax to agent ϕ_{1i} during period $i = 1, 2$ (the component of welfare resulting from service provision).

$$Q(\phi_2, a_1, \eta) = p \max\{0, \phi_2 - f(a_1, \eta)\}$$

$$l = \begin{cases} 1, & \text{if } f(a_1, \eta) \geq \phi_3 \\ 0, & \text{otherwise} \end{cases}$$

The agent's effort strategy $a = (a_1, a_2)$ is found as solution of the problem

$$\max_{a_1, a_2} \left(-L_0 + W_A(\phi_{11}, a_1) + \alpha \mathbb{E}_\eta \{l(-Q(\phi_2, a_1, \eta) + W_A(\phi_{12}, a_2)) - (1 - l)P\} \right) \tag{44}$$

$$g_A(\phi, a) \geq 0$$

Constraint (42) represents the participation constraint whereas (43) indicates that the penalty for low QoS level should be more severe when the license is withdrawn than when further service provision is granted.

The penalty to be paid by the agent can be determined first at the end of the first stage when both the environment state η and the agent's actual decision a_1 are revealed. Therefore the principal's decision problem (41)–(43) has the structure of a two-stage stochastic programming problem with recourse. Due to the discrete nature of the fine P the agent's objective function is generally not continuous in a_1. It may become continuous by letting the size of P depend on the deviation of the actual QoS level from the threshold ϕ_3. □

4 Solution Approaches

While the previous section focused on the utilization of the stochastic programming framework for modeling various aspects of agency problems in the presence of uncertainty, in this section we shall indicate its capabilities for the solution of models with the discussed features.

The research on the methods for solving decision making problems with several actors comprises different approaches such as stochastic games as in Filar and Vrieze (1997), or stochastic bilevel programming problems (SBLP), see Wynter (2001), and their generalizations, stochastic mathematical programs with equilibrium constraints (SMPEC) as in Evgrafov and Patriksson (2004); Shapiro (2006).

However, typically these approaches interpret the considered problem type as an extension of deterministic problems by uncertain parameters. In this paper we focus directly on the modeling and processing of uncertainty. Therefore we aim at the nontrivial adaptation of known stochastic programming methods, see Birge and Louveaux (1997); Ermoliev and Wets (1988); Shapiro and Ruszczynski (2003) and by developing new such methods tailored to the problem domain considered here.

The problems considered here possess some features that are challenging for the numerical methods. Since the agent's response on a given incentive schedule has to be taken into account, the principal's objective function is generally not convex and neither differentiable. The participation constraint represents a so-called connecting upper level constraint. It is located in the principal's subproblem but its feasibility depends also on the agent's response. As a consequence the set of the principal's feasible decisions may not be connected or convex. Also a possible existence of non-unique responses of the agent to some principal's decisions complicates the evaluation of the principal's problem. Therefore the stability of obtained optimal decisions should be investigated, for example by sensitivity analysis, as in Patriksson and Wynter (1997).

The numerical approach that we take here consists in the further development of the methods from the stochastic quasi-gradient (SQG) class, see Ermoliev (1988); Gaivoronski (1988, 2004). These methods have been developed for the solution of optimization problems with complex objective functions and constraints. This makes them especially applicable to stochastic programming problems with bilevel structure and nonlinear constraints as represented by agency relationships. The following example illustrates the main ideas and the potential of this framework.

Example 8. Solving principal-agent problem with stochastic quasi-gradient (SQG) method. We present here a test example of the agency model of the type (15)–(16). Let us take

$$U_P(a, \phi, \omega) = \phi_1(1 + y)$$

where similar to Example 1 $y = a + \epsilon$ with ϵ being some random variable with known distribution and zero mean. Additionally, there exist the lower and upper bounds on the decisions of both actors. We assume that the principal controls the lower bound ϕ_2 on the agent's decision a and that, in addition to the participation constraint, his decisions $\phi = (\phi_1, \phi_2)$ are subject to further constraint $\omega_1^2 \phi_1 - \phi_2 \geq 0$ where ω_1 is a normally distributed random parameter, $\omega_1 \sim N(0.5, 1)$. Thus, the principal's policy $\phi = (\phi_1, \phi_2)$ is obtained from the solution of the problem

$$\max_{\phi_1, \phi_2} \phi_1(1 + a) \tag{45}$$

$$\omega_1^2 \phi_1 - \phi_2 \geq 0 \tag{46}$$

$$W_A(\phi_1, a) \geq u \tag{47}$$

$$\phi_1 \in [\phi_{1L}, \phi_{1U}], \quad \phi_2 \in [\phi_{2L}, \phi_{2U}]$$

where the agent's response a is the optimal solution of the decision problem

Stochastic Programming Perspective on the Agency Problems Under Uncertainty 161

$$\max_{a} W_A(\phi_1, a) \tag{48}$$

$$a \in [\phi_2, a_U]$$

The agent's welfare $W_A(\phi_1, a)$ is determined similar to (37),

$$f_A(\phi_1, a, \omega_2) = -(\omega_2 a)^2 - (\phi_1 + \omega_2)a + 20\phi_1$$

We assume that the principal has imperfect information also about the parameter $\omega_2 \in \Omega_2 \subseteq \mathbb{R}$ and assumes that it has the normal distribution, $\omega_2 \sim N(0, 1)$. The numerical values of the model parameters are the following: $\phi_{1L} = \phi_{2L} = 0$, $\phi_{1U} = \phi_{2U} = a_U = 20$, $u = 5$.

Let us now compare solutions obtained by two approaches. The first approach replaces the random variables by their means and solves the resulting bilevel deterministic optimization problem. The second approach applies SQG method to the original stochastic problem, details of this method are described in the Appendix A.

The first approach results in the following deterministic equivalent of the stochastic programming problem (45)–(47)

$$\max_{\phi_1, \phi_2} \phi_1(1 + a) \tag{49}$$

$$0.25\phi_1 - \phi_2 \geq 0 \tag{50}$$

$$\phi_1(20 - a) \geq 5 \tag{51}$$

$$\phi_1 \in [0, 20], \ \phi_2 \in [0, 20]$$

with the response a being the optimal solution of

$$\max_{a} \phi_1(20 - a)$$

$$a \in [\phi_2, 20]$$

This problem yields the agent's response $a(\phi) = \phi_2$ for $\phi_1 > 0$ whereas the participation constraint (51) is violated for any response if $\phi_1 = 0$. This approach results in the optimal principal policy $\phi_T^* = (\phi_{1T}^*, \phi_{2T}^*) = (20, 5)$ with the agent's response $a_T^*(\phi_T^*) = 5$. With this strategy, the principal's and agent's welfares are $W_P(\phi_T^*, a_T^*) = 120$ and $W_A(\phi_T^*, a_T^*) = 100$, respectively.

This approach utilizes only the information about the means of the uncertain parameters ω_1 and ω_2. Employing the approach of stochastic programming allows to utilize more of the available information about these parameters and a better (for the principal) policy can be found.

Suppose that the constraint (46) is required to hold in average and, likewise, the average of the agent's welfare is considered as her utility. This results in the stochastic programming problem with bilevel structure

$$\max_{\phi_1,\phi_2} \phi_1(1+a) \tag{52}$$

$$\mathbb{E}_{\omega_1}\left\{\omega_1^2\phi_1 - \phi_2\right\} \geq 0 \tag{53}$$

$$\mathbb{E}_{\omega_2}\left\{-\omega_2^2 a^2 + (\phi_1 + \omega_2)a + 20\phi_1\right\} \geq 5 \tag{54}$$

$$\phi_1 \in [0, 20], \phi_2 \in [0, 20]$$

and the response a being the optimal solution of

$$\max_a \mathbb{E}_{\omega_2}\left\{-\omega_2^2 a^2 + (\phi_1 + \omega_2)a + 20\phi_1\right\} \tag{55}$$

$$a \in [\phi_2, 20]$$

This problem was solved by the SQG method described in Appendix A, see more detailed description of this method in Gaivoronski and Werner (2007). For the considered problem (52)–(54) we have found the optimal principal's strategy $\phi_S^* = (\phi_{1S}^*, \phi_{2S}^*) = (20, 12.249)$ with the agent's response $a_S^*(\phi_S^*) = 12.249$. Hence, the principal's welfare is $W_P(\phi_S^*, a_S^*) = 264.97$ whereas the agent's welfare is $W_A(\phi_S^*, a_S^*) = 5$.

A comparison of both solution methods reveals that actually two different problems are solved. The substitution of the average values leads to a linear deterministic equivalent formulation and much of the information about the agent's decision behavior and the random parameters ω_1 and ω_2 is lost. In contrast, the utilization of the stochastic programming methodology with bilevel features preserves the nonlinear structure of the original problems (45)–(48) and more of the available information is employed. For the considered model it leads to the policy that yields much higher optimal value of the principal's welfare. □

Somewhat more detailed analysis of the numerical properties of this approach was considered in Gaivoronski and Werner (2007).

5 Conclusions

In this paper we have discussed models and methods for the treatment of several types of uncertainty present in the agency problems. For this purpose we have considered the principal agency relationship between a regulator and a service provider in a liberalized telecom environment and demonstrated the utilization of stochastic programming concepts enhanced by the methods of bilevel programming. We have derived models for determination of incentive schedules under different aspects of imperfect information and outlined solution approaches, illustrated by an application example. The studies were restricted to the case of one agent. However, the models were quite general and provide the base for an extension to a multilateral formulation with several agents.

More research is needed into agency problems considered as stochastic programming problems with bilevel structure. Such studies will allow the computation of the optimal contracts, relaxing the traditional assumptions about information available to actors present in the current economic literature.

Acknowledgements The authors are grateful to two anonymous referees whose comments have contributed considerably to the improvement of the paper.

Appendix: Algorithm for Solution of Example 8

In the following we describe the main details of the SQG algorithm for Example 8. A more comprehensive analysis of the method is given in Gaivoronski and Werner (2007).

We characterize the optimal response $a = a(\phi)$ of the agent by Karush-Kuhn-Tucker optimality conditions of the agent's problem (48). Then a nonlinear one-level stochastic programming problem can be formulated which is equivalent to the original problem (52)–(55). The decision variables of this problem consist of the principal's and the agent's decision variables $\phi = (\phi_1, \phi_2)$ and a, respectively, and of the Lagrange multipliers $\lambda = (\lambda_1, \lambda_2)$ associated to the optimal response a.

$$\max_{\phi, a, \lambda} \phi_1(1 + a) \tag{56}$$

$$\mathbb{E}_{\omega_2}\left\{-2\omega_2^2 a - (\phi_1 + \omega_2) - \lambda_1 + \lambda_2\right\} = 0$$

$$\lambda_1(a - \phi_2) = 0 \tag{57}$$

$$\lambda_2(20 - a) = 0 \tag{58}$$

$$\mathbb{E}_{\omega_1}\left\{\omega_1^2 \phi_1 - \phi_2\right\} \geq 0$$

$$\mathbb{E}_{\omega_2}\left\{-\omega_2^2 a^2 - (\phi_1 + \omega_2)a + 20\phi_1 - 5\right\} \geq 0$$

$$\phi_2 \leq a \leq 20, \ 0 \leq \phi_1 \leq 20, \ 0 \leq \phi_2 \leq 20$$

$$\lambda_1, \lambda_2 \geq 0$$

This problem is ill-posed since it contains the complementarity constraints (57) and (58). For example, there exists no feasible solution, which strictly satisfies all inequality constraints. Therefore the usual constraint qualifications of nonlinear programming are violated at every feasible point. In order to deal with this difficulty we propose a decomposition of the complementarity constraints, resulting in a partition of problem (56)–(58) into a family of subproblems. Each subproblem is a convex stochastic programming problem with the random parameters $\omega = (\omega_1, \omega_2)$ and, given an initial point $x^0 = (\phi^0, a^0, \lambda^0)$, it is formulated as follows.

$$\max_{\phi,a,\lambda} \phi_1(1+a) \tag{59}$$

$$\mathbb{E}_\omega f_1(\phi,a,\lambda,\omega) \geq 0 \tag{60}$$

$$\mathbb{E}_\omega f_2(\phi,a,\lambda,\omega) = 0 \tag{61}$$

where

$$g_A(\phi,a) = \begin{pmatrix} a - \phi_2 \\ 20 - a \end{pmatrix}$$

$$f_1(\phi,a,\lambda,\omega) = \begin{pmatrix} g_{A,i}(\phi,a), & i \in I_C \\ \lambda_i, & i \in I_L \\ \omega_1^2\phi_1 - \phi_2 \\ -\omega_2^2 a^2 - (\phi_1 + \omega_2)a + 20\phi_1 - 5 \\ 20 - \phi_1 \\ 20 - \phi_2 \\ \phi_1 \\ \phi_2 \end{pmatrix}$$

$$f_2(a,\phi,\lambda,\omega) = \begin{pmatrix} g_{A,i}(a,\phi), & i \in \{1,2\} \setminus I_C \\ \lambda_i, & i \in \{1,2\} \setminus I_L \\ -2\omega_2^2 a - (\phi_1 + \omega_2) - \lambda_1 + \lambda_2 \end{pmatrix}$$

and

$$I_C = \{i \in \{1,2\} : g_{A,i}(\phi^0,a^0) > 0\}$$

$$I_L = \{i \in \{1,2\} : \lambda^0 > 0\}$$

The stochastic programming problem (59)–(61) can be solved iteratively by a SQG method. At each iteration step k a sample $\omega_k = (\omega_{1,k}^1, \ldots, \omega_{1,k}^{N_k}; \omega_{2,k}^1, \ldots, \omega_{2,k}^{N_k})$ of the random parameters ω is determined. Then new values for the iterates $x^k = (\phi_1^k, \phi_2^k, a^k, \lambda_1^k, \lambda_2^k)$, u^k and v^k are found where x^k denotes the decision variables, u^k the Lagrange multipliers of the inequality constraints (60) and v^k the Lagrange multipliers of the equality constraints (61). These variables are updated according to the rules

$$x^{k+1} = x^k - \alpha_x^k \xi_x^k$$

$$u^{k+1} = \max\{0, u^k + \alpha_u^k \xi_u^k\}$$

$$v^{k+1} = v^k + \alpha_v^k \xi_v^k$$

with the step sizes α_x^k, α_u^k and α_v^k satisfying the SQG conditions, see Gaivoronski (1988). Utilizing the Lagrangian function of problem (59)–(61) for a given observation $\omega_k^j, j = 1, ..., N_k$ of the random parameter and for the current iterates $x^k = (\phi^k, a^k, \lambda^k), u^k, v^k$

$$L\left(\phi^k, a^k, \lambda^k, u^k, v^k, \omega_k^j\right) = \phi_1^k\left(1 + a^k\right) + \left(u^k\right)^T f_1\left(\phi^k, a^k, \lambda^k, \omega_k^j\right) +$$
$$\left(v^k\right)^T f_2\left(\phi^k, a^k, \lambda^k, \omega_k^j\right)$$

the current search directions ξ_x^k, ξ_u^k and ξ_v^k can be determined by means of statistical estimates of the gradients of this Lagrangian, for example by

$$\xi_x^k = \frac{1}{N_k} \sum_{j=1}^{N_k} \nabla_x L\left(\phi^k, a^k, \lambda^k, u^k, v^k, \omega_k^j\right)$$

$$\xi_u^k = \frac{1}{N_k} \sum_{j=1}^{N_k} f_1\left(a^k, \phi^k, \lambda^k, \omega_k^j\right)$$

$$\xi_v^k = \frac{1}{N_k} \sum_{j=1}^{N_k} f_2\left(a^k, \phi^k, \lambda^k, \omega_k^j\right)$$

The iteration arrives in the vicinity of the optimal point when a stopping criterion such as (possibly relaxed) optimality conditions is satisfied. The number of the subproblems is finite. However, it grows exponentially with the numbers of constraints and decision variables of the agent's problem. More details about applications of the SQG algorithms in the Lagrangian context can me found in Ermoliev (1983); Kushner and Yin (2010).

References

Aghion, P., & Tirole, J. (1994). The management of innovation. *Quarterly Journal of Economics, 109*(4), 1185–1209.

Arifovic, J., & Karaivanov, A. (2010). Learning by doing vs. learning from others in a principal-agent model. *Journal of Economic Dynamics & Control, 34*, 1967–1992.

Armstrong, M. (1998). Network interconnection in telecommunications. *Economic Journal, 108*(448), 545–564.

Arnott, R., & Stiglitz, J. (1991). Moral hazard and nonmarket institutions: Dysfunctional crowding out of peer monitoring? *The American Economic Review, 81*(1), 179–190.

Audestad, J., Gaivoronski, A., & Werner, A. (2006). Extending the stochastic programming framework for the modeling of several decision makers: pricing and competition in the telecommunication sector. *Annals of Operations Research, 142*(1), 19–39.

Audestad, J.A., Gaivoronski, A., & Werner, A. (2002). Modeling market uncertainty and competition in telecommunication environment: Network providers and virtual operators. *Telektronikk, 97*(4), 46–64.

Baron, D., & Holmström, B. (1980). The investment banking contract for new issues under asymmetric information: Delegation and the incentive problem. *Journal of Finance*, *35*(5), 1115–1138.

Baron, D. P., & Myerson, R. B. (1982). Regulating a monopolist with unknown costs. *Econometrica*, *50*(4), 911–930.

Ben-Tal, A., Ghaoui, L. E., & Nemirovski, A. (2009). *Robust Optimization*. Princeton, NJ: Princeton University Press.

Birge, J. R., & Louveaux, F. (1997). *Introduction to Stochastic Programming*. New York: Springer.

Bond, P., & Gomes, A. (2009). Multitask principal-agent problems: Optimal contracts, fragility, and effort misallocation. *Journal of Economic Theory*, *144*, 175–211.

Cambini, C., & Jiang, Y. (2009). Broadband investment and regulation: A literature review. *TelecommunicationsPolicy*, *33*, 559–574.

Carlier, G. (2001). A general existence result for the principal-agent problem with adverse selection. *Journal of Mathematical Economics*, *35*, 129–150.

Carlier, G., & Dana, R. A. (2005). Existence and monotonicity of solutions to moral hazard problems. *Journal of Mathematical Economics*, *41*, 826–843.

Colson, B., Marcotte, P., & Savard, G. (2005). Bilevel programming: A survey. *4OR*, *3*, 87–107.

Curwen, P., & Whalley, J. (2007). Tele2 and the strategic role of virtual operations. *Info: The Journal of Policy, Regulation and Strategy for Telecommunications, Information and Media*, *9*(4/5), 55–69 (2007)

Dempe, S. (2002). *Foundations of Bilevel Programming*. Dordrecht: Kluwer.

Dutta, S., & Zhang, X. (2002). Revenue recognition in a multiperiod agency setting. *Journal of Accounting Research*, *40*, 67–83.

Ermoliev, Y. (1983). Stochastic quasigradient methods and their application to system optimization. *Stochastics*, *9*, 1–36.

Ermoliev, Y. (1988). Stochastic quasigradient methods. In Y. Ermoliev, & R. B. Wets (Eds.), *Numerical Techniques for Stochastic Optimization*, pp. 141–186. New York: Springer.

Ermoliev, Y., & Wets, R. B. (Eds.) (1988). *Numerical Techniques for Stochastic Optimization*. New York: Springer.

Evgrafov, A., & Patriksson, M. (2004). On the existence of solutions to stochastic mathematical programs with equilibrium constraints. *Journal of Optimization Theory and Applications*, *121*(1), 65–76.

Filar, J., & Vrieze, K. (1997). *Competitive Markov Decision Processes*. New York: Springer.

Gaivoronski, A. (1988). Stochastic quasigradient methods and their implementation. In Y. Ermoliev, & R. B. Wets (Eds.), *Numerical Techniques for Stochastic Optimization*, pp. 313–351. New York: Springer.

Gaivoronski, A. (2004). SQG: stochastic programming software environment. In S. Wallace, & W. Ziemba (Eds.), *Applications of Stochastic Programming, MPS-SIAM Series in Optimization*, pp. 637–670.

Gaivoronski, A., & Werner, A. (2007). A solution method for stochastic programming problems with recourse and bilevel structure (Under revision)

Grossman, S. J., & Hart, O. D. (1983). An analysis of the principal-agent problem. *Econometrica*, *51*(1), 7–45.

Gruber, H. (2002). Endogenous sunk costs in the market for mobile telecommunications: The role of licence fees. *The Economic and Social Review*, *33*(1), 55–64.

Guangzhou Hu, A. (2003). R&D organization, monitoring intensity, and innovation performance in chinese industry. *Economics of Innovation and New Technology*, *12*(2), 117–144.

Holmström, B. (1999). Managerial incentive problems: A dynamic perspective. *Review of Economic Studies*, *66*, 169–182.

Holmstrom, B., & Milgrom, P. (1987). Aggregation and linearity in the provision of intertemporal incentives. *Econometrica*, *55*(2), 303–328.

Holmstrom, B., & Milgrom, P. (1991). Multitask principal-agent analyses: Incentive contracts, asset ownership, and job design. *Journal of Law, Economics and Organisation*, *7*, 24–52.

Kall, P., & Wallace, S. W. (1994). *Stochastic Programming*. New York: Wiley.

Karni, E. (2008). Agency theory: Choice-based foundations of the parametrized distribution formulation. *Economic Theory*, *36*, 337–351.

Kushner, H. J., & Yin, G. G. (2010). *Stochastic Approximation and Recursive Algorithms and Applications*. New York: Springer.

Laffont, J. J. (1994). The new economics of regulation ten years after. *Econometrica*, *62*(3), 507–537.

Laffont, J. J., & Marimort, D. (2002). *The Theory of Incentives: The Principal-Agent Model*. Princeton, NJ: Princeton University Press.

Louta, M., Roussaki, I., & Anagnostou, M. (2003). Implications of 3G licensing to mobile telecommunications market dynamics. In *ConTEL 2003. Proceedings of the 7th International Conference on Telecommunications*, pp. 113–120.

Meyer, M., & Vickers, J. (1997). Performance comparisons and dynamic incentives. *Journal of Political Economy*, *105*, 547–581.

Mirrlees, J. (1999). The theory of moral hazard and unobservable behaviour: Part I. *The Review of Economic Studies*, *66*(1), 3–21.

Noam, E. M. (2010). Regulation 3.0 for telecom 3.0. *Telecommunications Policy*, *34*, 4–10.

Patriksson, M., & Wynter, L. (1997). Stochastic nonlinear bilevel programming. Tech. rep., PRISM, Université de Versailles – Saint Quentin en Yvelines, Versailles, France.

Patriksson, M., & Wynter, L. (1999). Stochastic mathematical programs with equilibrium constraints. *Operations Research Letters*, *25*(4), 159–167.

Pettersen, E. (2004). Managing end-user flexibility in electricity markets. Ph.D. thesis, Department of Industrial Economy and Technology Management, NTNU Trondheim, Norway.

Rochet, J. C., & Choné, P. (1998). Ironing, sweeping, and multidimensional screening. *Econometrica*, *66*(4).

Shapiro, A. (2006). Stochastic programming with equilibrium constraints. *Journal of Optimization Theory and Applications*, *128*(1), 223–243.

Shapiro, A., & Ruszczynski, A. (2003). *Stochastic Programming. Handbooks in Operations Research and Management Science*, vol. 10. Amsterdam: Elsevier.

Shavell, S. (1979). Risk sharing and incentives in the principal and agent relationship. *The Bell Journal of Economics*, *10*(1), 55–73.

Spence, M., & Zeckhauser, R. (1971). The allocation of social risk. Insurance, information, and individual action. *American Economic Review*, *61*(2), 380–387.

Stiglitz, J. (1974). Incentives and risk sharing in sharecropping. *Review of Economic Studies*, *41*(2), 219–255.

Strausz, R. (2006). Deterministic versus stochastic mechanisms in principal-agent models. *Journal of Economic Theory*, *128*, 306–314.

Tirole, J. (1999). Incomplete contracts: Where do we stand? *Econometrica*, *67*(4), 741–781.

Verikoukis, C., Mili, Z., Konstas, I., & Angelidis, P. (2004). Overview on telecommunications regulation framework in southeastern europe. In *MELECON 2004. Proceedings of the 12th IEEE Mediterranean Electrotechnical Conference*, vol. 2, pp. 607–610.

Wolf, D.D., & Smeers, Y. (1997). A stochastic version of a stackelberg-nash-cournot equilibrium model. *Management Science*, *43*(2), 190–197.

Wynter, L. (2001). Stochastic bilevel programs. In C. Floudas, & P. Pardalos (Eds.), *Encyclopedia of Optimization*. Dordrecht: Kluwer.

Sustainable Agriculture, Food Security, and Socio-Economic Risks in Ukraine

Oleksandra Borodina, Elena Borodina, Tatiana Ermolieva, Yuri Ermoliev, Günther Fischer, Marek Makowski, and Harrij van Velthuizen

Abstract In Ukraine, the growth of intensive agricultural enterprises that focus on fast profits contribute considerably to food insecurity and increasing socio-economic and environmental risks. Ukraine has important natural and labor resources for effective rural development; more than 50% of food production is still contributed by small and medium farms, despite the difficulties associated with economic instabilities and the lack of proper policy support. Currently, the main issue for the agro-policy is to use these resources in a sustainable way, enforcing robust long term development of rural communities and agriculture. In this chapter, we introduce a stochastic, geographically explicit model for designing forward-looking policies regarding robust resources allocation and composition of agricultural production, in order to enhance food security and rural development. In particular, we investigate the role of investments into rural facilities to stabilize and enhance the performance of the agrofood sector in view of uncertainties and incomplete information. The security goals are introduced in the form of multidimensional risk indicators.

1 Introduction

In Ukraine, production intensification with a focus on fast profits is one of the main drivers that restructure food markets and distribute resource management rights in an imbalanced way. Intensification is advantageous for large producers, while small

O. Borodina (✉) · E. Borodina
Institute of Economics and Forecasting, P. Mirnogo 26, 01011 Kiev, Ukraine
e-mail: oleksandra.borodina@gmail.com; oborodina@ief.org.ua

T. Ermolieva · Y. Ermoliev · G. Fischer · M. Makowski · H. van Velthuizen
International Institute for Applied Systems Analysis, Schlossplatz 1, A-2361 Laxenburg, Austria
e-mail: ermol@iiasa.ac.at; ermoliev@iiasa.ac.at; fisher@iiasa.ac.at; marek@iiasa.ac.at; velt@iiasa.ac.at

Y. Ermoliev et al. (eds.), *Managing Safety of Heterogeneous Systems*, Lecture Notes in Economics and Mathematical Systems 658, DOI 10.1007/978-3-642-22884-1_8, © Springer-Verlag Berlin Heidelberg 2012

and medium agricultural businesses abandon the market due to an inability to compete for scarce and costly resources without a proper policy support. As a result, a lack of producers diversification increases risks associated with food and water security, environmental pollution, loss of food diversity, deterioration of socio-economic conditions in rural areas, rural-urban migration, and loss of cultural heritage.

Investigating the dilemma between the economic growth and the degradation of rural areas in Ukraine requires the development of integrated approaches specifying interdependent socio-economic, demographic and environmental criteria of long-term sustainable rural community development. A set of such criteria has already been identified and implemented in the USA, as well as in the EU, see e.g., the LEADER I, LEADER II, LEADER+ programs (Agriholdings in Ukraine 2008; Leader European Observatory 2010). In Ukraine, similarly to the LEADER programs, rural development planning includes goals of stimulating investments into improving quality of life and social conditions; protection and friendly use of environmental and cultural values; introduction, utilization, and expansion of new technologies and markets of local producers and services.

The aim of this chapter is, first, to analyze implications of recent agricultural reforms and trade liberalization on agriculture and rural areas development in Ukraine. Secondly, according to this analysis, develop a decision theoretic framework for designing forward looking national and subnational agricultural policies. The focus is to support policy choice regarding optimal agricultural production structure with a specific concern to revive and consolidate small and medium scale producers and services in rural areas.

There exist different approaches to the analyses of optimal production structure and resources allocation in agriculture. Studies involving trade liberalization often rely on the concept of general equilibrium (GE). While GE models may provide useful information on several economic aspects of policy reforms, it may be inappropriate, and in some cases misleading, to rely extensively only on their use for planning sustainable development strategies (Scrieciu 2006). There exists vast literature summarizing the limitations of the GE analysis (Cramon-Taubadel von et al. 2010; Scrieciu 2006). Two main concerns dominate the discussion. The first is that GEs are too aggregate to include appropriate sustainability indicators with safety/security constraints and horizons of planning. The second raises the issue about "demand-price-supply" relations which are often largely driven by inherent uncertainties and current policies (Ermolieva et al. 2010), e.g., weather conditions or export-import quotas, and thus can differ from ideal "demand-price-supply" dependencies. The main risk of using the advice from ideal and aggregate GE models without accounting for possible alternative paths is that this may cause various unexpected economic and production shocks such as bankruptcy, non-payments, prices increase, noncompliance to market agreements, etc.

The main task of planning sustainable agriculture in Ukraine is to design necessary resources allocation and regulations for rehabilitation of rural areas (Borodina 2009, 2007; Christev et al. 2005; Libanova 2006; Pantyley 2009; Prokopa and Popova 2008). Therefore, in this chapter, we introduce an optimization model following general ideas of economic modeling outlined in Nobel Memorial Lecture by Koopmans (1975). He admits (pp. 240-243) that according to a frequently cited definition, economics is the study of "... best use of scarce resources ... ". Because of the existence of "... alternative ways of achieving the same end result that a genuine optimization problem arises" that may have different efficiency allocation criteria and constraints regarding available resources, capital, equipment, etc. Yet, "... with an optimal solution of the given problem, whether of cost minimization or output maximization, one can associate ... shadow prices, one for each resource, intermediate commodity or end-product". Koopmans further acknowledges that these shadow prices or dual variables can be used as a price system for the decentralization of decision, either through the operation of competitive markets, or as an instrument of national planning (Dantzig 1963; Kantorovich 1939; Kantorovich and Gavurin 1940; Kantorovich 1959; Koopmans 1947, 1975). In other words, although the optimization model may be the same, the institutional framework supported by the model can be fundamentally different.

In this chapter, we consider only a pre-institutional optimization framework, i.e., primal stochastic optimization resource allocation model. The analysis of dual problem, emerging pricing system, and decentralized solutions would require considerable extension of the chapter. In particular, important issues concern relations among emerging spot prices and safety/security constraints. In this chapter we also don't consider issues connected with data analysis, which are vital for proper treatment of inherent uncertainties. We simply assume that the data can always be characterized by scenarios. Therefore, the main issue is the design of policies (decisions) robust with respect to all potential scenarios. This framework assumes the existence of a policy analyst who may perform efficient allocation of resources. In general, the analyst may consider alternative objective functions that incorporate or emphasize various aspects of sustainability and security concepts.

In presence of uncertainties and resource (financial, land, water) constraints, irreversibility of deterministic solutions may incur high sank costs (Arrow and Fisher 1974). Therefore, there is a need for a two-stage decision making framework with anticipative and adaptive decisions (Arrow and Fisher 1974; Ermoliev and Wets 1988). The strategic (ex-ante) first-stage decisions taken before the uncertainties become known cannot be altered. In order to ensure the flexibility of the system under such decisions, they are supplemented by a set of corrective (ex-post) decisions implemented after the uncertainties are resolved. Thus, in the presence of uncertainty, e.g., climatic variability, markets shocks, demand and price fluctuations, etc., the strategic decisions are only partially implemented in the first stage, and can then be corrected in the second stage by learning from experience and further observations. Within the same modeling framework, the optimal combination of adaptive and anticipative decisions can be derived only by methods of two-stage stochastic optimization (STO). The two-stage STO model proposed in this chapter is geographically explicit. The application of the model is illustrated with an example of optimal investments into expansion of agricultural activities and rural services to employ potential workers migrating between Ukrainian regions as a result of job losses or financial/production instabilities.

The structure of the chapter is as follows. Section 2 summarizes main structural changes and current agricultural development trends in Ukraine and identifies key factors contributing to the worsening situation in rural communities. Section 3 outlines main criteria of rural community development as formulated in LEADER programs (Agriholdings in Ukraine 2008; Leader European Observatory 2010). It formulates a model that employs these criteria in the context of Ukraine. Section 4 discusses model application with selected numerical calculations and Sect. 5 concludes.

2 Structural Changes in Ukrainian Agricultural Sector

Agricultural enterprises in Ukraine are being actively restructured and integrated forming large agro-holdings. During 2005 and 2006, the number of the enterprises, which operate more than 10 thousand hectares of land, has increased by 27%; the average size of the total area in these enterprises has increased by 7% to more than 20 thousand hectares. Large agricultural farms may rather freely choose among the commodities to produce and in what amounts. This freedom induces specialization in more profitable products. As a result, agro-holdings concentrate primarily on intensive profitable production such as raw-materials for biofuels, which increases socio-economic and environmental risks in rural areas. Decreasing production diversity and diversion of land and water resources from direct food production undermines food security. It also worsens environmental quality through high fertilization rates and absence of necessary crop rotations. Without adequate regulations, these trends may lead to further land degradation, loss of fertile soils, water, air, soil pollution (Agriholdings in Ukraine 2008; Shnyrkov et al. 2006).

Apart from mono cropping which disturbs the supply of grains for direct consumption, food security problem has been exacerbated by inadequate import-export quotas and weather uncertainties. Imbalanced and unstable grains production affects, in particular, livestock sector, foremost, large animals and cows (see discussion in Sect. 3). Reasons for the decreasing number of animals in Ukraine are different for different locations and years. At the beginning of agricultural reforms, the loss of state subsidies following the collapse of the Soviet Union increased feed and production costs and reduced profitability of livestock enterprises. Further, in 2003, 2004 and 2005, the majority of large animals were slaughtered because of insufficient feeds due to low yields and intensive international trade (Tarassevych 2004). Decreased number of animals and declined meat production resulted in a substantial increase of meat prices. From March 2004 to March 2005, the price for meat increased by 56.8%. Due to the high share of meat in goods' basket (about 13%), meat deficit contributed about a 15% increase to the yearly inflation rate (Giucci and Bilan 2005). This shows how inadequate policies in the agricultural sector may produce dramatic effects within the sector with a spillover into the whole national economy. Large systems of bovine meat production turned to be most prone to frequent reforms and governmental regulations. Currently, large animals (among them cows) and bovine meat production in Ukraine concentrate primarily in household systems. Because of high production costs and risks, these producers will not invest into larger scale production. Livestock production is one of the most labor intensive agricultural activities, which may provide employment and social protection for many out of work in rural areas. However, without targeted investments this is unlikely to happen given high risks and strict quality norms imposed by the WTO accession (Heyets 2008; Shnyrkov et al. 2006).

Production intensification and land concentration led to many adverse problems, but most harmful are impacts on demographic and socioeconomic situation in rural areas. Intensive large scale enterprises and agro holdings require much fewer workers than soviet-type agro businesses. They make use of qualified labor force from cities, better educated with special skills and experience. This has released a rather substantial part of rural workers and inspired rural – urban migration in strive for short-term jobs (primarily in construction sector), what led to rural area depopulation and degradation (Libanova 2006; Pantyley 2009; Prokopa and Popova 2008). Depopulation and deterioration of living conditions and infrastructure in rural areas are also due to the fact that unlike the Soviet times when almost all expenses on the development, social security, health and fiscal provision of rural areas were taken by the state and local collective agrarian enterprises, during and after the reform "market" rules were introduced, i.e. agrarian enterprises make profits while local communities have to develop rural areas. It should be noted that a majority of large scale producers are registered in cities and rarely pay taxes into local budgets.

Most likely the agro-holdings will dominate the agricultural sector of Ukraine in the future. Considering their rash emergence and the increasing risks they cause to food security and rural development, new approaches for organization and planning need to be properly designed in order to enable agriculture and rural development with a multitude of farming activities. The government may impose regulations that provide equal and transparent financial support for doing business by all forms of enterprises in agricultural production and service sectors. This measure may reduce unequally distributed opportunities for subsidies and replace them by direct governmental/public investments, such as investment in practical education of rural community members, creation of market information systems, support of farm advisory services, and – most important – investments in rural infrastructure (roads, energy and water supply, health care, schools). Furthermore, it is necessary to put more emphasis on the impact of fiscal support measures to agriculture. Currently, the bias is strongly in favor of agro-holdings and urban areas.

3   Analysis of Pathways Towards Sustainable Rural Area Development

The trends highlighted in previous sections are alarming, and therefore call for adequate approaches for organization and forward-looking agricultural policies. There exists encouraging experience in planning rural development within

"LEADER"[1] programs. The programs implement incentives to encourage integrated, high-quality and original strategies for sustainable development, have a strong focus on partnership and networks for exchange of experience. In Ukraine, similar programs focus on revival of old and introduction of new rural activities to create rural jobs and enhance food security.

In this section we propose a two-stage stochastic optimization model to support policy-making on sustainable agricultural development under inherent risks, incomplete information, and resource constraints. Optimal adjustments of production and services by geographical locations are derived as a tradeoff between costs minimization, food security goals, targeted level of rural jobs, and the suitability criteria. The security goals are introduced in the form of multidimensional risk measures having direct connections (see remark in Sect. 5) with Value-at-Risk (VaR) and Conditional Value-at-Risk (CVaR) or expected shortfalls type indicators (Rockafellar and Uryasev 2000). For planning livestock production expansion, the suitability criteria include feeds and pastures requirements per unit livestock. The model is temporally explicit. In the current two-stage setting, it involves two stages (periods), contemporary and future. Each of these stages may include many time intervals. In other words, it may be easily expanded to a multi-period dynamic framework. The model is also geographically detailed. For now, it is implemented at the level of 25 Ukrainian regions, but may be disaggregated to finer resolutions. The model comprises three main modules with respective parameters, technical coefficients, criteria, and risks – socio-economic, environmental, and agricultural. The socio-economic module defines a balance between costs minimization and social goals including additional production to ensure jobs and food security; the environmental module controls pressure stemming from agricultural production in locations; the agricultural module imposes technical coefficients of agronomically sound practices. The model distinguishes producers of different agricultural commodities i in regions l and by production systems j. Production systems are characterized by different intensification levels, say, traditional (household), medium or intensive large scale producers. In general, there are considerable data requirements which cannot be fulfilled by traditional estimation procedures. The lack of repetitive observations of the same phenomenon raises important issue about using different sources and generators of data, explicit treatment of uncertainties and designing decisions robust with respect to inherent uncertainties.

Food security and rural development goals require allocating targeted production and respective rural workers by regions. Food targets include direct demand for food and feeds and indirect demand, e.g., international export obligations and inter-regional trades. Let $x_{ijl} \geq 0$ denote potential production of commodity i in region l and management system j. Increased production creates additional rural agricultural and nonagricultural (service) jobs. Define β_{ijl} as a number of workers to produce a unit of commodity x_{ijl}, and L_l – a targeted level of rural employment

---

[1] The acronym comes from *Liaison Entre les Actions de Development Rural*, i.e., *Links between Actions of Rural Development*.

in location l. Ignoring so far uncertainties, the goal to ensure required employment in location l is defined by the following constraints:

$$\sum_{ij} \beta_{ijl} x_{ijl} \geq L_l. \tag{1}$$

In general, L_l may not be known with certainty as it is difficult to predict, for example, how many people are likely to return from short-term urban jobs to rural areas. Therefore, constraint (1) as well as the following constraints (2) can be defined in terms of probabilistic constraints (7)–(8) or, within general two-stage stochastic optimization framework defined by functions (11)–(18). Migration of labor force between rural-urban areas and within regions depends on various factors, including availability of infrastructure, such as schools, trade centers, health and social provisions, transportation networks, entertaining and cultural centers, etc. The model may account for the behavioral components similarly to the model developed for the analysis of agricultural development in China (Ermolieva et al. 2005; Fischer et al. 2008, 2007) where behavioral criteria are combined with strictly planned governmental policies. In general, variable L_l may be characterized by alternative scenarios.

Data (Borodina 2009) on employment rates in rural services per unit of produce x_{ijl} permit to estimate the demand for jobs S_l by region l. Values S_l may be treated as random, i.e., defined either by probability distribution functions or by a set of potential scenarios. The willingness to work in infrastructure, for example in schools, depends on gender, age, educational level, i.e., values S_l can also be characterized by behavioral criteria. Thus, in addition to (1), x_{ijl} need to satisfy the condition on necessary expansion and employment in rural infrastructure:

$$\sum_{ij} \gamma_{ijl} \beta_{ijl} x_{ijl} \geq S_l. \tag{2}$$

Expansion of production and services requires investments. Their limitations are included in our model either as an overall budget constraint or as minimization of total costs and investments:

$$\sum_{il} V_{il} \left(\sum_{j} x_{ijl} \right) + \sum_{ijl} c_{ijl} x_{ijl} + \sum_{l} C_l(y_l) + \sum_{kl} c_{kl} y_{kl}, \tag{3}$$

where c_{ijl} are expenditures associated with production costs and wages of employees involved in production x_{ijl}. Investments V_{il} depend on the current level of regional development, i.e., depressive regions require higher investments. Cost functions $C_l(\cdot)$ and unit costs c_{kl} may be associated with trades agreements and transportation of feeds between regions, as explained below. Uncertainties of criterion (3) are associated, first of all, with market prices.

Food security and environmental constraints are introduced by equations (4), (5), (6), respectively:

$$\sum_{jl} x_{ijl} \geq d_i, \tag{4}$$

$$\sum_{ij} \delta_i x_{ijl} \leq a_l + y_l + \sum_k y_{kl} - \sum_k y_{lk}, \tag{5}$$

$$\sum_{ij} \sigma_i x_{ijl} \leq b_l. \tag{6}$$

Constraint (4) ensures that production levels x_{ijl} satisfy targeted national demand d_i of i-th commodity, which reflects food security goals; (5) ensures that allocations x_{ijl} satisfy availability of feeds in locations l, where δ_i is a technical coefficient defining feed requirements per unit of livestock. Variables $y_i \geq 0$ reflect possibility to expand feeding capacity a_l at cost $C_l(y_l)$; variables y_{kl} represent feed trading between different regions at cost c_{kl}. The same type of additional decision variables can be introduced in equations (4) for trading production commodities. Equation (6) allows production expansion only in locations with sufficient resources, such as pastures or cultivated land, thus ensuring efficient recycling of wastes and manure associated with new x_{ijl} units of production, σ_i is an ambient coefficient reflecting diverse recycling capacities (e.g., manure storage and processing facilities). Constraints (5) and (6) comprise the environmental module that safeguards environmental targets, land use, and agronomic norms.

Uncertainties, in particular, stochastic variables S_l, L_l require further model specification. We admit that information on S_l, L_l may be uncertain, and therefore variables x_{ijl} need to satisfy constraints (1)–(2) with some guaranteed certainty level for all possible scenarios of $S_l(\omega)$, $L_l(\omega)$ of S_l, L_l, where ω indicates uncertain events (scenarios) which may affect S_l, L_l, e.g., $\omega \in \{1, 2, \ldots, N\}$. Say, chances that constraints (1)–(2) are satisfied (under derived x_{ijl}) must be higher than the imposed levels $0 \leq p_l \leq 1, 0 \leq q_l \leq 1$. This requirement is expressed in terms of probabilistic constraints:

$$P\left[\sum_{ij} \beta_{ijl} x_{ijl} \geq L_l(\omega)\right] \geq p_l, \tag{7}$$

$$P\left[\sum_{ij} \gamma_{ijl} \beta_{ijl} x_{ijl} \geq S_l(\omega)\right] \geq q_l, \tag{8}$$

$0 \leq p_l \leq 1, 0 \leq q_l \leq 1$, which are similar to the well-known in engineering safety or reliability constraints. In the insurance business, they reflect the solvency constraints of insurance companies or banks, and are often defined by p_l, q_l in the range of $[0.001, 0.03]$, which corresponds to regulating the frequency of insolvency as an event that may occur once in 300-1000 years.

Constraints (7)–(8) describe a stochastic supply-demand relations of the employment: the demand $\beta_{ijl}x_{ijl}$ may be not completely satisfied by the random supply $L_l(\omega)$; similar relates to $\gamma_{ijl}\beta_{ijl}x_{ijl}$ and $S_l(\omega)$. If the analytical distributions of $L_l(\omega)$, $S_l(\omega)$ are known, then equations (7), (8) are reduced to linear equations defined by quantiles of the corresponding $L_l(\omega)$, $S_l(\omega)$. Generally, accounting for potential uncertainties of β_{ijl}, γ_{ijl} requires specific methods; in particular, (7), (8) may represent discontinuous constraints. To account for possibly highly discontinuous equations (7)–(8), we convert them into expected imbalances defined by convex functions:

$$E \max \left\{ 0, L_l(\omega) - \sum_{ij} \beta_{ijl} x_{ijl} \right\}, \tag{9}$$

$$E \max \left\{ 0, S_l(\omega) - \sum_{ij} \gamma_{ijl} \beta_{ijl} x_{ijl} \right\}. \tag{10}$$

Minimization of functions (9)–(10) implies costs π_l, ψ_l of decreasing the gaps or expected deficits of the employment in agriculture and services. Therefore, functions (9), (10) are modified to the following cost functions:

$$\pi_l E \max \left\{ 0, L_l(\omega) - \sum_{ij} \beta_{ijl} x_{ijl} \right\} \tag{11}$$

$$\psi_l E \max \left\{ 0, S_l(\omega) - \sum_{ij} \gamma_{ijl} \beta_{ijl} x_{ijl} \right\}. \tag{12}$$

In order to analyze the goals (3) and (11)–(12), the problem is formulated as follows: find production x_{ijl} minimizing the cost function

$$\sum_{il} V_{il} \left(\sum_{j} x_{ijl} \right) + \sum_{ijl} c_{ijl} x_{ijl} + \sum_{l} C_l(y_l) + \sum_{kl} c_{kl} y_{kl}$$

$$+ \sum_{l} \pi_l E \max \left\{ 0, L_l(\omega) - \sum_{ij} \beta_{ijl} x_{ijl} \right\} \tag{13}$$

$$+ \sum_{l} \psi_l E \max \left\{ 0, S_l(\omega) - \sum_{ij} \gamma_{ijl} \beta_{ijl} x_{ijl} \right\}$$

subject to constraints (4)–(6).

Function (13) can be considered as a stochastic version of scalarization functions used in multicriteria analysis. Formally, function (13) corresponds to a multicriteria stochastic minimization model with cost function (3) and risk functions (11)–(12). As analyzed in (Ermoliev and Wets 1988; Fischer et al. 2008), an appropriate choice of values π_l and ψ_l enables controlling the safety/security constraints (7), (8). We may also formulate a robust stochastic optimization model with an alternative scalarization function:

$$
\sum_{il} V_{il} \left(\sum_j x_{ijl} \right) + \sum_{ijl} c_{ijl} x_{ijl} + \sum_l C_l(y_l) + \sum_{kl} c_{kl} y_{kl}
$$

$$
+ E \max_l \pi_l \max \left\{ 0, L_l(\omega) - \sum_{ij} \beta_{ijl} x_{ijl} \right\} \tag{14}
$$

$$
+ E \max_l \psi_l \max \left\{ 0, S_l(\omega) - \sum_{ij} \gamma_{ijl} \beta_{ijl} x_{ijl} \right\}
$$

i.e., instead of the aggregate "expected" deficit defined by (13) as the sum of functions (11), (12), function (14) focuses on extreme random deficits (events) of the most suffering regions. The advantage of such an optimization problem is its focus on country-wide extreme events (scenarios) regarding demand-supply relations defined by $L_l(\omega), S_l(\omega)$, and $\sum \beta_{ijl} x_{ijl}, \sum \gamma_{ijl} \beta_{ijl}$. Minimization of functions (13), (14) corresponds to an important two-stage decision making formulation. To illustrate this, let us consider the case when parameters of the model do not depend on x_{ijl}. In this case, minimization of functions (13), (14) may be reduced to a linear programming (LP) problem using ex-ante decisions of the model defined by equations (4)–(6), (13) or (14), and additional second-stage ex-post decisions emerging after observations of random variables.

Let us consider the LP problem corresponding to minimization of (13) subject to constraints (4), (5), and (6). In general, ex-ante decisions x_{ijl}, y_{ijl} may lead to deficits defined by (9), (10). Let us consider a finite number of scenarios

$$
L_l^s, s = \overline{1 : N_l}, S_l^t, t = \overline{1 : M_l}
$$

of random variables $L_l(\omega)$ and $S_l(\omega)$. Two-stage model assumes that after the observation L_l^s and S_l^t of real random variables L_l and S_l, the arising deficit can be corrected by the second stage (ex-post) decisions Z_l^s and U_l^t. In our model, the second stage decisions Z_l^s in constraint (1) and U_l^t in constraint (2) may be associated with the use of better technologies or more qualified employees with higher wages. Decision variables Z_l^s and U_l^t ensure satisfaction of constraints

$$
\sum_{ij} \beta_{ijl} x_{ijl} + Z_l^s \geq L_l^s, \tag{15}
$$

$$\sum_{ij} \gamma_{ijl}\beta_{ijl}x_{ijl} + U_l^t \geq S_l^t \qquad (16)$$

for all possible random scenarios L_l^s, S_l^t, $s = \overline{1:N_l}$, and $t = \overline{1:M_l}$. Therefore, the second-stage feasible variables Z_l^s and U_l^t are, in general, random variables $Z_l(\omega)$ and $U_l(\omega)$ depending on random observations L_l^s, and S_l^t. The two-stage stochastic programming problem is formulated as minimization of the following function:

$$\sum_{il} V_{il}\left(\sum_{j} x_{ijl}\right) + \sum_{ijl} c_{ijl}x_{ijl} + \sum_{l} C_l(y_l) + \sum_{kl} c_{kl}y_{kl} \qquad (17)$$

$$+ \sum_{l} \pi_l EZ_l(\omega) + \sum_{l} \psi_l EU_l(\omega)$$

subject to constraints (4), (5), (6), (15), (16). If costs V_{il} and c_l, c_{kl} are linear (or piecewise linear convex function), then (17) may be solved by linear programming methods. Assume that scenarios L_l^s, $s = \overline{1:N_l}$, and S_l^t, $t = \overline{1:M_l}$, have probabilities $\vartheta_l^1, \ldots, \vartheta_l^{N_l}$ and $\mu_l^1, \ldots, \mu_l^{M_l}$, respectively. This is a natural assumption since results of questionnaires are usually quantified by likelihoods, e.g., with equal probabilities. Let us denote by Z_l^s and U_l^t the ex-post decision under scenarios L_l^s and S_l^t. Then, the proposed model can be formulated as the following linear programming problem in the space of ex-ante and ex-post decisions: minimize

$$\sum_{i.l} V_{il}\left(\sum_{j} x_{ijl}\right) + \sum_{ijl} c_{ijl}x_{ijl} + \sum_{l} C_l(y_l) + \sum_{kl} c_{kl}y_{kl} \qquad (18)$$

$$+ \sum_{l} \pi_l \sum_{s} \vartheta_l^s Z_l^s + \sum_{l} \omega_l \sum_{t} \mu_l^t U_l^t$$

subject to constraints (4), (5), (6), (15), and the constraints (15)–(16). It is easy to see that optimal decisions Z_l^s and U_l^t are calculated as

$$Z_l^s = \max\left\{0, L_l^s - \sum_{ij} \beta_{ijl}x_{ijl}\right\}, \qquad U_l^s = \max\left\{0, S_l^s - \sum_{ij} \gamma_{ijl}\beta_{ijl}x_{ijl}\right\},$$

for all scenarios $s = \overline{1:N_l}$ and $t = \overline{1:M_l}$. Therefore, the model defined by equations (4), (5), (6), (18) is indeed equivalent to the model defined by equations (4), (5), (6), (13), (15), (16) under random scenarios L_l^s and S_l^t.

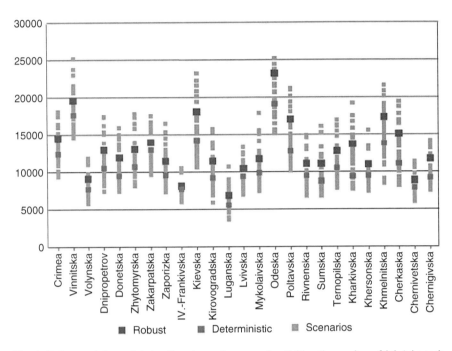

Fig. 1 Robust and deterministic allocations of new rural activities (in number of jobs) in each region

4 Numerical Application

In this section we summarize selected results (Borodina 2009) of collaboration between the IEF[2] and IIASA.[3] The application of the presented model at regional levels is illustrated with a case of livestock sector expansion and rural services development. Scenarios of migrants L_l^s and S_l^t in (15)–(16) are derived in (Borodina 2009) from experts opinions and national surveys. About 100 alternative scenarios are identified by ranges, and presented in Fig. 1. Other model parameters are also summarized in (Borodina 2009). Costs per animal operations, the ranking regions by depreciation level, transportation and production costs are available from Statistical Year Books of Ukraine.

The model operates in two modes: deterministic and stochastic. The solution of the deterministic model is optimal with respect to one scenario of migrants, e.g., expected values of L_l^s and S_l^t. In the stochastic mode, the number of migrants is not known in advance, and therefore the model derives a solution robust with respect to all scenarios.

---

[2] Institute of Economics and Forecasting of National Academy of Sciences, Ukraine.
[3] International Institute for Applied Systems Analysis.

To illustrate why two-stage STO produces robust risk-focusing solutions, we summarize the main differences between the deterministic (solution of the deterministic model) and the two-stage solutions. Deterministic model assumes complete information about agents, and therefore creates activities for the known number of migrants, which formally restrains the analysis to a trivial case in which $s = 1$ and $t = 1$ in (15)–(18). In the reality, however, often jobs are created for an expected or targeted number of migrants, while the real number of them is either lower or higher. Both cases, i.e., deficit and surplus, lead to direct and indirect costs. If activities are expanded (this also applies to the infrastructure, such as roads, schools, medical and cultural facilities), but the number of workers is overestimated, the investments will be lost. The situation may be improved by offering higher incomes and privileges in order to attract workers. Conversely, if jobs and facilities are in deficit, this will either cause regret situations among population or require up-front investments to immediately accommodate newcomers.

In contrast to the deterministic model, the two-stage solution is calculated assuming that the number of migrants is not known in advance. The costs and risks associated with situations of deficit and surplus described above are controlled by the second stage decision. Thus, the main idea of robust two-stage solution is to choose first-stage decisions x_{ijl} before knowing the true number of migrants such that the total expenses incurred by implementations of x_{ijl} and the costs of their possible corrections determined by second-stage decisions Z_l^s and U_l^t are minimized. In the event of "more-than-expected" migrants, the costs of second-stage decisions Z_l^s and U_l^t may reflect foreseen at stage 1 feasible adjustments of infrastructure, houses, farms, roads, etc. In the "less-than-expected" case, they may correspond to foreseen at stage 1 feasible increases of incomes or social benefits to attract more workers. In fact, for the simplicity of model formulation, functions (13), (14) ignore costs associated with the underestimation of migrants. Adjustments of the model for general case are trivial, and the discussion of the dual model is easy (see next section) for functions (13), (14).

According to expert estimates, it is anticipated that the number of migrants will exceed expected values (Fig. 1) of the deterministic model. Total costs (13) of optimal solution of the deterministic model and the robust solutions are illustrated in Fig. 2. For the solution of the deterministic model, the costs include costs of optimal single scenario solution and additional costs associated with the corrections of these solutions with respect to other potential scenarios. Costs of robust two-stage solutions are optimal with respect to both stages. Total costs of deterministic and robust solutions are about 7 and 5.5 millions of monetary units, respectively.

Figure 1 shows solutions in terms of rural work-places. Robust solution suggests creating activities accounting for percentiles of outcome with respect to all scenarios, while the deterministic model solution accounts only for expected scenario. These results so far provide only an aggregate region-level perspectives regarding agricultural expansion, which may be down-scaled to finer levels (i.e., villages, communities) applying technique developed in (Fischer et al. 2007).

In Fig. 1, alternative scenarios of migrants are depicted with grey color. Regarding financial support for additional livestock production allocation, the model

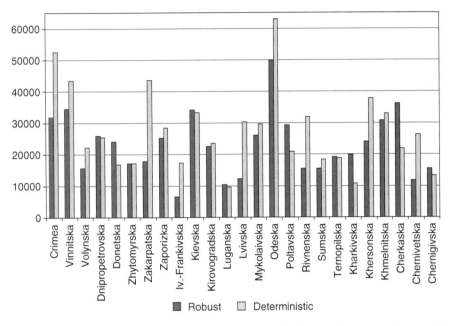

Fig. 2 Total costs associated with robust solution and optimal solution of the deterministic model

estimates that the support may come either in the form of voluntarily contributions or taxation of the intensive enterprises and part of the investments may be covered by governmental support or through other investments. The analysis of these alternatives requires formulation of the dual model and optimality/equilibrium conditions.

5 Concluding Remarks

This chapter summarizes agricultural developments in Ukraine in the period from 1990 to 2010. It identifies diverse risks induced by production intensification and concentration, in particular, risks associated with food security, environment pollution, worsening socio-economic and demographic conditions in rural areas of Ukraine. The problem of sustainable rural development and necessary agriculture expansion is formulated as a two-stage STO, which permits to account for inherent complex interactions and to derive forward-looking policies.

Numerical results review recent joint studies between IEF and IIASA on planning new activities and jobs in agricultural sector and rural services at the level of Ukrainian regions. In Ukraine it is expected that large number of short-term urban workers will migrate between regions and from urban to rural areas. Robust solution suggested by the two-stage STO model identifies levels of rural activities optimal with respect to a majority of possible migrants' scenarios. We illustrate the advantages (e.g. cost effectiveness) of robust solution in contrast to optimal solution of the deterministic model. Costs and risks associated with the deterministic model solution are much higher than costs and risks associated with robust solution derived by two-stage STO.

According to the general discussion in Sect. 1, the main purpose of this chapter is to develop only an integrated optimization model allowing a policy analyst to identify robust paths of future agriculture development in Ukraine improving socio-economic and environmental aspects of rural life, enhancing food security of the country.

Important remaining issue is the analysis of the dual problem, emerging optimality conditions, pricing system and decentralized solutions. The following example illustrates the type of important conclusions which can be derived from such analysis of a STO model. Risk functions (9), (10) embedded in cost function (13) define systemic risks of the whole food supply system; similar interpretation has the scalarizing function (14). It is unclear a priory, that minimization of cost-function (13) imposes implicit regional risk measures. This becomes clear only from analysis of the dual model and optimality conditions. Consider a slight modification of risk functions (9), (10) that reflects the discussion in the previous section. Let us introduce for each location l new decision variables h_l and g_l as risk reserves which have to be prepared ex-ante for making ex-post adjustments in the case of "less-than-expected" migrants flow. Then function (13) takes the form

$$\sum_{il} V_{il} \left(\sum_j x_{ijl} \right) + \sum_{ijl} c_{ijl} x_{ijl} + \sum_l C_l(y_l) + \sum_{kl} c_{kl} y_{kl} + \sum_l (\rho_l g_l + \varepsilon_l h_l)$$

$$+ \sum_l \pi_l E \max \left\{ 0, L_l(\omega) - \sum_{ij} \beta_{ijl} x_{ijl} - g_l \right\} \qquad (19)$$

$$+ \sum_l \psi_l E \max \left\{ 0, S_l(\omega) - \sum_{ij} \gamma_{ijl} \beta_{ijl} x_{ijl} - h_l \right\}$$

where ρ_l, ε_l are unit costs in l-th region associated with creation at stage 1 a unit of the risk reserve. The optimality conditions with respect to h_l and g_l lead to VaR and CVaR type risk measures with respect to decision variables h_l and g_l:

$$Prob \left[L_l(\omega) - \sum \beta_{ijl} x_{ijl} - g_l \geq 0 \right] = \rho_l / \pi_l,$$

$$Prob \left[S_l(\omega) - \sum \gamma_{ijl} \beta_{ijl} x_{ijl} - h_l \geq 0 \right] = \varepsilon_l / \psi_l,$$

jointly with other optimality conditions, possibly defined by diverse probability functions and dual variables. This becomes clear by taking the partial derivative with respect to x_{ijl} of function (19), assuming this derivative exists. In general, this requires the use of non-differentiable optimization techniques as in (Ermoliev and Leonardi 1981), or by formulating the dual problem for discrete approximation model similar as in (Koenker and Bassett 1978) defined by (4)–(6), (15), (16) and function (17).

Acknowledgements The authors are grateful to the participants of the IIASA/GAMM workshop on Coping with Uncertainty, and to the anonymous referees of this chapter, for their critical suggestions that led to improvements of this chapter.

References

Agriholdings in Ukraine: Good or Bad? German–Ukrainian Policy Dialogue in Agriculture. Institute for Economic Research and Policy Consulting, Policy Paper Series AgPP No 21, 2008.

Arrow, K. J., & Fisher, A. C. (1974). Preservation, uncertainty and irreversibility. *Quarterly Journal of Economics, 88*, 312–319.

Borodina, O. (2009). Food Security and Socioeconomic Aspects of Sustainable Rural Development in Ukraine. Interim Report IR-09-053, International Institute for Applied Systems Analysis (IIASA), Laxenburg, Austria.

Borodina, O. (2007). Applying Historical Precedent to a New Conventional Wisdom on Public Sector Roles in Agriculture and Rural Development: Case of Ukraine /FAO Research Report, Rome, 2007.

Christev, A., Kupets, O., & Lehmann, H. (2005). Trade Liberalization and Employment Effects in Ukraine. IZA Discussion Paper No. 1826, Institute for the Study of Labor (Forschungsinstitut zur Zukunft der Arbeit, Bonn).

Cramon-Taubadel von, S., Hess, S., & Brümmer, B. (2010). A Preliminary Analysis of the Impact of a Ukraine-EU Free Trade Agreement on Agriculture. Policy Research Working Paper 5264. The World Bank Development Research Group, Agriculture and Rural Development Team & Europe and Central Asia Region, April 2010.

Dantzig, G. B. (1963). *Linear Programming and Extensions*. Princeton: Princeton University Press.

Ermoliev, Y., & Leonardi, G. (1981). Some proposals for stochastic facility location models. *Mathematical Modeling, 3*, 407–420.

Ermoliev, Y., & Wets, R. (Eds.). (1988). Numerical techniques of stochastic optimization. *Computational Mathematics*. Berlin: Springer.

Ermolieva, T., Ermoliev, Y., Fischer, G., Jonas, M., Makowski, M., & Wagner, F. (2010). Carbon emission trading and carbon taxes under uncertainties. *Climatic Change, 103*(1-2).

Ermolieva T., Fischer, G., & van Velthuizen, H. (2005). Livestock Production and Environmental Risks in China: Scenarios to 2030", *FAO/IIASA Research Report*. Laxenburg, Austria: International Institute for Applied Systems Analysis.

Fischer, G., Ermolieva, T., Ermoliev, Y., & Sun, L. (2008). Risk-adjusted approaches for planning sustainable agricultural development. *Stochastic Environmental Research and Risk Assessment*, 1–10.

Fischer, G., Ermolieva, T., Ermoliev, Y., & van Velthuizen, H. (2007). Sequential downscaling methods for Estimation from Aggregate Data. In K. Marti, Y. Ermoliev, M. Makowski, & G. Pflug (Eds.), *Coping with Uncertainty: Modeling and Policy Issue*. Berlin, StateNew York: Springer.

Giucci, R., & Bilan, O. (2005). High inflation in Ukraine: Roots and remedies. Interim Report U1, Institute for Economic Research and Policy Consulting, German Advisory Group on Economic Reform.

Heyets, V. M. (2008). Agricultural and food markets under the WTO accession.

Kantorovich, L. V. (1939). Mathematicheskie Metody Organizatsii i Planirovania Proizvodstva, Leningrad State University Publishers, translated as "Mathematical Methods in the Organization and Planning of Production" (1960) in Management Science 6, 4, pp, 366–422.

Kantorovich, L. V., & Gavurin, M. K. (1940). (first version 1940, publ. 1949), "Primenenie matematicheskikh metodov v voprosakh analiza gruzopotokov," in Problemy povysheniia effektivnosti raboty transporta (The Use of Mathematical Methods in Analyzing Problems of Goods Transport, in Problems of Increasing the Efficiency in the Transport Industry, pp. 110–138). Academy of Sciences, U.S.S.R.

Kantorovich, L. V. (1959), Ekonomicheskii raschst nailichshego ispolzovania resursov, Acad. of SC., USSR (translated (1965), *The Best Use of Economic Resources*. Harvard University Press: Cambridge, Mass.

Koenker, R., & Bassett, G. (1978). Regression quantiles. *Econometrica, 46*(1), 33–50.

Koopmans, T. (1947). Optimum utilization of the transportation system. In *Proc. Intern. Statis. Conf.*, Voll. V. Washington, D.C.

Koopmans, T. (1975). Concepts of optimality and their uses. Nobel Memorial Lecture, December 11. Yale University, New Haven, Connecticut, USA.

Leader European Observatory: Assessing the added value of the LEADER approach. Rural Innovation Dossiern 4 (Links Between Actions for the Development of the Rural Community).

Libanova, E. (2006). Demographic crisys in Ukraine: Research problems, conclusions, and actions. In Steschenko (Ed.), *Complex Demographic Forecasts to 2050 in Ukraine*. Interim Report. NAS of Ukraine: Institute of Economics and Forecasting (In Ukrainian).

Pantyley, V. (2009). Demographic situation of rural population in country-regionplaceUkraine in the period of intensive socio-economic transformation. *Europ. Countries, 1*, 34–52. DOI: 10.2478/v10091/009-0004-6.

Prokopa, I., & Popova, O. (2008). *Depopulation in Rural Areas of Ukraine: Destructive Changes and Threats* (in Ukrainian). Interim Report. NAS of Ukraine: Institute of Economics and Forecasting.

Rockafellar, T., & Uryasev, S. (2000). Optimization of conditional value-at-Risk. *The Journal of Risk, 2*, 21–41.

Scrieciu, S. (2006). How useful are computable general equilibrium models for sustainability impact assessment? In C. Kirkpatrick, & C. George (Eds.), *Impact Assessment and Sustainable Development: European Practice and Experience*. Edward Elgar: Cheltenham.

Shnyrkov, A., Rogach, A., & Kopystyra, A. (2006). Ukraine's joining the WTO: realities and challenges. *Transition Studies Review, 13*(3), 513–523 (DOI 10.1007/s11300-006-0121-0). New York: Springer.

Tarassevych, O. (2004). Ukraine: Livestock and Products. Annual GAIN Report, UP4014, (http://www.fas.usda.gov/psd).

Multiple-Criteria Decision Support System for Siemianówka Reservoir under Uncertainties

Adam Kiczko and Tatiana Ermolieva

Abstract This paper presents a Multiple Criteria Decision Support System for the optimal management of the Siemianówka reservoir. The reservoir is localized on the Narew River upstream the NNP. The river system under considerations consists of a storage reservoir and a 100 km long River Narew reach, at which end the NNP is located. The goal of the work is to provide decision makers with a tool that would allow the safety of the NNP environmental requirements within the reservoir management policy to be included. An important issue is the competition between many water-dependent systems and agents, e.g., agriculture, energy, wetlands, for limited water resources. The proposed system allows a trade-off between different reservoir users to be found, including protected wetland ecosystems of the Narew Nation Park. Unobserved inflows play an essential role in the river water balance and are dealt with use of k-NN technique. In addition, as the optimization problem requires numerous realizations of the river model, a numerically efficient Stochastic Linear Transfer Function was applied to flow routing.

1 Introduction

Mitigation of negative changes in natural ecosystems is one of the challenges of the water management. In case of riverine environments, the most noticeable matter is protection of wetlands, which are considered as the ecosystems of a very high biodiversity level. Therefore their existence is crucial for sustainable development

---

A. Kiczko (✉)
Institute of Geophysics, Polish Academy of Science, Warsaw, Poland
e-mail: akiczko@igf.edu.pl

T. Ermolieva
International Institute for Applied Systems Analysis, Laxenburg, Austria
e-mail: ermol@iiasa.ac.at

Y. Ermoliev et al. (eds.), *Managing Safety of Heterogeneous Systems*, Lecture Notes in Economics and Mathematical Systems 658, DOI 10.1007/978-3-642-22884-1_9,
© Springer-Verlag Berlin Heidelberg 2012

of society. In this case such negative changes are especially seen in disruptions of river hydrological regime: long periods of droughts, shorter and smaller freshets. In result wetland areas often suffer from serious water shortages. The causes might be directly related to the human impact, but as well to some other inferences, like climate changes.

In this paper we would like to focus on adaptation of a management strategy of single reservoir system to include requirements of protected areas as one of the reservoir's main goal. This leads to the problem of a control of a multi-purpose reservoir which always takes a form of a supply/demand problem for a set of different, usually colliding users. The main difficulty comes when costs or benefits of certain users cannot be easily compared especially when economic terms are considered. For example, the benefits from energy production can be directly assessed, while introducing any economical measures for ecological or social requirements is rather a problematic issue.

There are many different approaches to this problem in water management. The most successful ones were based on such techniques like *Goal Programming* I.a. (Can and Houck 1984; Gandolfi and Salewicz 1991; Goulter and Castensson 1988; Yang et al. 1992, 1993), where optimization was constrained to a desirable value for each criterion. This allowed to obtain required trade-off between different criteria. An extension of this concept was proposed in form of interactive decision support systems, allowing the user to choose any appropriative solution form *pareto-*optimal surface. Such applications were presented by (Agrell et al. 1995; Berkemer et al. 1993; Makowski et al. 1995).

However, taking into account the stochastic character of reservoir management problem within multi-criteria analysis it is still a difficult task (Labadie 2004). In this paper we consider demands in stochastic way, in form of required safety levels of supply for each user. It was achieved by introducing special criteria functions, being similar to well known Value-at-Risk measures.

We consider here an application of Decision Support System for the multi-purpose Siemianówka reservoir. The reservoir is localized on the Narew River in North–East Poland. Downstream to it rich wetland ecosystem, enclosed within Narew National Park (NNP), is situated. The goal of the proposed system is to provide decision makers with a tool that would allow to control safety of the NNP environmental requirements within the reservoir management policy to be included. Important issues concern competition among many water-dependent systems and agents, e.g., agriculture, energy, wetlands, for limited water resources. Accounting for inherent uncertainties is a challenging key task. The control was performed in accordance with the Receding Horizon Optimal Control (RHOC) concept, where release amount is computed each time for a present inflow forecast.

The problem of reservoir management and water supply-demand under uncertainties and risks is formulated as a stochastic multi-criteria problem for preserving water mass balances.

2 The Upper Narew Mathematical Model

The Narew National Park is situated in north-east Poland and encloses valuable water-peat ecosystems of the anastomosing Upper Narew River, making this region unique in Europe. The NNP's flora consists of more than 600 species of vascular plants, including many protected varieties. Park wetland areas provide habitats for about 200 bird species, being one of the most important stop-over points for migrating birds. Due to its unique features, the NNP is an important site in the European Network of Natura 2000 (Dembek and Danielewska 1996).

The river reach under consideration (Fig. 1) is a primary, semi-natural form of a lowland river system, with relatively small water slope values equal to 0.02%. The annual river discharge at Suraż is 15.50 m^3/s. At the beginning of this reach a relatively big lowland storage reservoir Siemianówka is situated, with total capacity of about 80 mln m^3.

Freshets which in most of the other regions might cause significant threat, in NNP are part of a natural hydrological cycle. It can be seen that the localization of towns and villages follows the inundation zone border. In recent years, alarming changes have been observed in the Upper Narew River hydrological regime, manifested in a reduction of mean flows and shorter flooding periods. This results in a serious threat to rich wetland ecosystems. Local climate changes are one of causes of those changes. Mild winters combined with a reduction in annual rain levels have resulted in a reduction of the valley's ground-water resources. However, recent human activities also have had a significant influence on the deterioration of wetlands water conditions. River regulation work performed in the lower river reach has lowered water

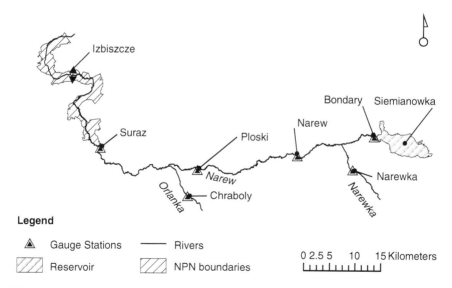

Fig. 1 Schematic map of the study area

levels. Additionally, a water storage reservoir constructed upstream of the NNP has had an important impact on water conditions, causing a reduction in flood peaks.

The idea for adaptation of reservoir management to improve water conditions at protect area comes from mid 1990's, when first symptoms of wetlands degradation become noticeable. For example (Okruszko et al. 1996) have even suggested that NNP should become the main reservoir recipient. Nevertheless requirements of other reservoir users cannot be neglected and in addition to fulfilling wetlands demands the following management goals have to be also included:

- Irrigation of grasslands for agriculture,
- Flood protection of adjacent to the reservoir areas, up to Narewka tributary,
- Fishery at reservoir,
- Energy production from water turbines.

To analyze the influence of reservoir on the river system and dependent subsystems, like wetlands, it was necessary to develop integrated model for Narew River. Because of complex structure of this multi-channel river system it was a challenging problem, broadly investigated by Kiczko et al. (2008). The Upper Narew river system can be represented by a diagram showed in the Fig. 2. It consist of two main subsystems: reservoir and river valley. Dynamics of Siemianówka reservoir was described with a simple balance equation:

$$\frac{dS(t)}{dt} = S(t) - U(t) + R(t) - P(t), \qquad (1)$$

where: $S(t)$ – reservoir storage $[m^3/s]$ at the time t, $U(t)$ – release amount $[m^3/s]$ to the river (control variable), $R(t)$ – inflow to the reservoir $[m^3/s]$, $P(t)$ – evaporation from the reservoir surface $[m^3/s]$. In addition reservoir storage $S(t)$ should not exceed minimal storage value S_{min} and maximal S_{max} for all t periods. Similar constraints are imposed on $U(t)$, which has to be higher than U_{min} and smaller than U_{max}:

$$S_{min} \leq S(t) \leq S_{max}$$
$$U_{min} \leq U(t) \leq U_{max} \qquad (2)$$

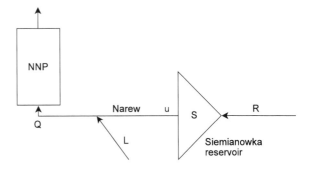

Fig. 2 The model of Upper Narew river system

Such simplification was possible because of its relatively small size, compare to used time scale, within the reservoir dynamics could be assumed to be linear.

River discharge through the NNP, Q is an effect of a flow transformation of reservoir's releases and lateral inflow to the river system L. Unfortunately this river reach is not well controlled and between the reservoir and NNP flow records only for one tributary were available. As a result of it more than the 30% of total inflow to the NNP comes from uncontrolled sub-basins. This increases a complexity of the reservoir control problem, as modeling of such external sources involves high uncertainty.

To identify flow conditions, a one dimensional flow routing model UNET (One-Dimensional Unsteady Flow Through a Full Network of Open Channels model, Barkau et al. (1989)) was applied. Despite this one-dimensional formulation, the UNET allows to include different conveyance conditions of river channel and flood-plains, which was necessary because of complex structure of the Narew river. The river reach was represented by 53 cross-sections at 2 km interval, obtained from the terrain survey. The model was calibrated by adjusting the Manning coefficients separately for the main channel and left, and right flood-plains. The water surface slope was used as a downstream boundary condition. As it was mentioned before, the model mass balance was a problematic issue because of the lack of data on lateral inflows. Therefore to take it into the account, it was assumed that unmeasured lateral inflows are linearly correlated with two known tributaries and can be described with the use of a linear regression model. Validation of the model was presented in the Fig. 3.

Distributed flow routing model was used to determine required flow conditions at protected areas. According to (Junk et al. 1989; Tockner et al. 2000) wetland ecosystems depend largely on river flow conditions, and particularly, on flooding. Because of complexity of ecological systems it is almost impossible to identify such demands directly. Nevertheless, (Banaszuk et al. 2002; Okruszko and Kiczko 2008) proposed

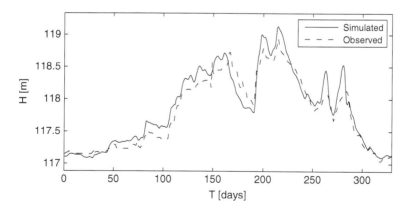

Fig. 3 Verification of UNET model for he Upper Narew river; water levels at Suraż river gauge during a spring freshet in 1983

a methodology for this region, based on analysis of past river flow patterns. In this case study UNET model was used to reconstruct flow conditions during a favorable from ecological point of view period. Such hydrological information was merged with data concerning localization of main plant communities. This allowed to estimate required magnitude of spring flooding and minimum admissible water flows during vegetation season. In a control problem this assumptions were introduced in a form of trajectory describing required inundation extent at the NNP $- \overline{A}_w(t)$.

In similar way flood risk for areas localized in reservoir's closest neighborhood, between Bondary river gauge and Narewka tributary (see Fig. 1), was estimated. Results obtained from flow routing methods were merged with Digital Terrain Model, giving flood extend for a certain discharges. As for this region only arable land could be affected, it was assumed that all flood loses are proportional to an inundation area. In addition to this, flow at the protected reach is directly affected by the reservoir and it was possible to describe attached costs with a simple relation to the release U. The criterion concerning flood loses was conditioned on admissible flooding area $- \overline{A}_f$.

Irrigation, as flood protection, applies to areas localized upstream to the Narewka tributary. Water mass is transferred through the river channel, so it is directly depended on the reservoir release U. Agriculture requires certain irrigation patterns of watering and draining during vegetation season. Fulfilling of these demands was considered previously as the main reservoir purpose and required values of release $- \overline{U}_a(t)$ were determined by the current reservoir's management instruction (BIPROMEL 1999).

Demands for fisheries and energy production could be easily described on the basis of reservoir's balance equation (1). Fisheries requires that certain reservoir storage $\overline{S}(t)$ is maintained and release $\overline{U}_s(t)$ does not exceed certain amounts during fishing times. While performance of hydro-power plant is in this case directly depended on release U and in the control problem it was assumed that losses occur when U is bellow the capacity of water turbines $- \overline{U}_e$.

The control system requires multiple realizations of the flow routing model. An implementation of the relatively complex UNET model would significantly reduce computational effectiveness. Moreover, so detailed representation of river flow behavior was indeed unnecessary. Because demands of NNP were conditioned by the inflow to NNP, only function linking the reservoir releases U and lateral inflows with discharge through NNP Q was needed. Such relation was evaluated in form of a Multiple Input Single Output (MISO) Transfer Function (TF) (Romanowicz et al. 2007):

$$Q_k = \frac{B\left(z^{-1}\right)}{A\left(z^{-1}\right)} U_{k-\delta_U} + \frac{C\left(z^{-1}\right)}{A\left(z^{-1}\right)} L^n_{k-\delta_L} \tag{3}$$

and an observation equation:

$$y_k = [S_k, Q_k]^T + \underline{e}_k \tag{4}$$

Q_k – water flow $[m^3/s]$ at protected and/or agricultural sites, L_k^n – lateral inflow to the river system from Narewka tributary $[m^3]$, y_k – observation vector, $\underline{e}_k$ – observation noise vector with normal distribution: $\sim N\left(0, \sigma^2\right)$, z^{-1} stands for the back-shift operator, $A\left(z^{-1}\right)$, $B_I\left(z^{-1}\right)$, $C\left(z^{-1}\right)$ are polynomials of n_a, n_b and n_c order, respectively:

$$A\left(z^{-1}\right) = 1 + a_1 z^{-1} + \ldots + a_{n_a} z^{-n_a}$$
$$B\left(z^{-1}\right) = b_0 + b_1 z^{-1} + \ldots + b_{n_b} z^{-n_b}$$
$$C\left(z^{-1}\right) = c_0 + c_1 z^{-1} + \ldots + c_{n_c} z^{-n_c} \tag{5}$$

This is a lumped, black-box model. It is important to note that in the presented formulation $\frac{C\left(z^{-1}\right)}{A\left(z^{-1}\right)}$ explains not only flow transformation of L^n but also unmeasured components of L, linearly correlated with L^n. Uncertainty of the model was described with the noise $\underline{e}_k$. River flow routing model, described by (3) was identified using *the Captain Toolbox* (Young et al. 2004).

3 Inflow Forecasts

In application of control system for receding optimization horizon all reservoir management decisions have to be based on the assumption concerning future values of inflows R_k and L_k^n. Therefore the estimation of the predictions uncertainties was essential in the formulation of a management problem. In typical applications, forecasts are based on Global Circulation Model realizations and flow predictions are evaluated with the use of downscaling methods combined with run-off modeling (Bates et al. 1998; Jones et al. 1995; Murphy 1999). For the Upper Narew basin this approach would provide reasonable predictions for approximately 10 days ahead. In the case of the Narew lowland river system, where freshet peak duration usually exceeds 1 month, a 10 day time horizon is insufficient. Therefore, for the purpose of control, inflow predictions were calculated from historical records, namely by means of the so-called nearest neighbor technique (k-NN). This is a non-parametric method, introduced to the hydrology by Karlsson and Yakowitz (1987). It has been widely used to forecast inflows from uncontrolled basins (Napiórkowski et al. 1999; Piotrowski et al. 2004).

In the proposed approach, one searches for the K points in Euclidean phase space, representing K trajectories of historical sequences of flows with an "embedding dimension" that equals p, that are the most similar (in the sense of the smallest Euclidean distance) to the point representing the current situation. Then the selected trajectories are applied to flow forecasting for the assumed time horizon.

Usually the future run-off is calculated as a mean of the trajectories K. However in this application a modification to this method was introduced. It was assumed that the selected trajectories were random sample from a distribution of possible future

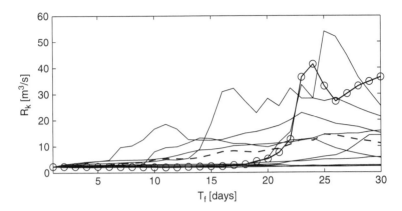

Fig. 4 An example forecast obtained with k-NN method for Bondary; solid lines stands for forecasted trajectories, dashed lines for an expected ones and lines with circles for observations

inflows, bringing this approach closer to the concept of GCM Ensemble Forecasts (Buizza et al. 1999). The number of points (trajectories) K and the embedding dimension were evaluated by trial and error.

Numerical experiments of the control model showed that this methodology improves the overall objective function value by 30% approximately, compared to the use of daily mean values of discharge as inflow predictions. In Fig. 4 an example forecast for flows at the Bondary river gauge is presented.

4 Optimization Problem

The reservoir control problem was formulated in accordance with the Receding Horizon Optimal Control, broadly described by Castelletti et al. (2008). The amount of release (in this case U) is computed at each decision step, taking into account present system state and inflow forecasts. The decision step length is determined by technical characteristics of the reservoir and availability of forecasts. For Siemianówka reservoir 1 day decision step, recommended in the reservoir management instruction (BIPROMEL 1999), was applied. Within the RHOC the optimization problem is solved for a chosen, finite time horizon, reflecting the information concerning future, possible inflows to the system (R and L). Usually it exceeds the duration of a single release disposal, which means that a solution of optimization problem provides also values of possible future releases. However, according to this concept only the first one is applied as the whole procedure, is repeated at next time step.

Application of a RHOC requires that an additional criterion for a system end-state is introduced. In this case it took a form of a penalty function for a deviation from desirable storage trajectory: $\overline{S}_{end}(t)$.

In general case, the reservoir control problem always takes a form of a stochastic control. It is because releases disposals are performed against highly uncertain system models and highly inaccurate inflow forecasts. For the Upper Narew Basin especially significant is the last element, which overwhelms uncertainty of flow routing. Therefore in an control system only uncertainty of the inflow forecast model was included. The result of it is that the control at each computation step was performed against stochastic realization of inflow R and L.

However, before the control problem formulation it was necessary to adopt particular form of management criteria. The main difficulty was in matching different factors, especially economical ones with demands of natural ecosystems. In the presented solution, certain demands were considered as required safety levels of supply. In other words, the control goal was to ensure the needs of a user within assumed probability. Such probabilistic constraints, for a single criterion, might be introduced through an optimization of specific piece-wise objective function. In this paper control criteria were formulated in manner, which allowed for such probabilistic interpretation of introduced measures. They took a form of cost functions, being minimized with respect to the control variable U and the given finite time horizon:

- Wetland demands

$$
y_1 = \max \begin{cases} \alpha_1 \left(\overline{A}_{W,k} - A_{W,k} - v_1 \right) & \text{if } \overline{A}_{W,k} \geq A_{W,k} + v_1 \\ (1 - \alpha_1) \left(A_{W,k} - \overline{A}_{W,k} - v_1 \right) & \text{else} \end{cases} \tag{6}
$$

- Irrigation:

$$
y_2 = \max \begin{cases} \alpha_2 \left(\overline{U}_{A,k} - U_{A,k} - v_2 \right) & \text{if } \overline{U}_{A,k} \geq U_{A,k} + v_2 \\ (1 - \alpha_2) \left(U_{A,k} - \overline{U}_{A,k} - v_2 \right) & \text{else} \end{cases} \tag{7}
$$

- Flood protection:

$$
y_3 = \max \begin{cases} 0 & \text{if } \overline{A}_{F,k} \geq A_{F,k} + v_3 \\ \left(A_{F,k} - \overline{A}_{F,k} - v_3 \right) & \text{else} \end{cases} \tag{8}
$$

- Reservoir storage:

$$
y_4 = \max \begin{cases} \alpha_4 \left(\overline{S}_k - S_k - v_4 \right) & \text{if } \overline{S}_k \geq S_k + v_4 \\ (1 - \alpha_4) \left(S_k - \overline{S}_k - v_4 \right) & \text{else} \end{cases} \tag{9}
$$

- Energy production:

$$
y_5 = \max \begin{cases} \left(\overline{U}_{E,k} - U_{E,k} - v_5 \right) & \text{if } \overline{U}_{E,k} \geq U_{E,k} + v_5 \\ 0 & \text{else} \end{cases} \tag{10}
$$

- Reservoir storage at the end of time horizon:

$$y_6 = \max \begin{cases} \alpha_6 \left(\overline{S}_{end} - S_{Th} - v_6\right) & \text{if } \overline{S}_{end} \geq S_{Th} + v_6 \\ (1 - \alpha_6) \left(S_{Th} - \overline{S}_{end} - v_6\right) \text{ else} \end{cases} \tag{11}$$

where y_1, y_2, y_3, y_4, y_5, y_6 stands for criteria values, v_1, v_2, v_3, v_4, v_5, v_6 – allowed deviations from goal trajectory (within which criterion value is equal to 0), α_1, α_2, α_4, α_6 – coefficients describing costs of not reaching goal value and respectively $1-\alpha_1$, $1-\alpha_2$, $1-\alpha_4$, $1-\alpha_6$ costs of exceedance. It is important to note that 8 and 10 are specific form of criterion function, where α coefficients are equal, respectively, to 0 and 1.

Criteria were aggregated with regard to a cost effectiveness concept i.e. with the weighted sum method. The crucial issue at this stage was: how to include stochastic (uncertain) character of the forecast to the optimization problem. There are two general solutions to this problem. First one, the most conservative, leads to the formulation of a stochastic optimization problem, within the control is aimed to find such U that ensures a maximum safety level against all possible inflow scenarios. In this case aggregation function might take the following form:

$$J = \frac{1}{N} \sum_{j=1}^{N} \sum_{I=1}^{6} \left(n_I y_I \left(U, R^j, L^j\right)\right) \tag{12}$$

where R^j and L^j stands for the jth inflow scenario (being a realization of the stochastic process) obtained from the nearest neighbor method (k-NN), N – number of such scenarios and n_I – Ith weighting coefficient for criteria scaling. Such approach allows to take into account of a transformation of an uncertainty through the union function, as the control was performed in the respect to the total expected cost of all subsystems.

The second way is to neglect the effect of the uncertainty transformation on the control process, then the optimization problem takes a deterministic form. Such solution is computationally cheaper, however it is expected that it would produce satisfactory results only for linear systems and in this paper it was considered only for comparison. The union function differs form the (12) in that the averaging was directly applied to the forecast ensemble:

$$J_D = \sum_{I}^{6} n_I y_I \left(U, \hat{R}, \hat{L}\right)$$

$$\hat{R}_k = \frac{1}{N} \sum_{j=1}^{N} R_k^j$$

$$\hat{L}_k = \frac{1}{N} \sum_{j=1}^{N} L_k^j \tag{13}$$

In the result the optimization was aimed to find optimal solution to the expected inflow forecast.

In case of the stochastic formulation, there is a strong connection with minimization of such convex functions and safety constraints, and so-called CVaR risk measures. As mentioned before, these fundamentally important connections allow to regulate safety constraints by adjusting the shape of presented piece-wise functions. In the following, this specific model was formulated without proofs, which can be derived in a similar manner from general results in (Ermoliev and Wets 1988; Ermoliev et al. 2000; Rockafellar and Uryasev 2000).

This concept was presented for wetlands demands, however it can be easily applied to the requirements of other subsystems. In case of single criterion problem the stochastic optimization problem takes the following form:

$$E\left\{\max\left[\alpha_W\left(\overline{A}_{W,k} - A_{W,k} - v_W\right), (1 - \alpha_W)\left(A_{W,k} - \overline{A}_{W,k} - v_W\right)\right]\right\} \qquad (14)$$

where $v_W \geq 0$ and $0 < \alpha \leq 1$. Such formulation (under rather general assumptions) allows to obtain the solution which is aimed to satisfy the demand for the inflow scenario of a given probability level. Particularly, in case when $v_W = 0$, this probability is equal to $1 - \alpha_W$. Thus, the minimization of aggregated criteria, for the stochastic case, composed of y_1 and other similar functions y_2, y_3, y_4, y_5 and y_6 with respect to decision variables S_K, U_K would yield a solution specifying the control variable U_K that satisfying required safety levels. If such a solution does not exist, a multi-criteria analyzing model would allow to find a compromise solution within desirable aspiration and reservation levels (Kiczko 2008). Of course such interpretation does not apply to the deterministic formulation, where α_l coefficients combined with n_l can be only considered in economical sense.

5 Results

Measuring performance of two different control formulations ((12) and (13)) was a problematic issue. Because of the multi-criteria character of the optimization it could not be done directly, on the basis of obtained criteria values. Therefore optimized trajectories of the model variables (U, S and A) were tested against the trajectories computed for the "perfect" forecast. Because such forecast was composed of real discharge data, the resulting trajectories were not affected by the uncertainty. Thus, the performance measures for two different control formulation was conditioned on similarity to this "perfect" trajectory.

Numerical experiments were performed for the spring and summer period of 1987. Values of α_l and n_l coefficients were presented in Table 1.

Computed trajectories were presented in the Fig. 5. It could be seen that the stochastic solution seems to be more conservative than deterministic one. It is especially obvious in case of the reservoir's storage. The stochastic model allowed

Table 1 Values of α_I and n_I coefficients used in computations

Criterion	α_I	n_I
Wetland demands (1)	0.9	4.5
Irrigation (2)	0.6	1
Flood Protection (3)	0	1
Fisheries (4)	0.5	$1 \cdot 10^{-6}$
Energy Production (5)	1	1
Final Storage (6)	0.5	$1 \cdot 10^{-6}$

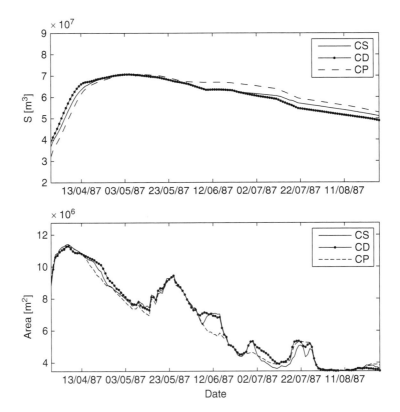

Fig. 5 Computed trajectories of reservoir storage S (upper plot) and inundation extend at wetland areas (lower plot); CS – trajectory for stochastic formulation, CD – trajectory for deterministic formulation; CP – trajectory for the "perfect" forecast

to maintain significantly higher water reserve. Of course, in the result better values for wetland criterion were achieved within the deterministic approach.

However the overall performance might be measured only in respect to the solution not affected by the uncertainty. Appropriative fit measures were showed in a Table 2. It can be seen that stochastic solution is much closer to the "perfect" one.

Table 2 Fit measures for the U, S and A trajectories obtained for the deterministic and stochastic control in respect to the "perfect" control; dU_{mean}, dS_{mean}, dA_{mean} – respectively mean deviation and dU_{max}, dS_{max}, dA_{max} – maximal deviation

Control type	dU_{mean}	dU_{max}	dS_{mean}	dS_{max}	dA_{mean}	dA_{max}
	$[m^3/s]$		$[10^6 m^3]$		$[10^5\ m^2]$	
Deterministic	1.12	8.49	3.36	7.84	3.0	16.7
Stochastic	0.89	6.64	2.04	5.39	2.6	16.7

6 Conclusions

In this paper we address a general task of regional reservoir control and water management. We formulate it as a stochastic and dynamic multi-criteria problem for preserving water mass balances under inherent risks and uncertainties in a region influenced by the reservoir policies. The management problem consists of optimizing several criteria: wetland water requirements, agricultural, energy production, flood protection, fishery and reservoir storage.

The proposed Decision Support System is a tool for decision makers to plan safe control of the environmental in the Narew National Park. The problem of water management was addressed in an integrated way. Proper expert knowledge, mathematical and modeling techniques were combined to provide relevant policy recommendations as to the best regulation of water supply to diverse systems characterized by different demand-supply priorities and costs associated with water shortages. The following issues were pointed out:

1. Formulation of the control problem according to the receding horizon optimal control,
2. The evaluation of the numerically efficient flow routing model was an essential point in this task. This was achieved with the use of the transfer function concept,
3. Representation of requirements of the reservoir users in a form of convex piece-wise cost functions. Such form allows to draw a link between stochastic optimization and probabilistic properties of the solution,
4. The inflow predictions were performed in accordance with the k-NN method. This approach is much less sophisticated than predictions based on GCM realizations, however, under significantly lower costs, it allows to obtain satisfactory results,
5. Comparison between stochastic and deterministic approaches. It has been shown that stochastic approach differs form the deterministic one. Stochastic formulation seems to be advantageous, as leads to the solution, which is significantly closer to the solution obtained for the "perfect" forecast.

The presented study requires future investigation. Efficiency of the control system could be improved with the use of the ensemble forecasts obtained form the GCM. In addition, the developed model does not take into account variability of natural system and therefore introduction of real time updating system could be considered.

References

Agrell, P., Lence, B., & Stam, A. (1995). An interactive multi-criteria decision model for reservoir management the shellmouth reservoir case. Tech. Rep. WP-95-090, International Institute for Applied Systems Analysis, Laxenburg, Austria.

Banaszuk, P., Banaszuk, H., Czubaszek, R., Jaros, H., Jekatierynczuk-Rudczyk, E., Kondratiuk, P., Micun, K., Roj-Rojewski, S., & Wysoka-Czubaszek, A. (2002). Narew National Park protection plan. Tech. rep., Narew National Park.

Barkau, R., Johnson, M., & Jackson, M. (1989) UNET: A Model of Unsteady Flow Through a Full Network of Open Channels. In *The proceedings of 1989 Hydraulics Conference, ASCE*, New Orleans, Louisiana, pp 1041–104.

Bates, B., Charles, S., & Hughes, J. (1998). Stochastic downscaling of numerical climate model simulations. *Environmental Modelling & Software 13*(3-4), 325–331.

Berkemer, R., Makowski, M., & Watkins, D. (1993). A prototype of a decision support system for river basin water quality management in Central and Eastern Europe. Tech. Rep. WP-93-049, International Institute for Applied Systems Analysis, Laxenburg, Austria.

BIPROMEL (1999). Siemianowka reservoir – water management rules (pol). Tech. rep., Bipromel, Warszawa.

Buizza, R., Hollingsworth, A., Lalaurette, F., & Ghelli, A. (1999). Probabilistic predictions of precipitation using the ECMWF ensemble prediction system. *Weather and Forecasting 14*(2), 168–189.

Can, E., & Houck, M. (1984). Real time reservoir operations by goal programming. *Journal of Water Resources Planning and Management 110*(3), 297–309.

Castelletti, A., Pianosi, F., & Soncini-Sessa, R. (2008). Water reservoir control under economic, social and environmental constraints. *Automatica 44*(6), 1595–1607.

Dembek, W., & Danielewska, A. (1996). Habitat diversity in the Upper Narew valley from the Siemianówka reservoir to Suraz. *Zeszyty Problemowe Postępów Nauk Rolniczych (Pol) (428)*, 25–38.

Ermoliev, Y., & Wets, R. (1988). Stochastic programming, an introduction. In *Numerical Techniques for Stochastic Optimization, Series in Computational Mathematics*, pp 1–32. New York: Springer.

Ermoliev, Y., Ermolieva, T., MacDonald, G., & Norkin, V. (2000). Stochastic optimization of insurance portfolios for managing exposure to catastrophic risks. *Annals of Operations Research, 99*, 207–225.

Gandolfi, C., & Salewicz, K. (1991). Water resources managment in the zambesi valley: Analysis of the lake kariba operation. In F. Van de Ven, D. Gutknecht, & K. Salewicz (Eds.), *Hydrology for the Management of Large Rivers*, vol. 201, pp 13–25. IAHS.

Goulter, I., & Castensson, R. (1988). Multi-objective analysis of boating and fishlife in lake sommen. *Water Resources Development 4*(3), 191–198.

Jones, R., Murphy, J., & Noguer, M. (1995). Simulation of climate change over Europe using a nested regional-climate model. I: Assessment of control climate, including sensitivity to location of lateral boundaries. *Quarterly Journal of the Royal Meteorological Society, 121*(526), 1413–1449.

Junk, W., Bayley, P., & Sparks, R. (1989). The flood pulse concept in river-floodplain system. *Canadian Special Publication of Fisheries and Aquatic Sciences*, pp. 110–127.

Karlsson, M., & Yakowitz, S. (1987). Nearest-neighbor methods for nonparametric rainfall-runoff forecasting. *Water Resources Research, 23*(7), 1300–1308.

Kiczko, A. (2008). Multi-criteria decision support system for Siemianówka reservoir under uncertainties. Tech. Rep. IR-08-026, International Institute for Applied Systems Analysis, Laxenburg, Austria.

Kiczko, A., Romanowicz, R., Napiórkowski, J., & Piotrowski, A. (2008). Integration of reservoir management and flow routing model – Upper Narew case study. *Publications of the Institute of Geophysics, Polish Academy of Sciences E-9*(405), 41–55.

Labadie, J. (2004). Optimal operation of multireservoir systems: State-of-the-Art Review. *Journal of Water Resources Planning and Management, 130*(2), 93–111.

Makowski, M., Somlyody, L., & Watkins, D. (1995). Multiple criteria analysis for regional water quality management: the Nitra River case. Tech. Rep. WP-95-022, International Institute for Applied Systems Analysis, Laxenburg Austria.

Murphy, J. (1999). An evaluation of statistical and dynamical techniques for downscaling local climate. *Journal of Climate, 12*(8), 2256–2284.

Napiórkowski, J., Kozlowski, A., & Terlikowski, T. (1999). Influence of inflow prediction on performance of water reservoir system. *Water Industry Systems: Modelling and Optimization Applications 2*, 39.

Okruszko, T., & Kiczko, A. (2008). Assessment of water requirements of swamp communities: the river Narew case study. *Publications of the Institute of Geophysics, Polish Academy of Sciences E-9*(405), 27–40.

Okruszko, T., Tyszewski, S., & Pusłowska, D. (1996). Water management in the Upper Narew valley. *Zeszyty Problemów Podstawowych Nauk Rolniczych (Pol), (428)*.

Piotrowski, A., Rowiński, P., & Napiórkowski, J. (2004). River flow forecast by means of selected black box models. In River Flow 2004, pp. 1375–1382.

Rockafellar, T., & Uryasev, S. (2000). Optimization of conditional-value-at-risk. *The Journal of Risk, 2*, 21–41.

Romanowicz, R., Kiczko, A., & Pappenberger, F. (2007). A state dependent nonlinear approach to flood forecasting. *Publications of the Institute of Geophysics, Polish Academy of Sciences E–7*(401), 223–230.

Tockner, K., Malard, F., & Ward, J. (2000). An extension of the flood pulse concept. *Hydrological Processes, 14*, 2861–2883.

Yang, Y., Burn, D., & Lence, B. (1992). Development of a framework for the selection of a reservoir operating policy. *Canadian journal of Civil Engineering 19*, 865–874.

Yang, Y., Bhatt, S., & Burn, D. (1993). Reservoir operating policies considering release change. *Civil Engineering Systems, 10*, 77–86.

Young, P., Taylor, C., Tych, W., Pedregal, D., & McKenna, C. (2004). The Captain Toolbox. Centre for Research on Environmental Systems and Statistics, Lancaster University (www.es.lancs.ac.uk/cres/captain).

Part III
Uncertainty and Optimization

A Deterministic Algorithm for Global Optimization

Yury Evtushenko and Mikhail Posypkin

Abstract An algorithm for solving global optimization problems is developed. The objective and constraints are required to have gradients satisfying Lipschitz condition. The problem may contain both continuous and integer variables and the objective may be non-convex and multimodal. Improved lower bounds and new techniques to reduce the number of algorithm steps by employing the gradient information are proposed for unconstrained optimization. Computational testing on different test problems demonstrate the efficiency of the proposed method in comparison with the state of the art approaches.

1 Introduction

Today there is a great variety of methods for solving global optimization problems (Pardalos et al. 2000; Pardalos and Resende 2002). These methods can be roughly divided into two big groups: *deterministic* and *non-deterministic* methods. The deterministic methods reach an approximate global minimum to the given accuracy. Non-deterministic methods use local search techniques, heuristics or their combination to locate good approximations for the global minimum but have no means to estimate the accuracy of the obtained results.

The main disadvantage of non-deterministic methods is the lack of certainty in the optimality of obtained solutions. Though for many problems the solution found by heuristic algorithms is satisfactory there are plenty of areas where the knowledge of accuracy of the obtained minima is mandatory. In such fields heuristics can't replace deterministic methods.

Y. Evtushenko · M. Posypkin (✉)
Institution of Russian Academy of Sciences Dorodnicyn Computing Centre of RAS, Moscow Russia
e-mail: evt@ccas.ru; mposypkin@mail.ru

Y. Ermoliev et al. (eds.), *Managing Safety of Heterogeneous Systems*, Lecture Notes in Economics and Mathematical Systems 658, DOI 10.1007/978-3-642-22884-1_10,
© Springer-Verlag Berlin Heidelberg 2012

Presently there is a great variety of deterministic methods. Many efficient methods were developed for convex optimization (Nesterov 2003). Such problems are relatively simple because of the following property: if the objective and constraints are convex functions and a local minimum exists then it is a global minimum. However in practice objective and constraints are often non-convex. For instance explicit introduction of uncertainties in linear and nonlinear mathematical programming models often leads to nonconvex/concave optimization model (Ermoliev and Norkin 2004).

The most successful deterministic methods for global optimization are based on interval analysis (Hansen 1992; Kearfott 1996), convexification (Tawarmalani and Sahinidis 2002), and Lipschitzian approaches (Pinter 1996; Strongin and Sergeyev 2000).

In this paper we propose a deterministic method for solving optimization problems with guaranteed accuracy. The algorithm requires the objective and constraints to have gradients satisfying Lipschitz conditions. The paper discusses the techniques to reduce the number of steps by employing the gradient information and handling discrete parameters in mixed-integer problems. The efficiency of the proposed approach is demonstrated on various test problems.

In the sequel the following notations are used:

- $\mathbb{Z}$ — the set of all integers,
- $\overline{1,n} = [1,n] \cap \mathbb{Z}$ — a set of all integers from 1 to n,
- Let $a,b \in \mathbb{R}^n$. Then $a = (\le,\ge,<,>)b$ if $a_i = (\le,\ge,<,>)b_i$ for all $i \in \overline{1,n}$,
- $[a,b] = \{x \in \mathbb{R}^n | a \le x \le b\}$ — a box with bounds $a,b \in \mathbb{R}^n$,
- $\mathbb{R}^n_+ = \{x \in \mathbb{R}^n | x_i \ge 0, i = \overline{1,n}\}$,
- $\mathbb{Z}_+$ — a set of all non-negative integers.

The paper is organized as follows. Section 2 describes the theoretical background. Section 3 considers different underestimations for objective function and constraints. The basic algorithm scheme is outlined in Sect. 4. Implementation details and experimental results are considered in Sect. 5.

2 Preliminaries

A global optimization problem can be formally stated as follows:

$$\text{Find } f_* = \min_{x \in X} f(x). \tag{1}$$

Where X is a set of *feasible points* or simply a *feasible set*. Without loss of generality assume that there is a box $[a,b]$ sufficiently large to contain at least one minimizer point. For problems with functional constraints the feasible set X is defined as follows

$$X = \{x \in [a,b] : g(x) \le 0\}, \tag{2}$$

A Deterministic Algorithm for Global Optimization 207

where function $g(x) : \mathbb{R}^n \to \mathbb{R}^m$ defines inequality constraints. Equality constraint $h(x) = 0$ can be replaced by a pair of inequality constraints $h(x) \le 0, -h(x) \le 0$. Mixed-integer problems imply additional restrictions on some variables:

$$X = \left\{ x \in [a,b] : g(x) \le 0, x_j \in \mathbb{Z}, j \in J \subseteq \{1,\dots,n\} \right\}. \tag{3}$$

Variables $x_j, j \in J$ are discrete, other variables are continuous. Definition (2) is a particular case of definition (3) with empty set of discrete variables. The rest of the section is devoted to the most general case (3).

The set X_* of optimal solutions of the problem (1) is defined as follows:

$$X_* = \{ x \in X : f(x) = f_* \}. \tag{4}$$

Exact optimal solutions for continuous problems seldom can be found numerically. In practice algorithms usually search *approximate solutions*. For $\varepsilon \in \mathbb{R}_+$ and $\delta \in \mathbb{R}_+^n$ we introduce the set of approximate ε, δ-optimal solutions defined as follows:

$$X_*^{\varepsilon, \delta} = \left\{ x \in X^\delta : f(x) \le f_* + \varepsilon \right\}, \tag{5}$$

where

$$X^\delta = \left\{ x \in [a,b] : g(x) \le \delta, x_j \in \mathbb{Z}, j \in J \subseteq \{1,\dots,n\} \right\} \tag{6}$$

is a δ-*feasible set*.

The numerical method for solving problem (1) considered in this paper is based on the *non-uniform covering approach* proposed in (Evtushenko 1971). This approach assumes processing sets $X^{(i)}, \dots, X^{(k)}$, $X^{(i)} \subseteq \mathbb{R}^n$ and points $x^{(1)}, \dots, x^{(k)}$, $x^{(i)} \in X^{(i)}$. For simplicity assume that processing is done in a sequential order and at i-th step the set $X^{(i)}$ and point $x^{(i)}$ are considered. The *current record* $u^{(i)}$ and the *best current solution* $\tilde{x}^{(i)}$ are defined as follows

$$u^{(i)} = f(\tilde{x}^{(i)}) = \min_{x \in N_i} f(x), \quad \tilde{x}^{(i)} \in N_i, \tag{7}$$

where $N_i = \{x^{(1)}, \dots, x^{(i)}\} \cap X_\delta$ is a sequence of δ-feasible points considered during first i steps of the algorithm.

Let $m^{(i)}(x)$ be an *underestimation* for a function $f(x)$ over set $X^{(i)} \cap X_*$, i.e. $f(x) \ge m^{(i)}(x)$ for all $x \in X^{(i)} \cap X_*$. Let $\varepsilon > 0$ and let $S^{(i)}$ be any such set that

$$S^{(i)} \subseteq \left\{ x \in X^{(i)} : m^{(i)}(x) \ge u^{(i)} - \varepsilon \right\}. \tag{8}$$

Then the *covering condition* is defined as follows:

$$X_* \cap \cup_{i=1}^k S^{(i)} \ne \emptyset. \tag{9}$$

The following theorem provides sufficient conditions for global optimality of the best current solution.

Theorem 1. *Let $X^{(1)}, \ldots, X^{(k)}$ be a sequence of sets and N_k be a sequence of δ-feasible points satisfying condition (9). Then $\tilde{x}^{(k)}$ defined as in (7) is an ϵ, δ-optimal solution of problem (1) i.e.*

$$f_* \geq u^{(k)} - \epsilon. \tag{10}$$

Proof. Consider the global minimizer x_* of problem (1). From (9) it follows that there is $i \in \overline{1,k}$ such that $x_* \in S^{(i)}$. Then $f_* = f(x_*) \geq m^{(i)}(x_*)$. According to (8) $m^{(i)}(x_*) \geq u^{(i)} - \epsilon \geq u^{(k)} - \epsilon$ by definitions of $S^{(i)}$ and $u^{(k)}$. Therefore inequality (10) holds. Since $\tilde{x}^{(k)} \in X^\delta$ and $f_* \geq u^{(k)} - \epsilon$ we conclude (according to (6)) that $\tilde{x}^{(k)}$ is an ϵ, δ-optimal solution of problem (1). $\qquad\square$

Theorem 1 is valid for arbitrary sequences of sets $\{X^{(i)}\}$ and points N_i satisfying property (9). The way of constructing sets $\{X^{(i)}\}$ and points from N_i is defined by an algorithmic implementation. In Sect. 4 we demonstrate how these sequences are constructed by bisection procedure.

3 Lower Bounds

Underestimations and lower bounds are essential for the proposed algorithm. In the rest of the paper we restrict our discussion to the case where sets $X^{(i)} = [a^{(i)}, b^{(i)}]$ are n-dimensional boxes. Consider a function $f(x) : \mathbb{R}^n \to \mathbb{R}$ and its underestimation $m^{(i)}(x) : \mathbb{R}^n \to \mathbb{R}$ over $X^{(i)} \cap X_*$. The *lower bound* $v^{(i)}$ for a function $f(x)$ over $X^{(i)}$ is computed as a solution of the following *relaxed* optimization problem:

$$v^{(i)} = \min_{x \in X^{(i)}} m^{(i)}(x). \tag{11}$$

Underestimations are constructed in a way to simplify the resolution of the problem (11). Various underestimations for Lipschitzian functions are considered in (Evtushenko et al. 2009). In this paper we focus on functions with Lipschician gradients. Section 3.1 describes the standard way of constructing Lipschitzian lower bounds for a general case. In Sect. 3.2 we demonstrate that these standard bounds can be improved for unconstrained optimization.

3.1 Lipschitzian Lower Bounds in a General Case

Let $f(x) : \mathbb{R}^n \to \mathbb{R}$ be a differentiable scalar function with gradient satisfying Lipschitz condition over a box $X^{(i)} = [a^{(i)}, b^{(i)}]$:

$$\|\nabla f(x) - \nabla f(y)\| \leq L^{(i)} \|x - y\|,$$

A Deterministic Algorithm for Global Optimization 209

for all $x, y \in X^{(i)}$ where $L^{(i)} \in \mathbb{R}$ is Lipschitz constant. The following underestimation is proposed in (Nesterov 2003):

$$m^{(i)}(x) = f\left(c^{(i)}\right) + \langle \nabla f\left(c^{(i)}\right), x - c^{(i)}\rangle - \frac{1}{2}L^{(i)}\left\|x - c^{(i)}\right\|^2, \quad (12)$$

The minimum $v^{(i)}$ of $m^{(i)}(x)$ over box $X^{(i)}$ can be found analytically:

$$v^{(i)} = f\left(c^{(i)}\right) + \langle \left|\nabla f\left(c^{(i)}\right)\right|, a^{(i)} - c^{(i)}\rangle - \frac{1}{2}L^{(i)}\left\|c^{(i)} - a^{(i)}\right\|^2, \quad (13)$$

where $c^{(i)} = (a^{(i)} + b^{(i)})/2$ is the center of box $X^{(i)}$.

Notice that if the function $f(x)$ is convex the quadratic part of the introduced bounds (12), (13) can be omitted. Thus for convex $f(x)$ the underestimation and the lower bound look as follows:

$$\begin{aligned} m^{(i)}(x) &= f\left(c^{(i)}\right) + \langle \nabla f\left(c^{(i)}\right), x - c^{(i)}\rangle, \\ v^{(i)} &= f\left(c^{(i)}\right) + \langle \left|\nabla f\left(c^{(i)}\right)\right|, a^{(i)} - c^{(i)}\rangle. \end{aligned} \quad (14)$$

3.2 Lipschitzian Lower Bounds for Unconstrained Optimization

Inequality (12) holds for the whole $X^{(i)}$. However it is sufficient to underestimate the objective only over a set $X_* \cap X^{(i)}$. For unconstrained problems all elements of X_* are stationary points of the objective $f(x)$, i.e. $\nabla f(x) = 0$. Consider such a stationary point x_s, $x_s \in X^{(i)}$. First notice (Nesterov 2003) that function

$$M^{(i)}(x) = f(x_s) + \langle \nabla f(x_s), x - x_s\rangle + \frac{1}{2}L^{(i)}\|x - x_s\|^2,$$

overestimates $f(x)$ over $X^{(i)}$. Since x_s is a stationary point $\nabla f(x_s) = 0$ and $M^{(i)}(x) = f(x_s) + \frac{1}{2}L^{(i)}\|x - x_s\|^2$. Thus the following inequality holds for any stationary point $x_s \in X^{(i)}$:

$$f(x_s) \geq f(c^{(i)}) - \frac{1}{2}L^{(i)}\|c^{(i)} - x_s\|^2.$$

The latter means that function

$$\hat{m}^{(i)}(x_s) = f(c^{(i)}) - \frac{1}{2}L^{(i)}\|c^{(i)} - x_s\|^2 \quad (15)$$

is an underestimation for objective $f(x)$ over a set of all stationary points in $X^{(i)}$ and hence over set $X^{(i)} \cap X_*$. The respective lower bound is given by the formula:

$$\hat{v}^{(i)} = f(c^{(i)}) - \frac{1}{2}L^{(i)}\|c^{(i)} - a^{(i)}\|^2. \quad (16)$$

Clearly this bound is more accurate than (13).

3.3 Computing Lipschitz Constants

There are different ways to calculate the value of Lipschitz constant. Various approximations are considered in (Strongin and Sergeyev 2000). The minimal value of valid Lipschitz constant $L_*^{(i)}$ can be obtained by solving the following optimization problem:

$$L_*^{(i)} = \max_{x \in X^{(i)}} \|H(x)\|, \tag{17}$$

where $H(x)$ is the Hessian of $f(x)$. Obtaining Lipschitz constants by directly solving problem (17) is senseless because this problem is at least as complex as the initial one (1). Fortunately inequalities (12), (15) remains valid for constants greater than the exact value $L_*^{(i)}$. Therefore it is sufficient to find any $L^{(i)} \geq \max_{x \in X^{(i)}} \|H(x)\|$. This overestimation can be found from the well-known inequality

$$\|H(x)\| \leq \max_{i \in \overline{1,n}} \sum_{j=1}^{n} |h_{ij}(x)| \tag{18}$$

by applying interval arithmetics (Hansen 1992).

4 Algorithm Details

4.1 Algorithm Overview

The algorithm for ensuring the covering condition (9) follows the general branch and bound scheme. The initial box $[a, b]$ is iteratively divided into smaller ones until all resulting boxes are discarded by *feasibility* or *optimality* tests. The feasibility test eliminates boxes not intersecting with X_*. The optimality test checks that searching inside the box can't improve the record for more than ϵ. Such boxes are added to the covering sequence $\{X_i\}$ and excluded from the further search. During the search the objective function is evaluated in a number of δ-feasible points. At i-th step the best solution $\tilde{x}^{(i)}$ with the lowest objective function value $u^{(i)} = f(\tilde{x}^{(i)})$ are saved.

The algorithm performs the following steps:

1. Setup a list of boxes $\mathbb{X} = \{[a, b]\}$ and the record $u_0 = \infty$.
2. If $\mathbb{X} = \emptyset$ then output the best current solution and exit otherwise take a box $X^{(i)}$ from the list $\mathbb{X}$.
3. Perform the feasibility test for $X^{(i)}$. If $X^{(i)}$ is eliminated then go to the step 2.
4. Update the record.
5. Perform the optimality test for $X^{(i)}$. If $X^{(i)}$ is eliminated then go to the step 2, otherwise divide $X^{(i)}$ along the longest edge, obtain two equal boxes and add them to the list $\mathbb{X}$;
6. go to the step 2.

A Deterministic Algorithm for Global Optimization

211

The outlined algorithm partitions the initial box into a number of smaller boxes. The produced boxes are either eliminated by the feasibility test or added to the covering sequence by the optimality test. Since boxes eliminated according to feasibility tests don't intersect with X_* the condition (9) holds for the resulting covering sequence $\{S^{(i)}\}$. Thus according to the Theorem 1 the best current solution $\tilde{x}^{(k)}$ is ϵ, δ-optimal. In practice the covering sequence is usually not constructed explicitly: the boxes failed feasibility or optimality tests are simply excluded from the further search.

4.2 Optimality Testing

We use the standard optimality test adopted in the majority of branch-and-bound algorithms:

1. evaluate the lower bound $v^{(i)} = \min_{x \in X^{(i)}} m^{(i)}(x)$ for the objective $f(x)$;
2. compare $v^{(i)}$ and $u^{(i)}$, if $v^{(i)} \geq u^{(i)} - \epsilon$ then add $X^{(i)}$ to the covering sequence and exclude it from the further search.

The lower bound is computed by formula (13) in a general mixed-integer non-linear programming case and by (16) for unconstrained optimization.

4.3 Feasibility Testing

Let $g(x) = (g_1(x), \ldots, g_m(x))$ be inequality constraint functions as defined in (2). Let $m_j^{(i)}(x)$ be an underestimation and $v_j^{(i)} = \min_{x \in X^{(i)}} m_j^{(i)}(x)$ be a lower bound for function $g_j(x)$, $j = \overline{1,m}$ calculated according to (13). If for at least one j from $\overline{1,m}$ holds $v_j^{(i)} > 0$ then the $X^{(i)}$ is infeasible and thus can be excluded from the further processing.

For problem (3) with integer variables the selected box $X^{(i)} = [a^{(i)}, b^{(i)}]$ can be reduced to the box $\hat{X}^{(i)} = [\hat{a}^{(i)}, \hat{b}^{(i)}]$ where

$$\hat{a}_j^{(i)} = \begin{cases} a_j^{(i)}, j \in N \setminus J, \\ \lceil a_j^{(i)} \rceil, j \in J, \end{cases} \qquad \hat{b}_j^{(i)} = \begin{cases} b_j^{(i)}, j \in N \setminus J, \\ \lfloor b_j^{(i)} \rfloor, j \in J \end{cases} \qquad \text{for all } j \in \overline{1,n},$$

If $\hat{b}_j^{(i)} < \hat{a}_j^{(i)}$ for at least one j, $1 \leq j \leq n$ then $X^{(i)}$ doesn't contain feasible points and thus can be safely excluded from the further search. Otherwise it is replaced by the box $\hat{X}^{(i)}$.

For unconstrained optimization we can introduce an additional feasibility test based on first-order optimality conditions. Let a box $X^{(i)} = [a^{(i)}, b^{(i)}]$ contains an optimal solution point x_*. From the first-order optimality condition $\nabla f(x_*) = 0$. From the Lipschitz condition

$$\|\nabla f(c_i)\| = \|\nabla f(c_i) - \nabla f(x_*)\| \leq L^{(i)} \|x_* - c_i\| \leq L^{(i)} \|b_i - c_i\|.$$

Thus if $\|\nabla f(c_i)\| > L^{(i)}\|b_i - c_i\|$ then $X^{(i)} \cap X_* = \emptyset$ and $X^{(i)}$ can be excluded from the further search. In what follows we call this assertion *gradient test*.

4.4 Updating the Record

At each iteration the objective function value is calculated in the new trial point $x^{(i)}$. For unconstrained optimization this point is the center $c^{(i)}$ of the box $X^{(i)}$. If $f(x^{(i)}) < u^{(i-1)}$ then $x^{(i)}$ and $f(x^{(i)})$ become the new best current solution and record respectively:

$$u^{(i)} = f(x^{(i)}), \tilde{x}^{(i)} = x^{(i)}.$$

For a problem with functional constraints $x^{(i)}$ is also taken equal to $c^{(i)}$ but the record is updated only if the point is δ-feasible, i.e. $g(c^{(i)}) \leq \delta$.

To update the record for mixed-integer problems the new trial point $x^{(i)}$ is defined as follows:

$$x_j^{(i)} = \begin{cases} c_j^{(i)}, j \in \overline{1,n} \setminus J, \\ \lfloor c_j^{(i)} \rfloor, j \in J. \end{cases}$$

If $g(x^{(i)}) \leq \delta$ and $f(x^{(i)}) < u^{(i-1)}$ then the best current solution and the record are updated.

5 Implementation and Experimental Results

It is worth to note that the proposed algorithm can exploit parallel processing because it generates new boxes per iteration. It has been implemented in the BNB-Solver framework (Evtushenko et al. 2009). This object-oriented framework for discrete and continuous parallel global optimization supports exact branch-bound algorithms, heuristic methods and hybrid approaches. BNB-Solver provides a support for distributed and shared memory architectures. The implementation for distributed memory machines is based on MPI (Snir et al. 1996) and thus can run on almost any computational cluster. In order to take advantages of multicore processors we provide a separate multi-threaded implementation for shared memory platforms. Details of parallel implementation, efficiency and speedup characteristics can be found in (Evtushenko et al. 2009). In the sequel we focus on serial implementation.

The goal of experiments was two-fold: to evaluate the efficiency of proposed new techniques (namely gradient test and lower bound (16)) and to compare our method to state-of-the-art solvers. Experiments were performed for unconstrained global optimization problems with polynomial objectives of the following form:

A Deterministic Algorithm for Global Optimization

$$f(x) = \sum_{i=1}^{m} a x_i^n + \sum_{d \in D} a_d x_{i_1}^{d_1} \ldots x_{i_m}^{d_m}, \tag{19}$$

where m is an even number of polynomial variables, n is a polynomial degree, $a > 0$ is a fixed real number, $D = \{(d_1, \ldots, d_m) : d_i \in \mathbb{Z}_+, \sum_{i=1}^{m} d_i \leq n - 1\}$ is a set of degree tuples of polynomial items and a_d are uniformly distributed random numbers, $a_d \in [0, a], d \in D$.

The search domain was a box $P = \{x \in \mathbb{R}^n : -B \leq x_i \leq B\}$, where $B = |D|$ and $|D|$ is a number of tuples in D. It is easy to show that $f(x) > f(0)$ for every $x \notin P$ and therefore $x_* \in P$. Upper bounds for Lipschitz constants used in computations were calculated according to (18).

Experiments were performed for five series of 10 problems each. Problems were randomly generated according to (19) with the following set of parameters:

- Series 1: $a = 10, m = 3, n = 4$;
- Series 2: $a = 10, m = 3, n = 6$;
- Series 3: $a = 10, m = 3, n = 8$;
- Series 4: $a = 10, m = 4, n = 4$;
- Series 5: $a = 10, m = 4, n = 6$.

In Table 1 we compare the average (AVR) , maximal (MAX) and minimal (MIN) running times for four different ways of optimality and feasibility testing:

- O1 — using lower bound (13);
- O2 — using lower bound (16);
- O1G — O1 coupled with the gradient test;
- O2G — O2 coupled with the gradient test.

We also used generated problems to test solvers from the GAMS[1] package. Only two solvers (BARON and LINDOGLOBAL) from this package claim to deterministically find a global solution. BARON (Tawarmalani and Sahinidis 2005) exploits constraint propagation, interval analysis, and duality bounds combined with a powerful branch-and-bound procedure. LINDOGLOBAL[2] also employs branch-and-bound approach and uses convexification of the objective and constraints to obtain lower bounds. In both solvers records are precalculated using local optimization and heuristic methods. Table 1 gives running time in seconds for these solvers in columns BR and LG respectively. For problems from Series 3,5 BARON terminates abnormally due to reaching memory limit (indicated by 'A' in the table). LINDOGLOBAL failed to find a correct solution for Series 2,3,5: it terminates after a few iterations and erroneously reports $x = 0_m$ as an answer (indicated by 'E' in the table). All experiments were run on the same PC Intel Core 2 Quad 2.83 GHz 4 Gb RAM and with the same absolute tolerance $\epsilon = 10^{-4}$. The O1 version of the

---

[1] http://www.gams.com/.

[2] http://www.lindo.com/.

Table 1 Running time in seconds for random polynomial unconstrained optimization problems

Series		O1	O2	O1G	O2G	BR	LG
1	AVR	1.28	0.15	0.38	0.14	1.07	5.42
	MAX	2.81	0.12	0.45	0.12	1.36	6.02
	MIN	0.58	0.2	0.31	0.21	0.65	3.43
2	AVR	32.73	1.81	4.67	1.79	4.86	E
	MAX	152.49	2.02	5.77	1.99	6.56	E
	MIN	9.89	1.63	4.27	1.66	3.68	E
3	AVR	T	11.02	28.06	11.24	A	E
	MAX	T	11.97	32.42	12.35	A	E
	MIN	T	10.16	25.55	10.46	A	E
4	AVR	97.724	2.41	9.13	2.32	3.51	23.63
	MAX	207.39	3.09	11.87	2.96	3.89	27.71
	MIN	35.93	2.00	7.82	1.94	3.02	20.54
5	AVR	T	72.63	243.39	69.05	A	E
	MAX	T	83.21	294.78	77.39	A	E
	MIN	T	65.36	228.57	61.89	A	E

coverage algorithm was not able to solve generated instances from Series 3,5 within 5 minutes time limit and was interrupted (indicated by 'T' in the table).

Results presented in Table 1 show that gradient test can remarkably improve the performance of the basic version O1 of the algorithm and gives a little improvement for O2. Best results are obtained using lower bound (16) combined with gradient test. Comparison with state-of-the-art solvers BARON and LINDOGLOBAL shows that for a considered class of unconstrained polynomial problems our method with lower bound (16) performs significantly better.

To test the proposed algorithm for the case of mixed-integer constrained optimization we selected a well-known pressure vessel design problem introduced in (Sandgren 1988). The objective is to minimize the total cost including the cost of the material, forming and welding. This problem has four design variables: thickness of the shell x_1, thickness of the head x_2, inner radius x_3 and length of the cylindrical section of the vessel x_4. Variables x_1 and x_2 are integer multiples of 0.0625 inch, which are the available thicknesses of rolled steel plates. Other two variables are continuous. The problem can be stated as follows:

Minimize:
$$f(x) = 0.6224x_1x_3x_4 + 1.7781x_2x_3^2 + 3.1661x_1^2x_4 + 19.84x_1^2x_3,$$
Subject to:
$$g_1(x) = -x_1 + 0.0193x_3 \leq 0,$$
$$g_2(x) = -x_2 + 0.00954x_3 \leq 0,$$
$$g_3(x) = -\pi x_3^2 x_4 - \tfrac{4}{3}\pi x_3^3 + 1296000 \leq 0,$$
$$g_4(x) = x_4 - 240 \leq 0,$$
$$x_1 = 0.0625z_1, z_1 \in \mathbb{Z},$$
$$x_2 = 0.0625z_2, z_2 \in \mathbb{Z}.$$

(20)

A Deterministic Algorithm for Global Optimization

Table 2 The comparison with other works

Source	Method	Minimum
Sandgren (1988)	branch & bound	8129.1036
Kannan and Kramer (1994)	augmented lagrange multipliers	7128.0428
Kalyanmoy (1997)	genetic algorithm	6410.3811
Coello Coello and Montes (2002)	genetic algorithm	6059.9463
Takahama and Sakai (2006)	particle swarm	6059.7143
Tahera et al. (2008)	genetic algorithm	6062.652
This paper	coverage approach	5850.3831

From the common sense it follows that all variables are positive. The constraint g_4 implies $x_4 \leq 240$. In (Coello Coello and Montes 2002) upper bounds for variables x_1, x_2, x_3 were set to 99 . We enlarged them to 200: $P = \{x \in \mathbb{R}^4 : 0 \leq x_1 \leq 200, 0 \leq x_2 \leq 200, 0 \leq x_3 \leq 200, 0 \leq x_4 \leq 240\}$. Finally variables x_1 and x_2 were replaced by integer variables z_1, z_2 and objective function and constraints were adjusted respectively.

We used underestimations (13) for constraints and the objective function. For the objective function $f(x)$ and the constraint $g_3(x)$ we calculated upper bounds for Lipschitz constants according to (18) for every new box $X^{(i)}$. For linear constraints $g_1(x), g_2(x)$ we used bound (14).

For the objective function $f(x)$ and constraints g_1, g_2, g_3 the precision was set to 10^{-7}. The algorithm processed 566227 boxes in 11.55 seconds on Intel Core 2 Quad 2.83 GHz. The obtained solution was

$$x_1 = 0.75, x_2 = 0.375, x_3 = 38.8601036266, x_4 = 221.365471361,$$
$$f(x) = 5850.3830518.$$

This solution provides the following values for constraints:

$$g_1(x) = -0.000000000007,$$
$$g_2(x) = -0.0042746114,$$
$$g_3(x) = -0.0000011429.$$

Table 2 compares the results obtained by our algorithm (the last line) and results obtained by other researchers. Our algorithm found the value 5850.3831 significantly better w.r.t. the best value 6059.7143 found before.

6 Related Works

First attempts to use Lipschitz property in optimization date back to early 70th. The seminal work (Evtushenko 1971) introduced Lipschitzian underestimations and described a deterministic algorithm for searching global optimum of a multivariate objective function. The algorithm is based on the non-uniform coverage of the feasible set by boxes. This paper also contained the Algol 60 recursive procedure implementing the proposed algorithm. Another approach for solving univariate global optimization problems was later independently proposed in works (Piyavskii 1972; Shubert 1972). In this approach the objective is supposed to satisfy Lipschitz condition with a known constant K. The "saw-tooth" like underestimation is iteratively constructed from the function values calculated at the ends of the processed intervals. The main limitation of this approach is that it can't be directly extended to the N-dimensional case.

These works gave rise to numerous research results in Lipschitzian optimization and software implementations see (Pinter 1996; Strongin and Sergeyev 2000) for survey. One of the most successful approaches was originally introduced in (Jones et al. 1993) for unconstrained black-box optimization. Later (Jones 1999) it was extended to handle functional and integer constrains. This algorithm called DIRECT (DIvide RECTangle) is based on biasing the search to most "promising" rectangles i.e. rectangles those are likely to contain global minima. The rectangles are selected on the basis of Lipschitzian lower bound where the Lipschitz constant is approximated at each step.

Most of these approaches are tailored to solving black-box problems and thus rely on estimations of Lipschitz constants rather than exact values. Though these search strategies converge to the minimum value at the limit they don't have means to measure the accuracy of the obtained solution. This leads to problems with selecting adequate stopping criteria. In contrast our approach uses strict upper bounds for Lipschitz constant given by formula (18) thereby guaranteeing the ϵ-optimality of the found point.

The absolute majority of works in Lipschitzian optimization assume that the objective function is Lipschitz-continuous. Few works dealing with Lipschitzian gradients (Evtushenko et al. 2009; Gergel 1997; Sergeyev 1998) use this property only to evaluate lower bounds and ignore the first-order optimality condition $\nabla f = 0$. We showed that this condition can be efficiently used in a simple gradient test and for calculating a lower bound (16) that doesn't require gradient evaluation. Experiments presented in Sect. 5 show that this bound tremendously improve the performance of the algorithm. Though the gradient test (some times also called *monotonicity test*) is widely used in interval approaches (Hansen 1992; Kearfott 1996) we pretend to be the first who applied this test to Lipschitzian optimization.

7 Conclusions

We have presented an algorithm for solving global optimization problems with guaranteed accuracy. The approach requires the objective and constraints to have gradients satisfying Lipschitz condition. In the paper the general algorithm scheme and its specialization for handling unconstrained and constrained mixed-integer problems were considered. Improved lower bounds and new techniques to reduce

A Deterministic Algorithm for Global Optimization

the number of algorithm steps by employing the gradient information were proposed for unconstrained optimization.

In order to test the approach we performed several experiments for unconstrained problems. Different optimality testing techniques for unconstrained optimization were compared. Experiments showed that the proposed novel lower bound could significantly (in several times) decrease the running time. We also compared our approach with production solvers on large set of random polynomials. This comparison showed that our approach is significantly faster for the considered set of problems. For the well known mixed-integer programming benchmark (pressure vessel design) the proposed algorithm was able to improve the best solution for this problem found so far.

Acknowledgements This work was supported by the Russian Foundation for Basic Research (Projects 08-01-00619-a, 09-01-12098-ofi-m) and by the Program P-14 of the President of the Russian Academy of Sciences.

References

Coello Coello, C.A., & Montes, E.M. (2002). Constraint-handling in genetic algorithms through the use of dominance-based tournament selection. *Advanced Engineering Informatics, 16,* 193–203.

Ermoliev, Y., & Norkin, V. (2004). Stochastic optimization of risk functions. In K. Marti, Y. Ermoliev, & G. Pflug (Eds.), *Dynamic Stochastic Optimization. Lecture Notes in Economics and Mathematical Systems,* pp. 225–249. New York: Springer.

Evtushenko, Yu. (1971). Numerical methods for finding global extreme (case of a non-uniform mesh). U.S.S.R. Comput. Maths. Math. Phys., Vol. 11, 6, pp. 38–54, Moscow.

Evtushenko, Yu., Malkova, V., & Stanevichyus, A. (2009). Parallel global optimization of functions of several variables. *Computational Mathematics and Mathematical Physics*, Vol. 49(2), pp. 246–260. MAIK Nauka. Doi:10.1134/S0965542509020055.

Evtushenko, Yu., Posypkin, M., & Sigal, I. (2009). A framework for parallel large-scale global optimization. *Computer Science: Research and Development*, Vol. 23(3), pp. 211–215. New York: Springer.

Gergel, V. P. (1997). A global optimization algorithm for multivariate functions with lipschitzian first derivatives. *Journal of Global Optimization, 10,* 257–281.

Hansen, E. (1992). *Global Optimization Using Interval Analysis*. New York: Dekker.

Jones, D. R., Perttunen, C. D., & Stuckman, B. E. (1993). Lipschitzian optimization without the Lipschitz constant. *Journal of Optimization Theory and Applications, 79*(1), 157–181.

Jones, D. R. (1999). The DIRECT global optimization algorithm. In A. Floudas, & P. Pardalos (Eds.), *Encyclopedia of Optimization*, pp. 725–735. New York: Springer.

Kalyanmoy, D. (1997). *GeneAS: A Robust Optimal Design Technique for Mechanical Component Design. Evolutionary Algorithms in Engineering Applications*, pp. 497–514. Berlin: Springer.

Kannan, B. K., & Kramer, S. N. (1994). An augmented lagrange multiplier based method for mixed integer discrete continuous optimization and its applications to mechanical design. *Journal of Mechanical Design, Transactions of the ASME, 116,* 318–320.

Kearfott, R. B. (1996). *Rigorous Global Search: Continuous Problems*. Dordrecht: Kluwer.

Nesterov, Y. (2003). *Introductory Lectures on Convex Optimization: A Basic Course (Applied Optimization)*. Netherlands: Springer.

Pardalos, P. M., Romeijn, E., & Tuy, H. (2000). Recent developments and trends in global optimization. *Journal of Computational and Applied Mathematics, 124*(1-2), 209–228.

Pardalos, P. M., & Resende, M. (Eds.) (2002). *Handbook of Applied Optimization*. Oxford: Oxford University Press.

Pinter, J. D. (1996). *Global Optimization in Action*. Dordrecht: Kluwer.

Piyavskii, S. A. (1972). An algorithm for finding the absolute extremum of a function. *USSR Computational Mathematics and Mathematical Physics, 12*, 57–67.

Snir, M., Otto, S., Huss-Lederman, S., Walker, D., & Dongarra, J. (1996). *MPI: The Complete Reference*. Boston: MIT Press.

Sandgren, E. (1988). Nonlinear integer and discrete programming in mechanical design. *Proceedings of the ASME Design Technology Conference*, pp. 95–105. Florida: Kissimine.

Sergeyev, Ya. D. (1998). Global one-dimensional optimization using smooth auxiliary functions. *Mathematical Programming, 81*(1), 127–146.

Strongin, R. G., & Sergeyev, Ya. D. (2000). *Global Optimization with Non-Convex Constraints: Sequential and Parallel Algorithms*. Dordrecht: Kluwer.

Shubert, B. (1972). A sequential method seeking the global maximum of a function. *SIAM Journal on Numerical Analysis, 9*, 379–388.

Tahera, K., Ibrahim, R. N., Lochert, P. B. (2008). GADYM – A novel genetic algorithm in mechanical design problems. *Journal for Universal Computer Science, 14*(15), 2566–2581.

Takahama, T., & Sakai, S. (2006). *Solving Constrained Optimization Problems by the ϵ-Constrained Particle Swarm Optimizer with Adaptive Velocity Limit Control*, pp. 1–7, CIS.

Tawarmalani, M., & Sahinidis, N. V. (2005). A polyhedral branch-and-cut approach to global optimization. *Mathematical Programming, 103*(2), 225–249.

Tawarmalani, M., Sahinidis N. V. (2002). *Convexification and Global Optimization in Continuous and Mixed-Integer Nonlinear Programming: Theory, Algorithms, Software, and Applications*. Dordrecht: Kluwer.

Robust Optimization by Fuzzy Linear Programming

Masahiro Inuiguchi

Abstract In this paper, possibilistic linear programming approaches, i.e., linear programming approaches with fuzzy coefficients, are reviewed from the perspective of robust optimization. The ideas of optimizing approaches and satisficing approaches are described rather than their technical and methodological aspects. In the first part of the paper, optimizing approaches are introduced. Possibly and necessarily optimal solutions are described as solution concepts for fuzzy linear programming problems. It is shown that a necessarily optimal solution is a solution preserving optimality from the fluctuations of coefficients within a certain range. However, because in many cases a necessarily optimal solution does not exist, a weakened solution concept, i.e., necessarily soft-optimal solutions is added. In the second part, satisficing approaches are briefly introduced. The necessity measure optimization model, the necessity fractile optimization model and the symmetric models are described. The solutions of these models preserve feasibility or satisfaction from the fluctuations of coefficients in a certain range. Finally, some concluding remarks are made.

1 Introduction

In real-world problems, we may face cases where the parameters of linear programming problems are not known exactly. In such cases, parameters can be treated as random variables or fuzzy variables, which are also called possibilistic variables (Dubois and Prade 1988; Hisdal 1978; Yager 2001; Zadeh 1978). The probability distribution that random variables obey is usually not easy to obtain, as it is assumed to be obtained by strict measurement on a ratio scale. On the other

---

M. Inuiguchi (✉)
Graduate School of Engineering Science, Osaka University, Toyonaka, Osaka, Japan
e-mail: inuiguti@sys.es.osaka-u.ac.jp

Y. Ermoliev et al. (eds.), *Managing Safety of Heterogeneous Systems*, Lecture Notes
in Economics and Mathematical Systems 658, DOI 10.1007/978-3-642-22884-1_11,
© Springer-Verlag Berlin Heidelberg 2012

hand, the possibility distribution restricting possibilistic variables can be obtained rather easily, because it is assumed to be obtained from experts' perception by measurement on an ordinal scale. The properties of probability, such as additivity, require the cardinality of probability, while the properties of possibility, such as maxivity (Dubois and Prade 1998), require only the ordinality of possibility. Here we do not discuss quantitative possibility, which has some connections with subjective probability, but only qualitative possibility (Dubois and Prade 1998). For these reasons, possibilistic programming approaches treating possibility distributions would be more convenient for modeling real-world optimization problems that include uncertainty.

In this paper, we review possibilistic linear programming approaches (see, for example, Dubois 1987; Inuiguchi and Ramík 2000) to robust optimization. Robustness in decision aiding and robust optimization have been surveyed in the literature (Ben-Tal and Nemirovski 2002; Aissi and Roy 2010). The robustness to which we refer in this paper implies the preservation of feasibility, satisfaction or optimality from the fluctuations of parameters within a certain range. Possibilistic linear programming approaches can be classified into three cases: the optimizing approach, the satisficing approach and the two-stage approach. Because the third approach has not yet been substantially developed, we focus on the other two approaches. First we review the optimization approach. We describe necessarily optimal solutions as solutions preserving optimality from the fluctuations of parameters within a certain range. Because a necessarily optimal solution does not always exist, necessarily soft-optimal solutions have been proposed. In necessarily soft-optimal solutions, the optimality conditions are relaxed to approximate optimality conditions. The relation to the minimax regret solution is shown, and a solution procedure for obtaining a best necessarily soft-optimal solution is briefly described.

Next we describe about the satisficing approach. This approach was developed earlier and has been described often in the literature. (Dubois 1987; Slowinski 1986; Slowinski and Teghem 1990; Tanaka and Asai 1984; Tanaka et al. 1984). We introduce the approach briefly and concisely, focusing on the idea of robust treatments. Then, the necessity measure optimization model, the necessity fractile optimization model and the symmetric model are presented. The solutions of these models preserve feasibility or satisfaction from the fluctuations of parameters within a certain range. It is emphasized that the reduced problems of those models have linearity to some extent.

Finally, we end the paper with some concluding remarks.

2 Optimizing Approach

2.1 Statement of the Problem

To explain the concept of robustly optimal solutions, let us consider the following linear programming problems with uncertain objective function coefficients:

$$\text{maximize } \gamma^{\mathrm{T}}x,$$
$$\text{s.t. } Ax \le b, \tag{1}$$

where A is a constant $m \times n$ matrix, $b = (b_1, b_2, \ldots, b_m)^{\mathrm{T}}$ is a constant vector, $x = (x_1, x_2, \ldots, x_n)^{\mathrm{T}}$ is a decision variable vector and $\gamma = (\gamma_1, \gamma_2, \ldots, \gamma_n)^{\mathrm{T}}$ is an uncertain variable vector corresponding to the objective function coefficient vector. In real-world problems, we may face cases where components of A and b are ambiguous and/or the inequality $\le$ is generalized to a fuzzy inequality (Inuiguchi et al. 1993). However, we treat the simpler case where A and b are constant and no fuzzy inequality is included.

When γ is regarded as a random variable vector, Problem (1) becomes a stochastic linear programming problem (Stancu-Minasian 1984). In this case, estimating the probability distribution that the random variable vector obeys would not be very easy because the probability should be measured on a cardinal scale. Sensitivity analysis (Derhy 2010) has been developed to evaluate the influence of the fluctuations of coefficients. However, it is only a local analysis and analysis with multiple parameters is not very easy. In this paper, we focus on a case where γ is seen as a possibilistic variable vector restricted by a possibility distribution. A possibility distribution is defined by a membership function $\mu_\Gamma : \mathbf{R}^n \to [0, 1]$ of a fuzzy set Γ showing the possible range of γ. The membership values $\mu_\Gamma(c)$ for a vector $c \in \mathbf{R}^n$ of the fuzzy set Γ do not need to be cardinal, but only ordinal. By this weak assumption, estimating of fuzzy set Γ would be easier than estimating the probability distribution. Thus, fuzzy set Γ can be estimated approximately by human experts.

For example, we may estimate a largest possible range for γ as an approximation of the support of fuzzy set Γ, a smallest possible range for γ as the core of fuzzy set Γ and an appropriate possible range for γ as the 0.5-level set of fuzzy set Γ. Between those ranges, membership values can be determined by some interpolation. Here, we define the support of Γ by $Supp(\Gamma) = \{c : \mu_\Gamma(c) > 0\}$, the core of Γ by $Core(\Gamma) = \{c : \mu_\Gamma(c) = 1\}$ and the α-level set of Γ by $[\Gamma]_\alpha = \{c : \mu_\Gamma(c) \ge \alpha\}$ for $\alpha \in (0, 1]$. $Supp(\Gamma)$ can be approximated by the ϵ-level set $[\Gamma]_\epsilon$ with very small positive number $\epsilon > 0$. This estimate is illustrated in Fig. 1 when $n = 2$.

Remark 1. In the above explanation, we assume that the appropriate possible range corresponds to $[\Gamma]_{0.5}$ but the value 0.5 is not essential. It can be replaced with any value between ϵ and 1. The correspondence between membership values of multiple fuzzy sets is essential in the possibilistic programming described in this paper.

Fuzzy programming problems are mathematical programming problems with fuzzy parameters and/or fuzzy relations (Inuiguchi et al. 1993). Possibilistic programming problems are fuzzy programming problems when membership functions of fuzzy sets with respect to uncertain parameters are regarded as possibility distributions. Moreover, interval and inexact programming problems (Bitran 1981, Soyster 1979) can be seen as special cases of fuzzy/possibilistic programming problems, where all fuzzy sets involved in the problems degenerate to crisp sets.

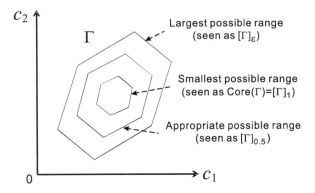

Fig. 1 Estimation of fuzzy set Γ

We interpret the membership function μ_Γ of fuzzy set Γ as a multivariate possibility distribution because Γ shows the possible range of γ. Then, Problem (1) can be seen as a possibilistic linear programming problem. We explain some optimizing approaches to such a problem related to robust optimization.

2.2 Possibility and Necessity

Now let us introduce possibility and necessity measures defined under possibility distributions in possibility theory. There are two different kinds of possibility theory (Dubois and Prade 1998): quantitative and qualitative. The quantitative possibility theory has some connections with subjective probability, while the qualitative possibility theory can be described either via a purely comparative approach, such as a partial ordering on events, or using set-functions ranging on an absolute, totally ordered scale (Dubois and Prade 1998). The possibility theory employed in this paper is the qualitative one.

To introduce the concepts of possibility and necessity, we consider a crisp case. Let A be a set of possible realizations and B a set of objectives satisfying a certain property $\mathscr{P}$. In other words, A shows a possible range, while B shows an event. Then we may say that the satisfaction of $\mathscr{P}$ is *possible* if and only if $A \cap B \neq \emptyset$, i.e., there exists an object z such that $z \in A$ and $z \in B$ (see Fig. 2). Moreover, we may say that the satisfaction of $\mathscr{P}$ is *necessary (certain)* if and only if $A \subseteq B$, i.e., for all objects $z \in A$, we have $z \in B$ (see Fig. 2). Then the possibility measure $\Pi_A(B)$ and necessity measure $N_A(b)$ are defined by

$$\Pi_A(B) = \begin{cases} 1, & \text{if } A \cap B \neq \emptyset, \\ 0, & \text{otherwise,} \end{cases} \quad N_A(B) = \begin{cases} 1, & \text{if } A \subseteq B, \\ 0, & \text{otherwise.} \end{cases} \quad (2)$$

Fig. 2 Possibility and necessity

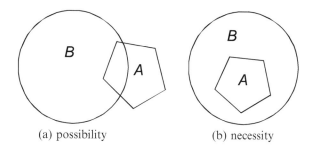

(a) possibility (b) necessity

When A and B are generalized to fuzzy sets, the possibility and necessity measures are defined by

$$\Pi_A(B) = \sup_z \min(\mu_A(z), \mu_B(z)), \tag{3}$$

$$N_A(B) = \inf_z \max(1 - \mu_A(z), \mu_B(z)), \tag{4}$$

where μ_A and μ_B are the membership functions of fuzzy sets A and B, respectively. Possibility measure $\Pi_A(B)$ and necessity measure $N_A(B)$ are depicted in Fig. 3. Note that $\Pi_A(B)$ in (3) and $N_A(B)$ in (4) degenerate to $\Pi_A(B)$ and $N_A(B)$ in (2) when A and B are crisp sets.

The Properties of possibility and necessity measures have been investigated in the literature (Dubois and Prade 1998), but we describe the following properties:

$$\Pi_A(B) > \alpha \Leftrightarrow (A)_\alpha \cap (B)_\alpha \neq \emptyset, \quad N_A(B) \geq \alpha \Leftrightarrow (A)_{1-\alpha} \subseteq [B]_\alpha, \tag{5}$$

where $(A)_\alpha$ is a strong α-level set defined by $(A)_\alpha = \{z : \mu_A(z) > \alpha\}$ for $\alpha \in [0, 1)$. In those properties, we can observe the relation between the possibility measure and non-empty intersection, as well as the relation between the necessity measure and inclusion.

Remark 2. Using a conjunction function $T : [0, 1] \times [0, 1] \to [0, 1]$ and an implication function $I : [0, 1] \times [0, 1] \to [0, 1]$ such that $T(0, 0) = T(0, 1) = T(1, 0) = 0$, $T(1, 1) = 1$, $I(0, 0) = I(0, 1) = I(1, 1) = 1$ and $I(1, 0) = 0$, the possibility and necessity measures can be generalized to

$$\Pi_A(B) = \sup_z T(\mu_A(z), \mu_B(z)), \tag{6}$$

$$N_A(B) = \inf_z I(\mu_A(z), \mu_B(z)). \tag{7}$$

Such generalized measures may work to express the variety of decision-maker's desires regarding possibility and necessity (Inuiguchi 2009).

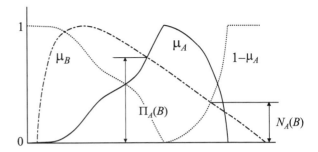

Fig. 3 Possibility and necessity measures

2.3 Possible and Necessary Optimalities

Two kinds of optimal solutions have been proposed to Problem (1) (Inuiguchi and Sakawa 1994). One is a possibly optimal solution, and the other is a necessarily optimal solution. Let X be the set of feasible solutions of Problem (1), i.e., $X = \{x : Ax \leq b\}$. Moreover, we define $S(c)$ as a set of optimal solutions to a linear programming problem $\max_{y \in X} c^T y$ with a constant objective function coefficient vector c, i.e.,

$$S(c) = \left\{ x \in X : c^T x = \max_{y \in X} c^T y \right\}. \tag{8}$$

Note that $S(c)$ is not always a singleton because a linear programming problem can have multiple optimal solutions. Then the possibly optimal solution set ΠS and the necessarily optimal solution set NS are defined by

$$\Pi S = \bigcup_{c \in \Gamma} S(c), \quad NS = \bigcap_{c \in \Gamma} S(c). \tag{9}$$

ΠS is a set of feasible solutions that is optimal for at least one objective coefficient vector $c \in \Gamma$. On the other hand, NS is a set of feasible solutions that is optimal for all objective coefficient vectors $c \in \Gamma$.

Let $V(x)$ be the set of objective function vectors c such that x is an optimal solution to $\max_{y \in X} c^T y$, i.e., $x \in S(c)$. Then the relations of ΠS and NS regarding possibility and necessity measures can be seen as

$$x \in \Pi S \Leftrightarrow \Pi_\Gamma(V(x)) = 1, x \in X \Leftrightarrow \Gamma \cap V(x) \neq \emptyset, x \in X$$
$$\Leftrightarrow \text{ there exists } c \in \Gamma \text{ such that } x \in S(c), \tag{10}$$
$$x \in NS \Leftrightarrow N_\Gamma(V(x)) = 1, x \in X \Leftrightarrow \Gamma \subseteq V(x), x \in X$$
$$\Leftrightarrow \text{ for all } c \in \Gamma, \text{ we have } x \in S(c). \tag{11}$$

Namely, a possibly optimal solution is a feasible solution optimal for at least one possible realization $c \in \Gamma$, whereas a necessarily optimal solution is a feasible solution optimal for all possible realizations $c \in \Gamma$. From this fact, we may see that

Robust Optimization by Fuzzy Linear Programming

a necessarily optimal solution is a robustly optimal solution because the solution remains optimal regardless of the fluctuation of c whithin the given range Γ.

The same solution set as ΠS was originally considered by Steuer (1981) but for a different purpose. Moreover, a solution set similar to NS was originally defined by Bitran (1981) in multiple objective cases. Those solution concepts were introduced to possibilistic programming by Luhandjula (1987) in multiple objective cases.

A few examples of possibly and necessarily optimal solutions are given in the following example.

Example 1. Let us consider Problem (1) with

$$
A = \begin{pmatrix} 3 & 4 \\ 3 & 1 \\ 0 & 1 \\ -1 & 0 \\ 0 & -1 \end{pmatrix}, \qquad b = \begin{pmatrix} 42 \\ 24 \\ 9 \\ 0 \\ 0 \end{pmatrix}, \tag{12}
$$

and

$$
\Gamma = \{(c_1, c_2)^{\mathrm{T}} : 3.5 \leq 2c_1 + c_2 \leq 5.5,\ 3.4 \leq c_1 + 2c_2 \leq 6, \\
1 \leq c_1 - c_2 \leq 1.3,\ 1 \leq c_1 \leq 2,\ 0.8 \leq c_2 \leq 2.2\}. \tag{13}
$$

Consider a feasible solution $(x_1, x_2)^{\mathrm{T}} = (6, 6)^{\mathrm{T}}$. The feasible region X is depicted in Fig. 4. Set the c_1-c_2 coordinate system so that its origin is located at $(x_1, x_2)^{\mathrm{T}} = (6, 6)^{\mathrm{T}}$. The set $V((6, 6)^{\mathrm{T}})$ of objective function coefficient vectors to which $(6, 6)^{\mathrm{T}}$ is optimal is a closed convex cone whose border lines are $4c_1 = 3c_2$ and $c_1 = 3c_2$, as shown in Fig. 4 on the c_1-c_2 coordinate system. Γ is also shown in Fig. 4 on the c_1-c_2 coordinate system. We observe that $\Gamma \cap V((6, 6)^{\mathrm{T}}) \neq \emptyset$ and $\Gamma \nsubseteq V((6, 6)^{\mathrm{T}})$. Therefore, $(x_1, x_2)^{\mathrm{T}} = (6, 6)^{\mathrm{T}}$ is a possibly optimal solution, but not a necessarily optimal solution.

However, let us have Γ defined by

$$
\Gamma = \{(c_1, c_2)^{\mathrm{T}} : c_1 + c_2 \geq 3,\ c_1 \geq c_2,\ c_1 \leq 2c_2,\ c_1 \leq 2.5,\ c_2 \leq 2\}. \tag{14}
$$

instead of that defined by (13). The situation is changed as shown in Fig. 5. In this case, we have $\Gamma \subseteq V((6, 6)^{\mathrm{T}})$, which implies $\Gamma \cap V((6, 6)^{\mathrm{T}}) \neq \emptyset$. Therefore, $(x_1, x_2)^{\mathrm{T}} = (6, 6)^{\mathrm{T}}$ is both a possibly optimal solution and a necessarily optimal solution.

These solution concepts are extended to the case when Γ is a fuzzy set (Inuiguchi and Sakawa 1994). Let $\chi_{S(c)}(x)$ and $\chi_{V(x)}(c)$ be characteristic functions of $S(c)$ and $V(x)$, respectively, i.e.,

$$
\chi_{S(c)}(x) = \begin{cases} 1, & \text{if } x \in S(c), \\ 0, & \text{otherwise}, \end{cases} \qquad \chi_{V(x)}(c) = \begin{cases} 1, & \text{if } c \in V(x), \\ 0, & \text{otherwise}. \end{cases} \tag{15}
$$

Fig. 4 An example of possibly optimal solution

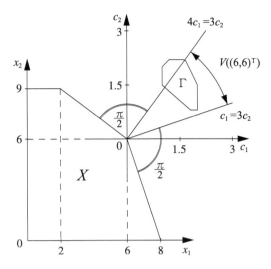

Fig. 5 An example of necessarily optimal solution

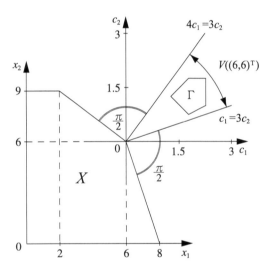

Then the possibly optimal solution set ΠS and the necessarily optimal solution set NS are defined by

$$\mu_{\Pi S}(x) = \sup_{c \in \Gamma} \min(\mu_\Gamma(c), \chi_{S(c)}(x))$$

$$= \sup_{c \in \Gamma} \min(\mu_\Gamma(c), \chi_{V(x)}(c)) = \Pi_\Gamma(V(x)), \quad (16)$$

$$\mu_{NS}(x) = \inf_{c \in \Gamma} \max(1 - \mu_\Gamma(c), \chi_{S(c)}(x))$$

$$= \inf_{c \in \Gamma} \max(1 - \mu_\Gamma(c), \chi_{V(x)}(c)) = N_\Gamma(V(x)). \quad (17)$$

Robust Optimization by Fuzzy Linear Programming

We may evaluate the degrees of possibility and necessity optimalities to feasible solutions using the membership function of a fuzzy set Γ.

From properties shown in (5), we have

$$\mu_{\Pi S}(x) > \alpha \Leftrightarrow (\Gamma)_\alpha \cap V(x) \neq \emptyset, \quad (18)$$

$$\mu_{NS}(x) \geq \alpha \Leftrightarrow (\Gamma)_{1-\alpha} \subseteq V(x). \quad (19)$$

Equation (18) implies that x such that $\mu_{\Pi S}(x) > \alpha$ is a feasible solution optimal for at least one $c \in (\Gamma)_\alpha$. On the other hand, (19) implies that x such that $\mu_{NS}(x) \geq \alpha$ is a feasible solution optimal for all $c \in (\Gamma)_{1-\alpha}$.

Because we are interested in robust optimization, we give an example of a necessarily optimal solution when Γ is a fuzzy set.

Example 2. Let us consider Problem (12) with fuzzy set Γ defined by the following membership function:

$$\mu_\Gamma(c_1, c_2) = \min\left(\mu_{\Gamma_1}(c_1), \mu_{\Gamma_2}(c_2)\right), \quad (20)$$

where μ_{Γ_1} and μ_{Γ_2} are membership functions of symmetric triangular fuzzy numbers $\Gamma_1 = ST(1.5, 1)$ and $\Gamma_2 = ST(1.5, 2)$. A symmetric triangular fuzzy number $M = ST(m^C, m^W)$ ($m^W > 0$) is defined by the following membership function:

$$\mu_M(r) = \max\left(0, 1 - \frac{|r - m^C|}{m^W}\right). \quad (21)$$

Consider a feasible solution $(x_1, x_2)^T = (6, 6)^T$. In the same way as Example 1, Fig. 6 shows the situation of the solution. In this example, the membership function of fuzzy set Γ draws a pyramid. In Fig. 6, the overhead view of fuzzy set Γ is

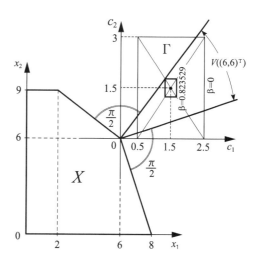

Fig. 6 An example of necessarily optimal solution with degree 0.176471

depicted as a rectangle with saltire lines, and the 0.823529-level set $[\Gamma]_{0.823529}$ is shown as a small box on the c_1-c_2 coordinate. As is shown in Fig. 6, we have $(\Gamma)_{0.823529} \subseteq V((6,6)^{\mathrm{T}})$ so that $(6,6)^{\mathrm{T}}$ is optimal for all $c \in [\Gamma]_{0.823529}$. Then $(6,6)^{\mathrm{T}}$ is a necessarily optimal solution to degree $0.176471 = 1 - 0.823529$.

In crisp cases, we may easily encounter a case where no necessarily optimal solution exists (see Example 1). By introducing fuzziness to Γ, i.e., several levels of possible range estimations, it becomes more probable that a necessarily optimal solution exists because we may set the smallest possible range as narrow as possible.

Inuiguchi and Sakawa (1994) investigated the possible and necessary optimality tests of a feasible solution when Γ is a vector of non-interactive fuzzy numbers (Inuiguchi et al. 2000). The possible optimality test problem of a given feasible solution is reduced to linear programming problems, while the necessary optimality test problem of a given feasible solution is reduced to a sequence of linear programming problems. Inuiguchi (2004) proposed an enumeration method of all possibly optimal extreme points with their possible optimality degrees. When the membership function of Γ is strictly quasi-concave, the necessarily optimal extreme points can be obtained by enumerating all possibly optimal extreme points. It is not necessarily to have enumerated all possibly optimal extreme points but to have the neighbors of those with possible optimality degree 1. For the case where Γ is a crisp box set, relations between solutions in multiple objective programming problems and possibly and necessarily optimal solutions have been investigated by Inuiguchi and Kume (1994). It has been shown that possibly optimal solutions are equivalent weakly efficient solutions of a multiple objective programming problems derived from Problem (1) and that necessarily optimal solutions are equivalent to completely optimal solutions.

2.4 Robust Soft-Optimal Solutions

Necessarily optimal solutions are the most rational solutions, but do not exist in many cases, while possibly optimal solutions are minimally rational solutions and often exist in large numbers. Therefore, intermediate solutions or relaxed necessarily optimal solutions have been investigated (Inuiguchi and Kume 1994; Inuiguchi and Sakawa 1995a, 1997, 1998). The proposed solution concept is called a necessarily soft-optimal solution or a robust soft-optimal solution.

To introduce the solution concept, we define the soft-optimal solution set $\tilde{S}(c)$ with respect to objective function coefficient vector c. Two definitions have been proposed, the difference-based definition $\tilde{S}_D(c)$ and the ratio-based definition $\tilde{S}_R(c)$:

$$\mu_{\tilde{S}_D(c)}(x) = \begin{cases} \mu_{Dif}\left(\max_{y \in X} c^T y - c^T x\right), & \text{if } x \in X, \\ 0, & \text{otherwise,} \end{cases} \quad (22)$$

$$\mu_{\tilde{S}_R(c)}(x) = \begin{cases} \mu_{Rat}\left(\dfrac{c^T x}{\max_{y \in X} c^T y}\right), & \text{if } x \in X, \\ 0, & \text{otherwise,} \end{cases} \quad (23)$$

where $\mu_{Dif} : \mathbf{R} \to [0, 1]$ and $\mu_{Rat} : \mathbf{R} \to [0, 1]$ are non-increasing and non-decreasing upper semi-continuous functions, respectively, such that $\mu_{Dif}(0) = \mu_{Rat}(1) = 0$ (see Fig. 7). The ratio-based definition is applicable only when $\forall c \in \Gamma$; $\max_{x \in X} c^T x > 0$. Members of $\tilde{S}_D(c)$ are solutions such that the difference between the objective function value and the optimal value is not very large. Namely, the optimality is relaxed based on the difference. On the other hand, members of $\tilde{S}_R(c)$ are solutions such that the ratio of the objective function value to the optimal value is close to 1. Namely, the optimality is relaxed based on the ratio. Members of both $\tilde{S}_D(c)$ and $\tilde{S}_R(c)$ are suboptimal solutions but the measures of their closeness to optimality are different.

Aissi and Roy (2010) mentioned similar measures. The difference-based definition corresponds to the absolute deviation, i.e., the value of the absolute regret in the worst case, while the ratio-based definition corresponds to the relative deviation, i.e., the value of the relative regret in the worst case.

Replacing $S(c)$ in (17) with $\tilde{S}(c)$, the necessarily soft-optimal solution set $\widetilde{NS}$ can be defined by the following membership function:

$$\mu_{\widetilde{NS}}(x) = \inf_c \max\left(1 - \mu_\Gamma(c), \mu_{\tilde{S}(c)}(x)\right). \quad (24)$$

When soft-optimal solution set $\tilde{S}(c)$ is defined based on the difference, $\tilde{S}(c)$ is substituted for $\tilde{S}_D(c)$. On the other hand, when soft-optimal solution set $\tilde{S}(c)$ is defined based on the ratio, $\tilde{S}(c)$ is substituted for $\tilde{S}_R(c)$.

Corresponding to $V(x)$, i.e., a set of objective function coefficient vectors c to which x is optimal, we may define a set $\tilde{V}(x)$ of objective function coefficient vectors c to which x is suboptimal by

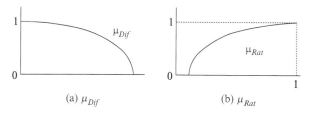

(a) μ_{Dif} (b) μ_{Rat}

Fig. 7 Functions μ_{Dif} and μ_{Rat}. (a) μ_{Dif} (b) μ_{Rat}

$$\mu_{\tilde{V}(x)}(c) = \mu_{\tilde{S}(c)}(x). \tag{25}$$

Then the necessarily soft-optimality solution set $\widetilde{NS}$ can also be characterized by

$$\mu_{\widetilde{NS}}(x) = \inf_c \max \left(1 - \mu_\Gamma(c), \mu_{\tilde{V}(x)}(c) \right). \tag{26}$$

From the property of the necessity measure in (5), we have

$$\mu_{\widetilde{NS}}(x) \geq \alpha \Leftrightarrow (\Gamma)_{1-\alpha} \subseteq [\tilde{S}(c)]_\alpha. \tag{27}$$

The right-hand side implies that solution x is sub-optimal at least to degree α for all possible objective function coefficient vectors c with degree more than $1 - \alpha$.

The following example shows to what extent the degree of robustness is increased by relaxing $S(c)$ to $\tilde{S}(c)$.

Example 3. Let us consider the same problem and the same solution as in Example 2. Namely, we discuss the necessary soft-optimality of $(6, 6)^T$. We use the difference-based soft-optimal solution set with μ_{Dif} defined by

$$\mu_{Dif}(r) = \begin{cases} 1, & \text{if } r \leq 0, \\ 1 - \dfrac{r}{5}, & \text{if } 0 < r \leq 5, \\ 0, & \text{if } r > 5. \end{cases} \tag{28}$$

The situation is shown in Fig. 8. As shown in Fig. 8, $\tilde{V}((6, 6)^T)$ is a fuzzy set such that the borders of α-level sets are depicted on the c_1-c_2 coordinate system. From (27), $\mu_{\widetilde{NS}}((6, 6)^T)$ is obtained by the supremum of $\{\alpha : (\Gamma)_{1-\alpha} \subseteq [\tilde{V}((6, 6)^T)]_\alpha\}$. From Fig. 8, we find that the supremum is 0.481481.

Comparing this with the result in Example 2, the degree is increased from 0.176471 to 0.481481 but the optimality is relaxed. Namely, we know that the difference of the objective function value of solution $(x_1, x_2)^T = (6, 6)^T$ from the optimal value is guaranteed to be at most $2.59260 \approx 5 \times (1 - 0.481481)$ as far as the fluctuation of the objective function coefficient vector c is in $(\Gamma)_{0.518519}$.

Necessarily soft-optimal solutions x can be ranked by their membership degrees $\mu_{\widetilde{NS}}(x)$. We may optimize the degree to obtain the best necessarily soft-optimal solution. From this point of view, we can formulate Problem (1) as the following optimization problem:

$$\underset{x \in X}{\text{maximize}} \ \mu_{\widetilde{NS}}(x). \tag{29}$$

Before describing the solution method for Problem (29), we give an example of the best necessarily soft-optimal solution as follows.

Example 4. Consider the same problem as in Example 3. The best necessarily optimal solution is obtained as $(x_1, x_2)^T = (4.69786, 6.97661)^T$. The situation of this solution is depicted in Fig. 9. As shown in Fig. 9, we observe that

Robust Optimization by Fuzzy Linear Programming 231

Fig. 8 An example of necessarily soft-optimal solution with degree 0.481481

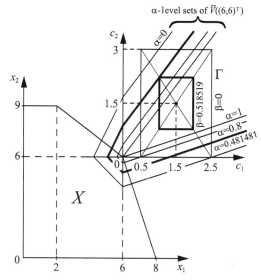

Fig. 9 An example of the best necessarily soft-optimal solution

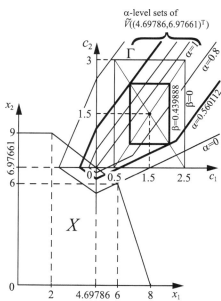

$(\Gamma)_{0.439888} \subseteq [\tilde{V}((4.69786, 6.97661)^T)]_{0.560112}$ holds. Then the degree is 0.560112, which is higher than the degree of the solution $(6, 6)^T$ shown in the previous example. Thus, we know that the objective function value of solution $(x_1, x_2)^T = (4.69786, 6.97661)^T$ from the optimal value is guaranteed to be at most $2.19944 \approx 5 \times (1 - 0.560112)$ as far as the fluctuation of the objective function coefficient vector c is in $(\Gamma)_{0.439888}$.

Under soft-optimality, the robustness of the solution is improved from the previous solution in Example 3. The improvement is not only in the largest difference from the optimal value but also in the range of fluctuation of c. The previous solution $(6, 6)^T$ was a unique necessarily optimal solution to the problem. The above fact implies that by relaxing the optimality (by permitting a small difference from the optimal value), we may obtain a better solution in the sense of both robustness and the worst regret (the largest difference from the optimal value).

Now let us describe the properties of Problem (29) when Γ is a crisp set. When Γ is a crisp set, Problem (29) with $\tilde{S}_D(c)$ is reduced to the following problem:

$$\underset{x \in X}{\text{minimize}} \ \underset{c \in \Gamma, \ y \in X}{\max} \ c^T y - c^T x, \tag{30}$$

where we introduce a natural assumption that the smaller the difference from the optimal value, the higher μ_{Dif} is. The difference from the optimal value, $\max_{y \in X} c^T y - c^T x$, can be seen as a regret of selecting x when the unknown objective function coefficient vector is c. Because we have

$$R(x) = \underset{c \in \Gamma, \ y \in X}{\max} \ c^T y - c^T x$$

$$= \underset{c \in \Gamma}{\max} \left(\max_{y \in X} c^T y - c^T x \right), \tag{31}$$

$R(x)$ can be understood as the maximum regret. Therefore, Problem (30) is a minimax regret problem, as investigated in (Inuiguchi and Sakawa 1995a; Inuiguchi and Tanino 2001a; Mausser and Laguna 1998, 1999).

On the other hand, when Γ is a crisp set, Problem (29) with $\tilde{S}_R(c)$ is reduced to the following problem:

$$\underset{x \in X}{\text{maximize}} \ \underset{c \in \Gamma}{\min} \ \frac{c^T x}{\max_{y \in X} c^T y}, \tag{32}$$

where we introduce a natural assumption that the larger the ratio to the optimal value, the higher μ_{Rat} is. Moreover, we assume $\forall c \in \Gamma; \ \max_{x \in X} c^T x > 0$. The ratio $c^T x / \max_{y \in X} c^T y$ can be understood as the achievement ratio of solution x to the optimal value when the unknown objective function coefficient vector is c. Then

$$WA(x) = \underset{c \in \Gamma}{\min} \ \frac{c^T x}{\max_{y \in X} c^T y} \tag{33}$$

can be seen as the worst achievement ratio. Thus, Problem (32) is a maximin achievement ratio problem, as investigated in (Inuiguchi and Sakawa 1997).

As are shown in (Inuiguchi and Kume 1994; Inuiguchi and Sakawa 1995a; 1997) those solutions have the following properties:

(a) $R(x) = 0$ (resp. $WA(x) = 1$) if and only if x is a necessarily optimal solution.
(b) $R(x) \geq 0, \forall x \in X$ (resp. $WA(x) \leq 1, \forall x \in X$).
(c) Any optimal solution of Problem (30) (resp. (32)) is a possibly optimal solution.

Robust Optimization by Fuzzy Linear Programming

Therefore, we may understand that optimal solutions to Problems (30) and (32) are possibly optimal solutions that minimize deviations from the necessary optimality.

The solution algorithms for Problem (30) were proposed by Inuiguchi and Sakawa (1995a), Mausser and Laguna (1998, 1999), and Inuiguchi and Tanino (2001a). Inuiguchi and Sakawa (1995a) proposed a two-phased approach. In the first phase, all possibly optimal extreme points are enumerated. Then, in the second phase, Problem (30) is solved by a relaxation procedure using the enumerated possibly optimal extreme points. This approach is proposed for problems in which Γ is a box set, but it is applicable for problems when Γ is a polytope (Inuiguchi and Tanino 2001a). Mausser and Laguna (1998) reformulated the problem as a mixed integer programming problem and applied a branch and bound method. This approach is also proposed for problems when Γ is a box set, but it is not applicable for problems with a polytope Γ. Moreover, Mausser and Laguna (1999) proposed a heuristic search approach to Problem (30). Inuiguchi and Tanino (2001a) use the same relaxation schema as Inuiguchi and Sakawa (1994), but they do not use the enumeration of all possibly optimal extreme points. They formulated the subproblem as a convex maximization problem and applied an outer approximation procedure together with a cutting hyperplane. This approach is proposed for cases when Γ is a polytope. It seems that in the case where Γ is a box set, Mausser and Laguna's approach (Mausser and Laguna 1998) is computationally efficient, and in the case where Γ is a polytope, Inuiguchi and Tanino's approach (Inuiguchi and Tanino 2001a) is computationally efficient.

Problem (32) is treated by Inuiguchi and Sakawa (1997). They modified Inuiguchi and Sakawa's two-phase approach to solve Problem (32) with a box set Γ. This approach can be extended to the case where Γ is a polytope. The other approaches to Problem (30) could also be modified for Problem (32).

When Γ is a fuzzy set, we need to introduce a bisection method on the membership degree. Inuiguchi and Sakawa (1998) extended the two-phase approach to the best necessarily soft-optimal solution problem (29) with $\tilde{S}_D(c)$ and a fuzzy set Γ. In this approach, they contrive the algorithm so as to converge the bisection method and the relaxation method simultaneously. Inuiguchi et al. (2001) proposed this solution approach to Problem (29) with $\tilde{S}_R(c)$ and a fuzzy set Γ.

3 Satisficing Approach

To explain satisficing approaches we consider the following possibilistic linear programming problem:

$$\begin{aligned}
&\text{maximize } \gamma^{\mathrm{T}}x, \\
&\text{subject to } \delta_i^{\mathrm{T}}x \precsim_i b_i, \ i = 1, \ldots, m,
\end{aligned} \tag{34}$$

where γ is an uncertain objective function coefficient vector whose possible range is known as a fuzzy set Γ. Similarly, δ_i is the i-th uncertain constraint function coefficient vector whose possible range is known as a fuzzy set Δ_i. $\lesssim_i$ is a fuzzy inequality relation depending on the i-th constraint. Combining $\lesssim_i$ with b_i, '$\lesssim_i b_i$' defines a fuzzy goal B_i with the linguistic expression "approximately smaller than b_i".

The satisficing approach employs the same idea regarding the treatment of the uncertainty as the conventional robust optimization approach (Ben-Tal and Nemirovski 2002). The history of the satisficing approach dates back more than three decades. The original treatment is known as robust programming (Negoita et al. 1976). Robust programming was proposed for Problem (34) with a singleton Γ, i.e., the objective function is a usual crisp function $c^\mathrm{T} x$. In robust programming, Problem (34) with a crisp objective function $c^\mathrm{T} x$ is formulated as

$$
\begin{aligned}
&\text{maximize } c^\mathrm{T} x, \\
&\text{subject to } \Delta_i^\mathrm{T} x \subseteq B_i, \ i = 1, \dots, m.
\end{aligned}
\tag{35}
$$

By this formulation, each possible range of the left-hand side values, $\Delta_i^\mathrm{T} x$, is managed so as to be included in a satisfactory range B_i. The fluctuations of the left-hand side values are bounded by B_i. In this sense, a solution of Problem (35) has robustness in the satisfaction of constraints.

After that, treatment by using fuzzy max (Tanaka et al. 1984) and treatment by using an inequality relations of fuzzy numbers (Orlovsky 1980; Tanaka and Asai 1984) were proposed. Since the proposal of possibility theory (Zadeh 1978), possibility and necessity measures have been used to treat constraints with fuzzy coefficient vectors and to treat objective functions with fuzzy coefficient vectors (Dubois 1987). Because we concentrate on robust treatments in possibilistic programming approaches, we describe only models using necessity measures.

Using necessity measures, Problem (34) can be addressed by

$$
\begin{aligned}
&\text{maximize } \psi(t, \alpha_0, \alpha_1, \dots, \alpha_m), \\
&\text{subject to } N_{\Gamma^\mathrm{T} x}([t, \infty)) \geq \alpha_0, \\
&\qquad\qquad N_{\Delta_i^\mathrm{T} x}(B_i) \geq \alpha_i, \ i = 1, \dots, m,
\end{aligned}
\tag{36}
$$

where t is a target value the of objective function, and $\psi : \mathbf{R} \times [0, 1]^{m+1} \to \mathbf{R}$ is a function non-decreasing with all arguments.

Because necessity measure $N_{\Delta_i^\mathrm{T} x}(B_i)$ is strongly related to the inclusion relation between $\Delta_i^\mathrm{T} x$ and B_i, it can be seen as a degree of the inclusion. The treatment by $N_{\Delta_i^\mathrm{T} x}(B_i) \geq \alpha_i$ corresponds to the treatment by $\Delta_i^\mathrm{T} x \subseteq B_i$ in robust programming. Then the treatment of constraints is in the same spirit as robust programming. However, the treatment by $N_{\Delta_i^\mathrm{T} x}(B_i) \geq \alpha_i$ can control degree α_i. The larger α_i is, the more robust in the satisfaction of the i-th constraint is the solution.

To treat the objective function with uncertain coefficients, a target value t is introduced, and the objective function value is required to be not less than the target value with an appropriate certainty level. This requirement is expressed by $N_{\Gamma^{\mathrm{T}}x}([t, +\infty)) \geq \alpha_0$. The larger t is, the better the solution. The larger α_0 is, the more robust is the satisfaction of the requirement to be not less than t.

Then we have multiple objectives, maximizing α_i, t and α_0. The objective of Problem (36) is an aggregation of those objectives. In the literature (Inuiguchi et al. 1993; Inuiguchi and Ramík 2000), the following models are proposed:

Necessity measure optimization model: In this model, α_i, $i = 1, 2, \ldots, m$ and target value t are predetermined by the decision-maker as $\bar{\alpha}_i$, $i = 1, 2, \ldots, m$ and $\bar{t}$. Then ψ is defined by

$$\psi(t, \alpha_0, \alpha_1, \ldots, \alpha_m) = \begin{cases} \alpha_0, & \text{if } t \geq \bar{t} \text{ and } \alpha_i \geq \bar{\alpha}_i, i = 1, 2, \ldots, m, \\ 0, & \text{otherwise.} \end{cases} \tag{37}$$

Then the problem is reduced to

$$\begin{aligned} &\text{maximize } N_{\Gamma^{\mathrm{T}}x}([\bar{t}, \infty)), \\ &\text{subject to } N_{\Delta_i^{\mathrm{T}}x}(B_i) \geq \bar{\alpha}_i, \ i = 1, \ldots, m. \end{aligned} \tag{38}$$

Necessity fractile optimization model: In this model, α_i, $i = 0, 1, \ldots, m$ are predetermined by the decision-maker as $\bar{\alpha}_i$, $i = 0, 1, \ldots, m$. Then ψ is defined by

$$\psi(t, \alpha_0, \alpha_1, \ldots, \alpha_m) = \begin{cases} t, & \text{if } \alpha_i \geq \bar{\alpha}_i, i = 0, 1, \ldots, m, \\ 0, & \text{otherwise.} \end{cases} \tag{39}$$

Then the problem is reduced to

$$\begin{aligned} &\text{maximize } t, \\ &\text{subject to } N_{\Gamma^{\mathrm{T}}x}([t, \infty)) \geq \bar{\alpha}_0, \\ &\qquad\qquad N_{\Delta_i^{\mathrm{T}}x}(B_i) \geq \bar{\alpha}_i, \ i = 1, \ldots, m, \end{aligned} \tag{40}$$

Symmetric model: In this model, t is predetermined by the decision-maker as $\bar{t}$. Then ψ is defined by

$$\psi(t, \alpha_0, \alpha_1, \ldots, \alpha_m) = \begin{cases} \min_{i=0,1,\ldots,m} \alpha_i, & \text{if } t \geq \bar{t}, i = 0, 1, \ldots, m, \\ 0, & \text{otherwise.} \end{cases} \tag{41}$$

Then the problem is reduced to

$$\begin{aligned} &\text{maximize } \alpha, \\ &\text{subject to } N_{\Gamma^{\mathrm{T}}x}([\bar{t}, \infty)) \geq \alpha, \\ &\qquad\qquad N_{\Delta_i^{\mathrm{T}}x}(B_i) \geq \alpha, \ i = 1, \ldots, m, \end{aligned} \tag{42}$$

The necessity measure optimization and necessity fractile optimization models are conceptually illustrated in Fig. 10. In Fig. 10, level curves of the fuzzy region

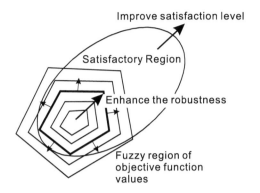

Fig. 10 Illustration of necessity measure optimization and necessity fractile optimization models

of objective function values $\varGamma^T x$ are depicted by pentagons, while the satisfactory region $[t, +\infty)$ is depicted by an ellipsoid (which is not the real shape, but will suffice for illustration). In the necessity measure optimization model, $\varGamma^T x$ is improved so that the fixed $[\bar{t}, +\infty)$ includes the lower α-level set of $\varGamma^T x$. In this way, the robustness is enhanced. On the other hand, in the necessity fractile optimization model, $\varGamma^T x$ and satisfaction level t are improved, keeping the condition that the fixed $\bar{\alpha}$-level set of $\varGamma^T x$ is included in the satisfactory region $[t, +\infty)$. The two models can be combined as in (Inuiguchi and Sakawa 1995b).

Considering the correspondence between probability measures and possibility/ necessity measures, constraints $N_{\varDelta_i^T x}(B_i) \geq \bar{\alpha}_i$, $i = 1, \ldots, m$ can be seen as counterparts of chance constraints in stochastic programming (Stancu-Minasian 1984). From this point of view, Problem (36) is called a modality constrained programming problem (Inuiguchi et al. 1993).

When $\varGamma$ is a vector of non-interactive fuzzy numbers, many papers (for example, Inuiguchi et al. 1993; Inuiguchi and Ramík 2000) show that Problems (38) and (40) are reduced to linear programming problems and that Problem (42) can be solved by a bisection method together with the linear programming technique. Inuiguchi and his collaborators (see Inuiguchi 2010) investigated more general $\varGamma$ without a great loss of linearity in the reduced problems.

Remark 3. In Problem (36), we only use necessity measures, but we can also introduce possibility measures. By introducing possibility measures, we can treat the decision-maker's wishes, for which we may seek only possibility rather than necessity. Moreover, by introducing generalized possibility and necessity measures, we may treat a variety of both the decision-maker's requirements on the constraints and objective function values while considering the uncertainty. For example, the necessity measure defined by (4) cannot treat the inclusion of fuzzy sets, $A \subseteq B$ defined by $\mu_A(z) \leq \mu_B(z)$, $\forall z$. However, using a suitable implication function, we can treat the inclusion by means of necessity measures (Inuiguchi and Tanino 2000).

Remark 4. To treat more general linear programming problems with fuzzy coefficients, we should consider constraints whose right-hand sides are also linear functions with fuzzy coefficients (Dubois 1987; Inuiguchi et al. 1993) because the transposition of fuzzy numbers from the right-hand side to the left-hand side is not always permitted. By such transposition, the meaning of the equation or the inequality constraint can be changed in some treatments of constraints. This phenomenon is related to the fact that equations with fuzzy numbers cannot always be solved (Dubois and Prade 1983).

4 Concluding Remarks

In this paper, we reviewed possibilistic linear programming approaches in view of robust optimization under fuzzy information. The optimizing approach has been investigated to obtain a feasible solution preserving optimality or sub-optimality from fluctuations of parameters within a certain range. On the other hand, the satisficing approach has been developed to obtain a feasible solution preserving feasibility or satisfaction from fluctuations of parameters within a certain range. The optimization approach does not require target values to objective functions, but does require expensive computational effort to solve the reduced problems. The satisficing approach often reduces problems to tractable ones but may require target values to objective functions.

The solution algorithms in the optimizing approach are not yet well developed. However, it is expected that global optimization techniques can solve the problems as proposed in discrete optimization cases (Kasperski 2008). In the satisficing approach, the solution algorithms are developing well. In the linear programming case of the satisficing approach, it would be a challenge to investigate to what extent we can generalize the fuzzy set and the necessity measure without loss of tractability. Other than those approaches, the two-stage approach (Inuiguchi and Tanino 2001b) is conceivable but not yet substantially developed.

References

Aissi, H., & Roy, B. (2010). Robustness in multi-criteria decision aiding. In M. Ehrgott, J. R. Figueira, & S. Greco (Eds.), *Trends in Multiple Criteria Decision Analysis*, pp. 87–122. New York: Springer.

Ben-Tal, A., & Nemirovski, A. (2002). Robust optimization – methodology and applications. *Mathematical Programming, Seriers B, 92*, 453–480.

Bitran, G. R. (1981). Linear multiple objective problems with interval coefficient. *Management Science, 26*, 694–706.

Derhy, M.-F. (2010). *Linear Programming, Sensitivity Analysis and Related Topics*. NJ: Prentice-Hall.

Dubois, D. (1987). Linear programming with fuzzy data, In J. C. Bezdek (Ed.), *Analysis of Fuzzy Information, Vol. III: Applications in Engineering and Science*, pp. 241–263. Boca Raton, FL: CRC Press.

Dubois, D., & Prade, H. (1983). Inverse operations for fuzzy numbers. In *Proc. of IFAC Symp. on Fuzzy Inform., Knowl. Represent. and Decis. Anal.*, pp. 399–404. Marseille.

Dubois, D., & Prade, H. (1988). *Possibility Theory: An Approach to Computerized Processing of Uncertainty*. New York and London: Plenum Press.

Dubois, D., & Prade, H. (1998). Possibility theory: Qualitative and quantitative aspects. In D. M. Gabbay, & P. Smets (Eds.), *Handbook of Defeasible Reasoning and Uncertainty Management Systems*, Vol. 1, pp. 169–226. Dordrecht: Kluwer.

Hisdal, E. (1978). Conditional possibilites independence and noninteraction. *Fuzzy Sets and Systems*, *1*, 299–309.

Inuiguchi, M. (2004). Enumeration of all possibly optimal vertices with possible optimality degrees in linear programming problems with a possibilistic objective function. *Fuzzy Optimization and Decision Making*, *3*, 311–326.

Inuiguchi, M. (2009). A semi-infinite programming approach to possibilistic optimization under necessity measure constraints, In *Proc. of 2009 IFSA World Congr. and 2009 EUSFLAT Conf. (IFSA/EUSFLAT 2009)*, Calouste Gulbenkian Foundation, pp.873–878 Lisbon, Portugal.

Inuiguchi, M. (2010). Approaches to linear programming problems with interactive fuzzy numbers. In W. A. Lodwick, & J. Kacprzyk (Eds.), *Fuzzy Optimization: Recent Developments and Applications*. New York: Springer.

Inuiguchi, M., Ichihashi, H., & Kume, Y. (1993). Modality constrained programming problems: A unified approach to fuzzy mathematical programming problems in the setting of possibility theory. *Information Sciences*, *67*, 93–126.

Inuiguchi, M., & Kume, Y. (1994). Minimax regret in linear programming problems with an interval objective function, In G. H. Tzeng, H. F. Wang, U. P. Wen, & P. L. Yu (Eds.), *Multiple Criteria Decision Making*, pp. 65–74. New York: Springer.

Inuiguchi, M., & Ramík, J. (2000). Possibilistic linear programming: A brief review of fuzzy mathematical programming and a comparison with stochastic programming in portfolio selection problem. *Fuzzy Sets and Systems*, *111*, 3–28.

Inuiguchi, M., Ramík, J., & Tanino, T. (2000). Oblique fuzzy vectors and their use in possibilistic linear programming. *Fuzzy Sets and Systems*, *135*, 123–150.

Inuiguchi, M., & Sakawa, M. (1994). Possible and necessary optimality tests in possibilistic linear programming problems. *Fuzzy Sets and Systems*, *67*, 29–46.

Inuiguchi, M., & Sakawa, M. (1995a). Minimax regret solutions to linear programming problems with an interval objective function. *European Journal of Operational Research*, *86*, 526–536.

Inuiguchi, M., & Sakawa, M. (1995b). A possibilistic linear program is equivalent to a stochastic linear program in a special case. *Fuzzy Sets and Systems*, *76*, 309–318.

Inuiguchi, M., & Sakawa, M. (1997). An achievement rate approach to linear programming problems with an interval objective function. *Journal of the Operational Research Society*, *48*, 25–33.

Inuiguchi, M., & Sakawa, M. (1998). Robust optimization under softness in a fuzzy linear programming problem. *International Journal of Approximate Reasoning*, *18*, 21–34.

Inuiguchi, M., & Tanino, T. (2000). A new class of necessity measures and fuzzy rough sets based on certainty qualifications, In *Proc. of the Second Int. Conf. on Rough Sets and Current Trends in Comput.* (RSCTC2000), pp. 223–230, Banff, Canada.

Inuiguchi, M., & Tanino, T. (2001a). On computation methods for a minimax regret solution based on outer approximation and cutting hyperplanes. *International Journal of Fuzzy Systems*, *3*, 548–557.

Inuiguchi, M., & Tanino, T. (2001b). Two-stage linear recourse problems under non-probabilistic uncertainty, In Y. Yoshida (Ed.), *Dynamical Aspects in Fuzzy Decision Making*, pp. 117–140. Heidelberg: Physica-Verlag.

Inuiguchi, M., Tanino, T., & Tanaka, H. (2001). Optimization approaches to possibilistic linear programming problems, In *Proc. of Joint 9th IFSA World Congr. and 20th NAFIPS Int. Conf.*, pp. 2724–2729. Vancouver.

Kasperski, A. (2008). *Discrete Optimization with Interval Data: Minimax Regret and Fuzzy Approach*. Berlin: Springer.

Luhandjula, M. K. (1987). Multiple objective programming problems with possibilistic coefficients. *Fuzzy Sets and Systems*, *21*, 135–145. (1987)

Mausser, H. E., & Laguna, M. (1998). A new mixed integer formulation for the maximum regret problem. *International Transactions in Operational Research, 5*, 389–403.

Mausser, H. E., & Laguna, M. (1999). A heuristic to minimax absolute regret for linear programs with interval objective function coefficients. European Journal of Operational Research, *117*, 157–174.

Negoita, C. V., Minoiu, S., & Stan, E. (1976). On considering imprecision in dynamic linear programming. *Economic Computation & Economic Cybernetics Studies & Research, 3*, 83–95.

Orlovsky, S. A. (1980). On formalization of a general fuzzy mathematical programming problem. *Fuzzy Sets and Systems, 3*, 311–321.

Słowinski, R. (1986). A multicriteria fuzzy linear programming method for water supply system development planning. *Fuzzy Sets and Syst. 19*, 217–237.

Słowinski, R. & Teghem, J. (eds.) (1990). *Stochastic versus Fuzzy Approaches to Multiobjective Mathematical Programming under Uncertainty*, Dordrecht: Kluwer Academic Publishers.

Soyster, A. L. (1979). Inexact linear programming with generalized resource sets. *European Journal of Operational Research, 3*, 316–321.

Stancu-Minasian, I. M. (1984). *Stochastic Programming with Multiple Objective Functions*. Dordrecht: D. Reidel Publishing Company.

Steuer, R. E. (1981). Algorithms for linear programming problems with interval objective function coefficients. *Mathematics of Operations Research, 6*, 333–348.

Tanaka, H., & Asai, K. (1984). Fuzzy linear programming probelms with fuzzy numbers. *Fuzzy Sets and Systems, 13*, 1–10.

Tanaka, H., Ichihashi, H., & Asai, K. (1984). A formulation of fuzzy linear programming problem based on comparison of fuzzy numbers. *Control and Cybernetics, 13*, 185–194.

Yager, R. R. (2001). Determining equivalent values for possibilistic variables. *IEEE Transactions on Systems, Man, and Cybernetics, Part B, Cybernetics, 31*(1), 19–31.

Zadeh, L. A. (1978). Fuzzy sets as a basis for a theory of possibility. *Fuzzy Sets and Systems, 1*, pp. 3–28.

Various Types of Objective Functions of Clustering for Uncertain Data

Yasunori Endo and Sadaaki Miyamoto

Abstract Whenever we classify a dataset into some clusters, we need to consider how to handle the uncertainty included into data. In those days, the ability of computers were very poor, and we could not help handling data with uncertainty as one point. However, the ability is now enough to handle the uncertainty of data, and we hence believe that we should handle the uncertain data as is. In this paper, we will show some clustering methods for uncertain data by two concepts of "tolerance" and "penalty-vector regularization". The both concepts are more useful to model and handle the uncertainty of data and more flexible than the conventional methods. By the way, we construct a clustering algorithm by putting one objective function. Hence, we can say that the whole clustering algorithm depends on its objective function. In this paper, we will thereby introduce various types of objective functions for uncertain data with the concepts of tolerance and penalty-vector regularization, and construct the clustering algorithms for uncertain data.

1 Introduction

Clustering methods are known as very useful tools in many fields for data mining and we can find the construction of datasets through the clustering methods.

Now, the more ability of computers increase, the more works for uncertainty have been studied. In the past, each datum handled by the computers was approximately represented as one point or value because of poor ability of the computers. However, the ability is now enough to handle the data with uncertainty called uncertain data and many researchers have tried to handle original data from the viewpoint that the datum should been represented as not one point approximately but certain

Y. Endo (✉) · S. Miyamoto
Department of Risk Engineering, Faculty of Systems and Information Engineering,
University of Tsukuba, Tsukuba, Japan
e-mail: endo@risk.tsukuba.ac.jp; miyamoto@risk.tsukuba.ac.jp

Y. Ermoliev et al. (eds.), *Managing Safety of Heterogeneous Systems*, Lecture Notes
in Economics and Mathematical Systems 658, DOI 10.1007/978-3-642-22884-1_12,
© Springer-Verlag Berlin Heidelberg 2012

distribution exactly in a data space. The goal of this paper is to show some clustering methods for uncertain data.

Whenever we construct the clustering methods for the uncertain data, we have one problem, that is, how should we represent the uncertainty of data?

The easiest way to represent the uncertain data is to use the concept of interval, i.e., $[\underline{x}, \overline{x}]$. This methodology is deterministic in the meaning that each interval datum is handled as one point by introducing a measure or a dissimilarity between intervals, e.g., minimum distance, maximum distance and Hausdorff distance. However, the way has the following problems.

1. The dissimilarity between data, which plays very important role to construct clustering algorithms, is defined on distance. There are many distances between intervals but we don't have any guide to select the suitable distance. Moreover, some distances between intervals don't satisfy the axiom of metric.
2. We often construct clustering methods through the methodology of optimization. If we represent each datum as an interval, it becomes difficult to formulate the problem in the framework of the optimization.

To solve the above problems, another way has been presented in which the uncertainty is represented by certain probabilistic density function (PDF). In this way, the clustering is regarded as a method to determine some parameters of the PDF and hence the way can be called parametric. However, the methodology has other problems as follows.

1. The type of PDF determines not only the shape of uncertainty of data but also the detail of the clustering algorithm. However, we don't have clear indication to select which type of PDFs to use.
2. Even if we select one type of PDF, we have no belief that the selected PDF is suitable to represent the uncertainty of data.

To solve the above problems, we have proposed "tolerance" of an convenient tool to handle uncertain data and applied some of clustering algorithms (Endo et al. 2005, 2008; Hasegawa et al. 2008; Kanzawa et al. 2007, 2008; Murata et al. 2006). In our proposed tolerance, tolerance vectors (Endo et al. 2005) and penalty ones (Hasegawa et al. 2008) play main role. Each uncertain datum is allowed to allocate any position by those vectors as far as the constraints for those vectors are satisfied and the position is derived as an optimal solution of a given objective function. Hence, we can say that this concept is in the framework of methodology of soft computing. Penalty vectors are similar to tolerance ones and the methods using penalty-vector regularization become more flexible than tolerance vectors because no constraint for the vectors is needed. Moreover, the concept has been developed by using kernel trick in (Kanzawa et al. 2008). The method can classify datasets which consist of clusters of uncertain data with nonlinear boundary.

In this paper, we will show some clustering methods for uncertain data with two concepts of tolerance and penalty-vector regularization. The procedures to construct these methods are based on fuzzy c-means (FCM), i.e,

Step 1 An objective function is defined.
Step 2 Some solutions which minimize the objective function are derived.
Step 3 An clustering algorithm is constructed using the solutions.

It means that the whole algorithm depends on the objective function, and we will thereby introduce various types of objective functions for uncertain data to construct the clustering algorithms. We notice that the discussion is mathematical and there are no numerical examples.

2 FCM-Based Clustering with Tolerance

Fuzzy c-means (FCM) clustering is one of the most typical and effective methods. In this section, we try to introduce the concept of tolerance into FCM and show some objective functions of FCM-based clustering with tolerance which handles uncertain data. First, we introduce basic concept of tolerance. Second, we explain the conventional FCM. Third, we show some objective functions of FCM-based clustering with tolerance.

2.1 Basic Concept of Tolerance

In general, a datum $x \in \Re^p$ with uncertainty is presented by some interval, i.e.,

$$[\underline{x}, \overline{x}] = [(\underline{x}_1, \ldots, \underline{x}_p)^T, (\overline{x}_1, \ldots, \overline{x}_p)^T] \subset \Re^p.$$

In our proposed tolerance, such a datum is represented by

$$x + \varepsilon = (x_1, \ldots, x_p)^T + (\varepsilon_1, \ldots, \varepsilon_p)^T \in \Re^p$$
$$= (x_1 + \varepsilon_1, \ldots, x_p + \varepsilon_p)^T$$

and a constraint for ε_j like that

$$|\varepsilon_j| \leq \xi_j.$$

A vector $\varepsilon = (\varepsilon_1, \ldots, \varepsilon_p)^T \in \Re^p$ is called tolerance vector. If we assume that

$$\begin{cases} x_j = \dfrac{\overline{x}_j + \underline{x}_j}{2}, \\ \xi_j = \dfrac{|\overline{x}_j - \underline{x}_j|}{2}, \end{cases}$$

the formulation is equivalent to the above interval.

This concept of tolerance is very useful in the reason that we can handle uncertain data in the framework of optimization to use the concept without introducing some particular measure between intervals. For example, let's consider calculation of distance $d(X, Y)$ between $X = [\underline{x}, \overline{x}]$ and $Y = [\underline{y}, \overline{y}]$. We have to introduce some measure between intervals to calculate it, e.g.,

$$\begin{cases} d_{\min}(X,Y) = \min\{\|\underline{y} - \overline{x}\|, \|\overline{y} - \underline{x}\|\}, & \text{(minimum distance)} \\ d_{\max}(X,Y) = \max\{\|\underline{y} - \overline{x}\|, \|\overline{y} - \underline{x}\|\}, & \text{(maximum distance)} \\ d_{\text{Hausdorff}}(X,Y) = \max\{\|\overline{y} - \overline{x}\|, \|\underline{y} - \underline{x}\|\}. & \text{(Hausdorff distance)} \end{cases}$$

However, if we use tolerance, we don't need any particular distance, that is, a distance $d(X, Y)$ between $X = x + \varepsilon_x$ ($\|\varepsilon_x\| \leq \xi_x$) and $Y = y + \varepsilon_y$ ($\|\varepsilon_y\| \leq \xi_y$) can be calculated as $\|(x - y) + (\varepsilon_x - \varepsilon_y)\|$. From the above, we know that this tool is useful when we handle the data, especially data with missing values of their attributes, in the framework of optimization like as fuzzy c-means clustering (Endo et al. 2008).

Here, we can consider two types of constraints of tolerance, that is,

$$\|\varepsilon\|^2 \leq \kappa^2 \tag{1}$$

and

$$|\varepsilon_j| \leq \kappa_j. \tag{2}$$

We call the above "hyper-sphere type" and the below "hyper-rectangle type". We show two examples of tolerance, an example of hyper-sphere type tolerance on two dimensional Euclidean space: $\|\varepsilon_k\|^2 \leq \kappa_k^2$ in Fig. 1 and one of hyper-rectangle type tolerance on two dimensional Euclidean space: $|\varepsilon_{kj}| \leq \kappa_{kj}$ in Fig. 2. If the data

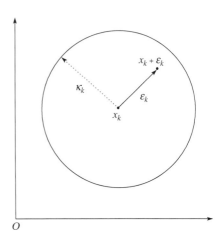

Fig. 1 An example hyper-sphere type tolerance on the two dimensional Euclidean space: $\|\varepsilon_k\|^2 \leq \kappa_k^2$

Various Types of Objective Functions of Clustering for Uncertain Data

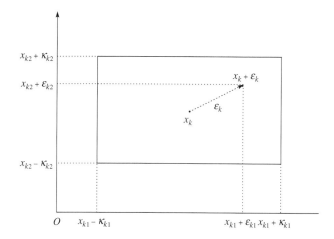

Fig. 2 An example of hyper-rectangle tolerance on the two dimensional Euclidean space: $|\varepsilon_{kj}| \leq \kappa_{kj}$

space is a color space, for example, one axis may be the color of red and the other be the color of blue.

Now, we define some symbols for the following discussion. n and c are numbers of data and clusters. x_k, ε_k and κ_k mean the k-th datum in $\Re^p$, the tolerance vector of x_k and the tolerance range of ε_k, respectively. v_i and u_{ki} mean the cluster center of the i-th cluster in $\Re^p$ and belongingness of x_k for v_i in $[0, 1]$. Moreover, $\|\bullet\|_2$ and $\|\bullet\|_1$ represent L_2- and L_1-norms.

2.2 Fuzzy c-Means

As mentioned above, clustering algorithms based on FCM are constructed according to the procedure in Sect. 1. Here, we show the procedure to construct the algorithm of FCM.

Step 1 An objective function is defined as follows:

$$J_{\text{FCM}} = \sum_{k=1}^{n}\sum_{i=1}^{c}(u_{ki})^m d_{ki} \quad (d_{ki} = (\|x_k - v_i\|_2)^2) \tag{3}$$

m ($m > 1$) is a fuzzification parameter. The constraint is as follows:

$$\sum_{i=1}^{c} u_{ki} = 1, \quad \forall k. \tag{4}$$

Step 2 The solutions which minimize the objective function are derived as follows:

$$\begin{cases} u_{ki} = \dfrac{(1/d_{ki})^{\frac{1}{m-1}}}{\sum_{l=1}^{c}(1/d_{kl})^{\frac{1}{m-1}}} \\[4mm] v_i = \dfrac{\sum_{k=1}^{n}(u_{ki})^m x_k}{\sum_{k=1}^{n}(u_{ki})^m} \end{cases}$$

Step 3   An clustering algorithm is constructed using the solutions as Algorithm 1.

Algorithm 1 (FCM)

FCM1   *Give the initial cluster centers $V = \{v_i\}$.*
FCM2   *Update $U = \{u_{ki}\}$ by the optimal solution of u_{ki} on fixing V.*
FCM3   *Update V by the optimal solution of v_i on fixing U.*
FCM4   *Stop if a convergence criterion satisfies. Otherwise go back to* **FCM2***.*

We can consider some convergence criteria, for example,

- an iteration time $L \geq L_0$,
- $\max_{k,i} |u_{ki}^{(L+1)} - u_{ki}^{(L)}| \leq \eta$,
- $\max_{k,i} \left(\dfrac{|u_{ki}^{(L+1)} - u_{ki}^{(L)}|}{|u_{ki}^{(L)}|} \right) \leq \eta$,
- $\sum_{i=1}^{c} \sum_{k=1}^{n} |u_{ki}^{(L+1)} - u_{ki}^{(L)}| \leq \eta$,
- $\sum_{i=1}^{c} \sum_{k=1}^{n} \left(\dfrac{|u_{ki}^{(L+1)} - u_{ki}^{(L)}|}{|u_{ki}^{(L)}|} \right) \leq \eta$,
- $\max_i \|v_i^{(L+1)} - v_i^{(L)}\| \leq \eta$.

From the above, we can see that the objective function plays essential role. We note that the calculation results depend on the initial values and this problem is essential and very difficult to solve.

2.3   Fuzzy c-Means with Tolerance

2.3.1   Standard Model on L_2-Norm

In this section, we show FCM clustering in which the concept of tolerance is introduced to handle the uncertain data. The clustering is called FCM with tolerance (FCMT). The objective function is as follows:

$$J_{\text{FCMT}} = \sum_{k=1}^{n} \sum_{i=1}^{c} (u_{ki})^m d_{ki}. \quad (d_{ki} = (\|x_k + \varepsilon_k - v_i\|_2)^2) \tag{5}$$

Various Types of Objective Functions of Clustering for Uncertain Data

This function is presented by replacing x_k of one of FCM by $x_k + \varepsilon_k$. The constraints are (4), and (1) or (2).

The optimal solutions which minimize the above function are derived by Lagrange multiplier. We show how to derive the solutions with the constraint (2). We note that the solutions with the constraint (1) can be derived in the same way.

We introduce the following Lagrange function to solve the optimization problem:

$$L_{\text{FCMT}} = J_{\text{FCMT}} + \sum_{k=1}^{n} \gamma_k \left(\sum_{i=1}^{c} u_{ki} - 1 \right) + \sum_{k=1}^{n} \sum_{j=1}^{p} \psi_{kj} \left(\varepsilon_{kj}^2 - \kappa_{kj}^2 \right).$$

From the Kuhn-Tucker condition, the necessary conditions are as follows:

$$
\begin{cases}
\dfrac{\partial L_{\text{FCMT}}}{\partial v_{ij}} = 0, \ \dfrac{\partial L_{\text{FCMT}}}{\partial u_{ki}} = 0, \ \dfrac{\partial L_{\text{FCMT}}}{\partial \varepsilon_{kj}} = 0, \\[2ex]
\dfrac{\partial L_{\text{FCMT}}}{\partial \gamma_k} = 0, \ \dfrac{\partial L_{\text{FCMT}}}{\partial \psi_{kj}} \leq 0, \ \psi_{kj} \dfrac{\partial L_{\text{FCMT}}}{\partial \psi_{kj}} = 0, \ \psi_{kj} \geq 0.
\end{cases}
\tag{6}
$$

From the convexity of J_{FCMT}, it is sufficient to consider the case of (6).

First, we derive the optimal solution of u_{ki}.

$$\frac{\partial L_{\text{FCMT}}}{\partial u_{ki}} = m(u_{ki})^{m-1} \| x_k + \varepsilon_k - v_i \|^2 + \gamma_k = 0. \tag{7}$$

Thus, we have

$$u_{ki} = \left(\frac{-\gamma_k}{m \| x_k + \varepsilon_k - v_i \|^2} \right)^{\frac{1}{m-1}}.$$

In addition, from the constraints (4), we have

$$\sum_{l=1}^{c} \left(\frac{-\gamma_k}{m \| x_k + \varepsilon_k - v_l \|^2} \right)^{\frac{1}{m-1}} = 1. \tag{8}$$

From (7) and (8), we have the following optimal solution of u_{ki}:

$$u_{ki} = \frac{(1/d_{ki})^{\frac{1}{m-1}}}{\sum_{l=1}^{c} (1/d_{kl})^{\frac{1}{m-1}}}. \tag{9}$$

Second, we derive the optimal solution of v_{ij}.

$$\frac{\partial L_{\text{FCMT}}}{\partial v_{ij}} = -\sum_{k=1}^{n} 2(u_{ki})^m (x_{kj} + \varepsilon_{kj} - v_{ij}) = 0.$$

Thus, we have the following optimal solution of v_{ij}:

$$v_{ij} = \frac{\sum_{k=1}^{n} (u_{ki})^m (x_{kj} + \varepsilon_{kj})}{\sum_{k=1}^{n} (u_{ki})^m}. \tag{10}$$

Third, we derive the optimal solution of ε_{kj}.

$$\frac{\partial L_{\text{FCMT}}}{\partial \varepsilon_{kj}} = \sum_{i=1}^{c} 2(u_{ki})^m (x_{kj} + \varepsilon_{kj} - v_{ij}) + 2\psi_{kj}\varepsilon_{kj} = 0.$$

Thus, we have

$$\varepsilon_{kj} = \frac{-\sum_{i=1}^{c} (u_{ki})^m (x_{kj} - v_{ij})}{\sum_{i=1}^{c} (u_{ki})^m + \psi_{kj}}. \tag{11}$$

On the other hand,

$$\psi_{kj} \frac{\partial L_{\text{FCMT}}}{\partial \psi_{kj}} = \psi_{kj}(\varepsilon_{kj}^2 - \kappa_{kj}^2) = 0.$$

Therefore, we have to consider two cases, $\psi_{kj} = 0$ and $\varepsilon_{kj}^2 = \kappa_{kj}^2$. First, we consider the case of $\psi_{kj} = 0$.

$$\frac{\partial L_{\text{FCMT}}}{\partial \varepsilon_{kj}} = \sum_{i=1}^{c} 2(u_{ki})^m (x_{kj} + \varepsilon_{kj} - v_{ij}) = 0.$$

Thus, we have

$$\varepsilon_{kj} = \frac{-\sum_{i=1}^{c} (u_{ki})^m (x_{kj} - v_{ij})}{\sum_{i=1}^{c} (u_{ki})^m}. \tag{12}$$

Next, we consider the case of $\varepsilon_{kj}^2 = \kappa_{kj}^2$. From (11) and $\varepsilon_{kj}^2 = \kappa_{kj}^2$, we have

$$\varepsilon_{kj}^2 = \left\{ \frac{-\sum_{i=1}^{c} (u_{ki})^m (x_{kj} - v_{ij})}{\sum_{i=1}^{c} (u_{ki})^m + \psi_{kj}} \right\}^2 = \kappa_{kj}^2.$$

Thus, we have

$$\sum_{i=1}^{c} (u_{ki})^m + \psi_{kj} = \pm \frac{\left| \sum_{i=1}^{c} (u_{ki})^m (x_{kj} - v_{ij}) \right|}{\kappa_{kj}}.$$

From $\psi_{kj} \geq 0$, the right side is positive. Therefore, from (11) we get

$$\varepsilon_{kj} = -\frac{\kappa_{kj} \sum_{i=1}^{c} (u_{ki})^m (x_{kj} - v_{ij})}{\left| \sum_{i=1}^{c} (u_{ki})^m (x_{kj} - v_{ij}) \right|}. \tag{13}$$

Various Types of Objective Functions of Clustering for Uncertain Data 249

Finally, we have the optimal solution of ε_{kj} which satisfies (12) and (13) as follows:

$$
\begin{cases}
\varepsilon_{kj} = -\alpha_{kj} \sum_{i=1}^{c} (u_{ki})^m (x_{kj} - v_{ij}), \\
\alpha_{kj} = \min \left\{ \dfrac{\kappa_{kj}}{|\sum_{i=1}^{c} (u_{ki})^m (x_{kj} - v_{ij})|}, \dfrac{1}{\sum_{i=1}^{c} (u_{ki})^m} \right\}.
\end{cases}
\tag{14}
$$

We can construct the algorithm of FCMT using the above optimal solutions as follows:

Algorithm 2 (FCMT)

FCMT1   *Give the initial cluster centers* $V = \{v_i\}$ *and initial tolerance vectors* $E = \{\varepsilon_k\}$.
FCMT2   *Update* $U = \{u_{ki}\}$ *by the optimal solution of* u_{ki} *on fixing* V *and* E.
FCMT3   *Update* V *by the optimal solution of* v_i *on fixing* E *and* U.
FCMT4   *Update* E *by the optimal solution of* ε_k *on fixing* U *and* V.
FCMT5   *Stop if a convergence criterion satisfies. Otherwise go back to* **FCMT2**.

2.3.2   Entropy-Based Model on L_2-Norm

Instead of (5), we can consider the following objective function:

$$
J_{\text{eFCMT}} = \sum_{k=1}^{n} \sum_{i=1}^{c} u_{ki} d_{ki} + \lambda^{-1} \sum_{k=1}^{n} \sum_{i=1}^{c} u_{ki} \log u_{ki}. \quad (d_{ki} = \|x_k + \varepsilon_k - v_i\|_2^2)
$$

This function is based on the objective function of entropy-based FCM (Miyamoto and Mukaidono 1997) and we call the clustering entropy-based FCMT (eFCMT). The constraints are (4), and (1) or (2). We have the following optimal solutions under the constraint (2) by the similar procedure as FCMT and we can use Algorithm 2.

$$
\begin{cases}
u_{ki} = \dfrac{e^{-\lambda \|x_k + \varepsilon_k - v_i\|_2^2}}{\sum_{l=1}^{c} e^{-\lambda \|x_k + \varepsilon_k - v_l\|_2^2}}, & (15) \\[3mm]
v_{ij} = \dfrac{\sum_{k=1}^{n} u_{ki} (x_{kj} + \varepsilon_{kj})}{\sum_{k=1}^{n} u_{ki}}, & (16) \\[3mm]
\begin{cases}
\varepsilon_{kj} = -\alpha_{kj} (x_{kj} - \sum_{i=1}^{c} u_{ki} v_{ij}), \\
\alpha_{kj} = \min \left\{ \dfrac{\kappa_{kj}}{|x_{kj} - \sum_{i=1}^{c} u_{ki} v_{ij}|}, 1 \right\}.
\end{cases} & (17)
\end{cases}
$$

2.3.3 Standard Model on L_1-Norm

Here, we consider FCM with tolerance on L_1-norm (FCMT-L_1). The objective function is given by replacing L_2-norm by L_1-norm in the objective function of FCMT (5) as follows:

$$J_{\text{FCMT-}L_1} = \sum_{k=1}^{n}\sum_{i=1}^{c}(u_{ki})^m d_{ki}. \quad (d_{ki} = \|x_k + \varepsilon_k - v_i\|_1) \tag{18}$$

The constraints are (4) and (2).

This function cannot be differentiated by ε_k and v_i so that we have to another way to find the optimal solutions.

First, we can derive the optimal solution of u_{ki} by using Lagrange function as follows:

$$u_{ki} = \frac{(1/d_{ki})^{\frac{1}{m-1}}}{\sum_{l=1}^{c}(1/d_{kl})^{\frac{1}{m-1}}}. \quad (d_{ki} = \|x_k + \varepsilon_k - v_i\|_1) \tag{19}$$

Second, we describe the method (based on (Jajuga 1991)) to obtain the optimal solution of v_{ij}. From (18), we can consider the following semi-objective function:

$$J_{ij}(v_{ij}) = \sum_{k=1}^{n}(u_{ki})^m |x_{kj} + \varepsilon_{kj} - v_{ij}|. \tag{20}$$

If the semi-objective function (20) is minimized, the whole objective function (18) is also minimized. According to the following procedures, the optimal solution of v_{ij} is calculated.

Algorithm 3 (Calculation of v_{ij})

Step 1 *All data are sorted in ascending order in each dimension.*

$$x_{1j} + \varepsilon_{1j}, \ldots, x_{nj} + \varepsilon_{nj}$$

$$\downarrow Sorting$$

$$x_{q(1)j} + \varepsilon_{q(1)j} \leq \cdots \leq x_{q(n)j} + \varepsilon_{q(n)j}$$

where $q(k)$ is substitution of $(1, \ldots, n)$.

Step 2 *We calculates as follows.*

$$S = -\frac{1}{2}\sum_{k=1}^{n}(u_{ki})^m.$$

Step 3 *It starts from $r = 0$ and the following calculations are repeated between $S < 0$.*

Various Types of Objective Functions of Clustering for Uncertain Data

$$r := r + 1,$$
$$S := S + (u_{q(r)i})^m.$$

Step 4　*From the above calculation, we obtain*

$$v_{ij} = x_{q(r)j} + \varepsilon_{q(r)j}. \tag{21}$$

Third, we consider the way to obtain the optimal solution of ε_{kj}. The procedure is as same as v_{ij}.

Algorithm 4 (Calculation of ε_{kj})

Step 1　*Data are sorted in ascending order in each dimension.*

$$v_{1j} - x_{kj}, \dots, v_{cj} - x_{kj}$$

$$\downarrow Sorting$$

$$v_{q(1)j} - x_{kj} \le \dots \le v_{q(c)j} - x_{kj}$$

where $q(i)$ is substitution of $(1, \dots, c)$.
Step 2　*We calculates as follows.*

$$S = -\frac{1}{2} \sum_{i=1}^{c} (u_{ki})^m.$$

Step 3　*It starts from $r = 0$ and the following calculations are repeated between $S < 0$.*

$$r := r + 1,$$
$$S := S + (u_{kq(r)})^m.$$

Step 4　*From the above calculation, we obtain*

$$\varepsilon_{kj} = \text{sign}(v_{q(r)j} - x_{kj}) \times \min\{|v_{q(r)j} - x_{kj}|, \kappa_{kj}\}. \tag{22}$$

We can construct the algorithm of FCMT-L_1 using the above way to calculate the optimal solutions. The algorithm is as same as Algorithm 2. However, we notice that the updating processes V and E need the above algorithms 3 and 4, respectively. We show the algorithm as follows:

Algorithm 5 (FCMT-L_1)

FCMT-$L_1$1   *Give the initial cluster centers $V = \{v_i\}$ and initial tolerance vectors $E = \{\varepsilon_k\}$.*
FCMT-$L_1$2   *Update $U = \{u_{ki}\}$ by the optimal solution of u_{ki} on fixing V and E.*
FCMT-$L_1$3   *Update V by the calculation algorithm of v_i on fixing E and U.*
FCMT-$L_1$4   *Update E by the calculation algorithm of ε_k on fixing U and V.*
FCMT-$L_1$5   *Stop if a convergence criterion satisfies. Otherwise go back to FCMT-$L_1$2.*

2.3.4   Entropy-Based Model on L_1-Norm

Instead of (18), we can consider the following objective function:

$$J_{\text{eFCMT-}L_1} = \sum_{k=1}^{n}\sum_{i=1}^{c} u_{ki} d_{ki} + \lambda^{-1}\sum_{k=1}^{n}\sum_{i=1}^{c} u_{ki} \log u_{ki}. \quad (d_{ki} = \|x_k + \varepsilon_k - v_i\|_1)$$

We call the clustering based on the above objective function entropy-based FCMT-L_1 (eFCMT-L_1). The constraints are (4) and (2).

The optimal solution can be obtained in the same way as FCMT-L_1, that is,

$$u_{ki} = \frac{e^{-\lambda\|x_k + \varepsilon_k - v_i\|_1}}{\sum_{l=1}^{c} e^{-\lambda\|x_k + \varepsilon_k - v_l\|_1}}. \tag{23}$$

Also, we use Algorithm 3 and Algorithm 4 with $m = 1$ to calculate the optimal solutions of v_{ij} and ε_{kj}, respectively. The algorithm of eFCMT-L_1 is as same as Algorithm 5.

3   FCM-Based Clustering Using Penalty-Vector Regularization

In the above section, we introduced the concept of tolerance to handle the uncertainty of data and some objective functions to construct clustering algorithms. However, there are many cases in which the uncertainty can not be estimated. When we use the tolerance, we have to determine the value of κ_k so that it is difficult to introduce the concept in these cases. Thus, we have to consider another way of soft computing to handle the uncertainty. First, we introduce basic concept of penalty-vector regularization to do that. Second, we show some objective functions of FCM-based clustering using penalty-vector regularization.

Various Types of Objective Functions of Clustering for Uncertain Data   253

3.1 Basic Concept of Penalty-Vector Regularization

Here, we introduce a new concept of penalty-vector regularization. The concept is similar to the concept of tolerance and penalty vectors play important role. However, the difference from the concept of tolerance is that there is no constraint of penalty vectors.

We define some symbols at the beginning. In addition to the symbols in the above section, we define penalty vector $\delta_k = (\delta_{k1}, \ldots, \delta_{kp})^T \in \Re^p$, and a set of penalty vectors $\Delta = \{\delta_1, \ldots, \delta_n\}$. The uncertain datum is represented as $x_k + \delta_k$. In addition, we define weighting coefficient w_{klj} $(w_{klj} \geq 0)$ and weighting matrix as follows:

$$W_k = \begin{pmatrix} w_{k11} & \cdots & w_{k1p} \\ \vdots & \ddots & \vdots \\ w_{kp1} & \cdots & w_{kpp} \end{pmatrix}. \tag{24}$$

One of the simplest form of the matrix is as follows:

$$W_k = \begin{pmatrix} w_{k1} & & 0 \\ & \ddots & \\ 0 & & w_{kp} \end{pmatrix}. \tag{25}$$

Now, we introduce the following penalty term:

$$\sum_{k=1}^{n} \delta_k{}^T W_k \delta_k = \sum_{j=1}^{p} \sum_{l=1}^{p} w_{klj} \delta_{kl} \delta_{kj}. \tag{26}$$

We assume that W_k is a symmetric matrix, i.e., $w_{kj} = w_{jk}$. In case that W_k is a diagonal matrix, the penalty term is represented as follows:

$$\sum_{k=1}^{n} \delta_k{}^T W_k \delta_k = \sum_{k=1}^{n} \sum_{j=1}^{p} w_{kjj} \left(\delta_{kj} \right)^2.$$

3.2 Fuzzy c-Means Using Penalty-Vector Regularization

3.2.1 Standard Model on L_2-Norm

We add the penalty term (26) to the objective function of FCM (3) and obtain the following objective function.

$$J_{\text{FCMP}} = \sum_{k=1}^{n} \sum_{i=1}^{c} (u_{ki})^m d_{ki} + \sum_{k=1}^{n} \delta_k{}^T W_k \delta_k. \quad (d_{ki} = \|x_k + \delta_k - v_i\|_2^2) \tag{27}$$

The more it approaches $(x_k + \delta_k)$ by v_i, the more the first term of (27) becomes small. On the other hand, the penalty term of (27) grows in proportion to squared δ_{kj}. Therefore, the bigger w_{kj} is, the smaller the optimal solution δ_{kj} which minimizes (27) becomes. Oppositely, the smaller w_{kj} is, the bigger the optimal solution δ_{kj} which minimizes (27) becomes. Hence, we know that W_k means the uncertainty of each datum x_k.

Our goal of this section is to derive optimal solutions U, V and Δ which are minimize the objective function (27) under the constraint (4).

First, we consider to derive optimal solutions of U. We introduce the following Lagrange function to do that.

$$L_{\text{FCMP}} = J_{\text{FCMP}} + \sum_{k=1}^{n} \gamma_k \left(\sum_{i=1}^{c} u_{ki} - 1 \right).$$

$$\frac{\partial L_{\text{FCMP}}}{\partial u_{ki}} = m(u_{ki})^{m-1} d_{ki} + \gamma_k = 0. \tag{28}$$

Thus, we have

$$u_{ki} = \left(\frac{-\gamma_k}{m d_{ki}} \right)^{\frac{1}{m-1}}.$$

In addition, from the constraint (4), we have

$$\sum_{l=1}^{c} \left(\frac{-\gamma_k}{m d_{ki}} \right)^{\frac{1}{m-1}} = 1. \tag{29}$$

From (28) and (29), we have

$$u_{ki} = \frac{(1/d_{ki})^{\frac{1}{m-1}}}{\sum_{l=1}^{c} (1/d_{kl})^{\frac{1}{m-1}}}.$$

Second, we consider to derive optimal solutions of V.

$$\frac{\partial L_{\text{FCMP}}}{\partial v_{ij}} = - \sum_{k=1}^{n} 2(u_{ki})^m (x_{kj} + \delta_{kj} - v_{ij}) = 0.$$

Thus, we have

$$v_{ij} = \frac{\sum_{k=1}^{n} (u_{ki})^m (x_{kj} + \delta_{kj})}{\sum_{k=1}^{n} (u_{ki})^m}.$$

Last, we consider to derive Δ.

$$\frac{\partial L_{FCMP}}{\partial \delta_{kj}} = \frac{\partial}{\partial \delta_{kj}} \left(\sum_{k=1}^{n} \sum_{i=1}^{n} \sum_{j=1}^{p} (u_{ki})^m (x_{kj} + \delta_{kj} - v_{ij})^2 + \sum_{l=1}^{p} \sum_{j=1}^{p} w_{klj} \delta_{kl} \delta_{kj} \right)$$

$$= 2 \sum_{i=1}^{c} (u_{ki})^m (x_{kj} + \delta_{kj} - v_{ij}) + 2 \sum_{l=1}^{p} w_{klj} \delta_{kl}$$

$$= 0.$$

Thus, we get the following equation:

$$\left(\sum_{i=1}^{c} (u_{ki})^m \right) \delta_{kj} + \sum_{l=1}^{p} w_{klj} \delta_{kl} + \sum_{i=1}^{c} (u_{ki})^m (x_{kj} - v_{ij}) = 0.$$

The above equation holds for any j $(1 \leq j \leq p)$. Hence,

$$\left(\sum_{i=1}^{c} (u_{ki})^m \right) \begin{pmatrix} \delta_{k1} \\ \vdots \\ \delta_{kp} \end{pmatrix} + \begin{pmatrix} w_{k11} & \cdots & w_{kp1} \\ \vdots & \ddots & \vdots \\ w_{k1p} & \cdots & w_{kpp} \end{pmatrix} \begin{pmatrix} \delta_{k1} \\ \vdots \\ \delta_{kp} \end{pmatrix}$$

$$+ \begin{pmatrix} \sum_{i=1}^{c} (u_{ki})^m (x_{k1} - v_{i1}) \\ \vdots \\ \sum_{i=1}^{c} (u_{ki})^m (x_{kp} - v_{ip}) \end{pmatrix} = 0. \qquad (30)$$

We put some symbols as follows.

$$\begin{cases} A_k = \left(\sum_{i=1}^{c} (u_{ki})^m \right) I + W_k^T, \\ \\ B_k = \begin{pmatrix} \sum_{i=1}^{c} (u_{ki})^m (x_{k1} - v_{i1}) \\ \vdots \\ \sum_{i=1}^{c} (u_{ki})^m (x_{kp} - v_{ip}) \end{pmatrix}. \end{cases}$$

Here, I is a unit matrix.

Using the symbol A_k and B_k, (30) can be rewritten as follows.

$$A_k \delta_k + B_k = 0.$$

Finally, we can get the following solution.

$$\delta_k = -(A_k)^{-1} B_k.$$

We need regularization of A_k for existence of optimal solutions Δ.
We can use Algorithm 2 as FCMP with the above optimal solutions.

3.2.2 Entropy-Based Model on L_2-Norm

Instead of (27), we can consider the following objective function:

$$J_{\text{eFCMP}} = \sum_{k=1}^{n}\sum_{i=1}^{c} u_{ki} d_{ki} + \lambda^{-1} \sum_{k=1}^{n}\sum_{i=1}^{c} u_{ki} \log u_{ki} + \sum_{k=1}^{n} \delta_k{}^T W_k \delta_k. \qquad (31)$$

$$(d_{ki} = \|x_k + \delta_k - v_i\|_2^2)$$

This function is based on the objective function of entropy-based FCM (Miyamoto and Mukaidono 1997) and we call the clustering entropy-based FCMP (eFCMP). The constraints are (4). We have the following optimal solutions by the similar procedure as FCMP and we can use Algorithm 2.

$$\begin{cases} u_{ki} = \dfrac{e^{-\lambda \|x_k + \delta_k - v_i\|_2^2}}{\sum_{l=1}^{c} e^{-\lambda \|x_k + \delta_k - v_l\|_2^2}}, & (32) \\[3ex] v_{ij} = \dfrac{\sum_{k=1}^{n} u_{ki}(x_{kj} + \delta_{kj})}{\sum_{k=1}^{n} u_{ki}}, & (33) \\[3ex] \delta_k = -(A_k|_{m=1})^{-1} B_k|_{m=1}. & (34) \end{cases}$$

3.2.3 Standard Model on L_1-Norm

Here, we consider FCM using penalty-vector regularization on L_1-norm (FCMP-L_1). The objective function is given by replacing L_2-norm by L_1-norm in the objective function of FCMP (27) as follows:

$$J_{\text{FCMT-}L_1} = \sum_{k=1}^{n}\sum_{i=1}^{c}(u_{ki})^m d_{ki} + \sum_{k=1}^{n} \delta_k{}^T W_k \delta_k. \quad (d_{ki} = \|x_k + \delta_k - v_i\|_1) \quad (35)$$

The constraints are (4) and W_k is (25).

First, we can derive the optimal solution of u_{ki} by using the Lagrange function as follows:

$$u_{ki} = \frac{(1/d_{ki})^{\frac{1}{m-1}}}{\sum_{l=1}^{c}(1/d_{kl})^{\frac{1}{m-1}}}. \qquad (36)$$

Second, we derive the optimal solution of v_{ij}. Same as the above, we consider the following semi-objective function:

$$J_{ij}(v_{ij}) = \sum_{k=1}^{n}(u_{ki})^m |x_{kj} + \delta_{kj} - v_{ij}|.$$

Various Types of Objective Functions of Clustering for Uncertain Data

This semi-objective function becomes equal to (20) by replacing ε_k by δ_k. Therefore, we can obtain the optimal solution of v_i using Algorithm 3.

Third, we describe the method to obtain the optimal solution of δ_{kj}. The semi-objective function of δ_{kj} is

$$J_{kj}(\delta_{kj}) = \sum_{i=1}^{c}(u_{ki})^m|x_{kj} + \delta_{kj} - v_{ij}| + w_{kj}\delta_{kj}{}^2.$$

We show Algorithm 6 to calculate the optimal solution of δ_{kj}.

Algorithm 6 (Calculation of δ_{kj})

Step 1   *Data are sorted in ascending order in each dimension.*

$$v_{1j} - x_{kj}, \ldots, v_{cj} - x_{kj}$$

$$\downarrow Sorting$$

$$v_{q(1)j} - x_{kj} \le \cdots \le v_{q(c)j} - x_{kj}$$

where $q(i)$ is substitution of $(1, \ldots, c)$.

Step 2   *Calculates as follows.*

$$S := -\sum_{i=1}^{c}(u_{kq(i)})^m + 2w_{kj}(v_{q(1)j} - x_{kj}),$$

$$r := 0.$$

Step 3   *If $S > 0$, δ_{kj} is obtained as follows and finish this algorithm*

$$\delta_{kj} = \frac{\sum_{i=1}^{c} u_{kq(i)}{}^m}{2w_{kj}}$$

Step 4   *Update as follows.*

$$r := r + 1,$$

$$S := S + 2u_{kq(r)}{}^m.$$

Step 5   *If $S > 0$, δ_{kj} is obtained as follows and finish this algorithm*

$$\delta_{kj} = v_{q(r)j} - x_{kj}$$

Step 6   *If $r = c$, δ_{kj} is obtained as follows and finish this algorithm*

$$\delta_{kj} = -\frac{\sum_{i=1}^{c} u_{kq(i)}{}^m}{2w_{kj}}$$

Step 7 *Update as follows.*

$$S := S + 2w_{kj}\left(v_{q(r+1)j} - v_{q(r)j}\right).$$

Step 8 *If $S > 0$, δ_{kj} is obtained as follows and finish this algorithm*

$$\delta_{kj} = \frac{\sum_{i=r+1}^{c}\left(u_{kq(i)}\right)^m - \sum_{i=1}^{r}\left(u_{kq(i)}\right)^m}{2w_{kj}}$$

Otherwise, go back to **Step 4.**

The algorithm of FCMP-L_1 is as same as Algorithm 5.

3.2.4 Entropy-Based Model on L_1-Norm

Instead of (35), we can consider the following objective function:

$$J_{\text{eFCMP-}L_1} = \sum_{k=1}^{n}\sum_{i=1}^{c} u_{ki}d_{ki} + \lambda^{-1}\sum_{k=1}^{n}\sum_{i=1}^{c} u_{ki}\log u_{ki} + \sum_{k=1}^{n}\delta_k{}^T W_k \delta_k. \tag{37}$$

$$(d_{ki} = \|x_k + \delta_k - v_i\|_1)$$

We call the clustering based on the above objective function entropy-based FCMP-L_1 (eFCMP-L_1). The constraints are (4) and (2).

The optimal solution can be obtained in the same way as FCMT-L_1, that is,

$$u_{ki} = \frac{e^{-\lambda\|x_k + \varepsilon_k - v_i\|_1}}{\sum_{l=1}^{c} e^{-\lambda\|x_k + \varepsilon_k - v_l\|_1}}. \tag{38}$$

Also, we use Algorithm 3 and Algorithm 6 with $m = 1$ to calculate the optimal solutions of v_{ij} and δ_{kj}, respectively. The algorithm of eFCMP-L_1 is as same as Algorithm 5.

4 Conclusion

In this paper, we showed some clustering methods for uncertain data with the concepts of tolerance and penalty-vector regularization. The whole clustering algorithm depends on the objective function, and we proposed various types of objective functions to construct some clustering algorithms for uncertain data. The purpose of this paper is to propose the objective functions with the concepts of tolerance and penalty-vector regularization and we didn't show numerical examples.

Various Types of Objective Functions of Clustering for Uncertain Data 259

In case to use the algorithms from the proposed objective functions, dependence on the initial values is a very difficult problem. This problem essentially lies in many clustering algorithms which is constructed using iterative optimization as well as our proposed algorithms.

We now focus two techniques. One is kernel trick and the other is pairwise constraints. Kernel trick is a useful tool to classify a dataset into some clusters with nonlinear boundaries. Pairwise constraints are informations whether a pair of data should be included into one cluster or not. The pairwise constraint that a pair of data should be in one cluster is called "must-link" and the constraint that a pair of data should not be in one cluster is called "cannot-link." We believe that the proposed concept of tolerance and penalty-vector regularization can be applied to the techniques and we will discuss the problems in the forthcoming paper with numerical examples.

In this paper, we didn't compare and evaluate the proposed algorithms. That's why the main purpose of the paper is to formulate the uncertain data in the framework of optimization and construct some fuzzy c-means clustering algorithms under the formulation. I hope the users of our proposed algorithms consider their properties to choose the adequate one from those.

Acknowledgements This study is partly supported by the Grant-in-Aid for Scientific Research (C) (Project No.21500212) from the Ministry of Education, Culture, Sports, Science and Technology, Japan.

References

Bezdek, J. C. (1981). *Pattern Recognition with Fuzzy Objective Function Algorithms*. New York: Plenum.

Endo, Y., Hasegawa, Y., Hamasuna, Y., & Miyamoto, S. (2008). Fuzzy c-means for data with rectangular maximum tolerance range. *Journal of Advanced Computational Intelligence and Intelligent Informatics, 12*(5), 461–466.

Endo, Y., Murata, R., Haruyama, H., & Miyamoto, S. (2005). Fuzzy c-means for data with tolerance. In *Proc. 2005 International Symposium on Nonlinear Theory and Its Applications*, pp. 345–348.

Hasegawa, Y., Endo, Y., & Hamasuna, Y. (2008). *On fuzzy c-means for data with uncertainty using spring modulus*, SCIS&ISIS 2008.

Jajuga, K. (1991). L_1-norm based fuzzy clustering. *Fuzzy Sets and Systems, 39*, 43–50.

Kanzawa, Y., Endo, Y., & Miyamoto, S. (2007). Fuzzy c-means algorithms for data with tolerance based on opposite criterions. *IEICE Transactions Fundamentals, E90-A*(10), 2194–2202.

Kanzawa, Y., Endo, Y., & Miyamoto, S. (2008). Fuzzy c-means algorithms for data with tolerance using kernel functions. *IEICE Transactions Fundamentals, E91-A*(9), 2520–2534.

Miyamoto, K., & Mukaidono, M. (1997). Fuzzy c-means as a regularization and maximum entropy approach. In *Proc. of the 7th International Fuzzy Systems Association World Congress* (IFSA'97), Vol. 2, pp. 86–92.

Murata, R., Endo, Y., Haruyama, H., & Miyamoto, S. (2006). On fuzzy c-means for data with tolerance. *Journal of Advanced Computational Intelligence and Intelligent Informatics, 10*(5), 673–681.

Part IV
Analysis and Optimization of Technical Systems Under Uncertainty

Stochastic Optimal Open-Loop Feedback Control of Dynamic Structural Systems under Stochastic Uncertainty

Kurt Marti and Ina Stein

Abstract In order to stabilize mechanical structures under dynamic applied loads, active control strategies are taken into account. The structures usually are stationary, safe and stable without external dynamic disturbances, such as strong earthquakes, wind turbulences, water waves, etc. Thus, in case of dynamic disturbances, additional control elements can be installed enabling active control actions. Active control strategies for mechanical structures are applied in order to counteract heavy applied dynamic loads, such as earthquakes, wind, water waves, etc. which would lead to large vibrations causing possible damages of the structure. Modeling the structural dynamics by means of a system of first order random differential equations for the state vector (displacement vector q and its time derivative $\dot{q}$), robust optimal controls are determined in order to cope with the stochastic uncertainty involved in the dynamic parameters, the initial values and the applied loadings.

1 Dynamic Structural Systems Under Stochastic Uncertainty

1.1 *Stochastic Optimal Structural Control: Active Control*

In order to omit structural damages and therefore high compensation (recourse) costs, active control techniques are used in structural engineering. The structures usually are stationary, safe and stable without considerable external dynamic disturbances. Thus, in case of heavy dynamic external loads, such as earthquakes, wind turbulences, water waves, etc., which cause large vibrations with possible damages, additional control elements can be installed in order to counteract applied dynamic loads, see (Block 2008; Soong 1988, 1990).

K. Marti (✉) · I. Stein
Federal Armed Forces University Munich, Aerospace Engineering and Technology,
Neubiberg/Munich, Germany
e-mail: kurt.marti@unibw-muenchen.de; ina.stein@online.de

Y. Ermoliev et al. (eds.), *Managing Safety of Heterogeneous Systems*, Lecture Notes
in Economics and Mathematical Systems 658, DOI 10.1007/978-3-642-22884-1_13,
© Springer-Verlag Berlin Heidelberg 2012

The structural dynamics is modeled mathematically by means of a linear system of second order differential equations for the m–vector $q = q(t)$ of displacements. The system of differential equations involves random dynamic parameters, random initial values, the random dynamic load vector and a control force vector depending on an input control function $u = u(t)$. Robust, i.e. parameter-insensitive optimal feedback controls u^* are determined in order to cope with the stochastic uncertainty involved in the dynamic parameters, the initial values and the applied loadings. In practice, the design of controls is directed often to reduce the mean square response (displacements and their time derivatives) of the system to a desired level within a reasonable span of time.

The performance of the resulting structural control problem under stochastic uncertainty is evaluated therefore by means of a convex quadratic cost function $L = L(t, z, u)$ of the state vector $z = z(t)$ and the control input vector $u = u(t)$. While the actual time path of the random external load is not known at the planning stage, we may assume that the probability distribution or at least the moments under consideration of the applied load and other random parameters are known. The problem is then to determine a robust, i.e. parameter-insensitive (open-loop) feedback control law by minimization of the expected total costs, hence, a stochastic optimal control law.

As mentioned above, in active control of dynamic structures, cf. (Block 2008; Nagarajaiah and Narasimhan 2007; Soong 1988, 1990; Soong and Costantinou 1994; Spencer and Nagarajaiah 2003; Yang and Soong 1988), the behavior of the m-vector $q = q(t)$ of displacements with respect to time t is described by a system of second order linear differential equations for $q(t)$ having a right hand side being the sum of the stochastic applied load process and the control force depending on a control n-vector function $u(t)$:

$$M\ddot{q} + D\dot{q} + Kq(t) = f(t, \omega, u(t)), t_0 \leq t \leq t_f. \tag{1a}$$

Hence, the force vector $f = f(t, \omega, u(t))$ on the right hand side of the dynamic equation (1a) is given by the sum

$$f(t, \omega, u) = f_0(t, \omega) + f_a(t, \omega, u) \tag{1b}$$

of the applied load $f_0 = f_0(t, \omega)$ being a vector-valued stochastic process describing e.g. external loads or excitation of the structure caused by earthquakes, wind turbulences, water waves, etc., and the actuator or control force vector $f_a = f_a(t, \omega, u)$ depending on an input or control n-vector function $u = u(t), t_0 \leq t \leq t_f$. Here, ω denotes the random element, lying in a certain probability space (Ω, A, P), used to represent random variations. Furthermore, M, D, K, resp., denotes the $m \times m$ mass, damping and stiffness matrix. In many cases the actuator or control force f_a is linear, i.e.

$$f_a = \Gamma_u u \tag{1c}$$

with a certain $m \times n$ matrix Γ_u.

Stochastic Optimal Open-Loop Feedback

By introducing appropriate matrices, the linear system of second order differential equations (1a,b) can be represented by a system of first order differential equations as follows:

$$\dot{z} = g(t, \omega, z(t\omega), u) := Az(t, \omega) + Bu + b(t, \omega) \tag{2a}$$

with

$$A := \begin{pmatrix} 0 & I \\ -M^{-1}K & -M^{-1}D \end{pmatrix}, \qquad B := \begin{pmatrix} 0 \\ M^{-1}\Gamma_u \end{pmatrix}, \tag{2b}$$

$$b(t, \omega) := \begin{pmatrix} 0 \\ M^{-1}f_0(t, \omega). \end{pmatrix} \tag{2c}$$

Moreover, $z = z(t)$ is the $2m$-state vector defined by

$$z = \begin{pmatrix} q \\ \dot{q} \end{pmatrix} \tag{2d}$$

fulfilling a certain initial condition

$$z(t_0) = \begin{pmatrix} q(t_0) \\ \dot{q}(t_0) \end{pmatrix} := \begin{pmatrix} q_0 \\ \dot{q}_0 \end{pmatrix} \tag{2e}$$

with given or stochastic initial values $q_0 = q_0(\omega), \dot{q}_0 = \dot{q}_0(\omega)$.

1.2 Robust (Optimal) Open-Loop Feedback Control

Assuming here that at each time point t the state $z_t := z(t)$ is available, the control force $f_a = \Gamma u$ is generated by means of a PD-controller. Hence, for the input n-vector function $u = u(t)$, we have

$$u(t) := \varphi\left(t, q(t), \dot{q}(t)\right) = \varphi\left(t, z(t)\right) \tag{3a}$$

with a feedback control law $\varphi = \varphi(t, q, \dot{q})$. Efficient approximate feedback control laws are constructed here by using the concept of *open-loop feedback control*. Open-loop feedback control is the main tool in *model predictive control*, cf. (Allgöwer 2000; Marti 2008; Richalet et al. 1978), which is very often used to solve optimal control problems in practice. The idea of *open-loop feedback control* is to construct a feedback control law quasi *argument-wise*, see cf. (Aoki 1967; Ku and Athans 1973).

A major issue in optimal control is the *robustness*, cf. (Dullerud and Paganini 2000), i.e. the insensitivity of the optimal control with respect to parameter variations. In case of random parameter variations, robust optimal controls can be obtained by means of stochastic optimization methods, cf. (Marti 2008). Thus, we introduce the following concept of an *stochastic optimal (open-loop) feedback control*.

Definition 1. In case of stochastic parameter variations, robust, hence, parameter-insensitive optimal (open-loop) feedback controls obtained by stochastic optimization methods are also called *stochastic optimal (open-loop) feedback controls*.

1.3 Stochastic Optimal Open-Loop Feedback Control

Finding a stochastic optimal open-loop feedback control, hence, an optimal (open-loop) feedback control law being insensitive as far as possible with respect to random parameter variations, means that besides optimality of the control law also its insensitivity with respect to stochastic parameter variations should be guaranteed. Hence, in the following sections we develop now a stochastic version of the (optimal) open-loop feedback control method, cf. (Marti 2008, 2010/11,/). An short overview on this novel stochastic optimal open-loop feedback control concept is given below:

At each intermediate time point $t_b \in [t_0, t_f]$, based on the observed state $z_b = z(t_b)$ at t_b, a stochastic optimal open-loop control $u^* = u^*(t) = u^*(t|(t_b, z_b)), t_b \le t \le t_f$, is determined first on the remaining time interval $[t_b, t_f]$, see Fig. 1, by stochastic optimization methods, cf. (Marti 2008).

Having a stochastic optimal open-loop control $u^* = u^*(t|(t_b, z_b)), t_b \le t \le t_f$, on each remaining time interval $[t_b, t_f]$ with an arbitrary starting time $t_b, t_0 \le t_b \le t_f$, a stochastic optimal open-loop feedback control law is then defined as follows:

Definition 2.

$$\varphi^* = \varphi(t_b, z(t_b)) := u^*(t_b) = u^*(t_b|(t_b, z_b)), t_0 \le t_b \le t_f. \quad (3b)$$

Hence, at time $t = t_b$ just the "first" control value $u^*(t_b) = u^*(t_b|(t_b, z_b))$ of $u^*(\cdot|(t_b, z_b))$ is used only. For each other argument $(t, z_t) := (t, z(t))$ the same construction is applied.

For finding stochastic optimal open-loop controls, on the remaining time intervals $t_b \le t \le t_f$ with $t_0 \le t_b \le t_f$, the stochastic Hamilton function of the control problem is introduced. Then, the class of H− minimal controls, cf. (Kalman et al. 1969), can be determined in case of stochastic uncertainty by solving a finite-dimensional stochastic optimization problem for minimizing the conditional expectation of the stochastic Hamiltonian subject to the remaining deterministic control constraints at each time point t. Having a H− minimal control, the related

Fig. 1 Remaining time interval

Stochastic Optimal Open-Loop Feedback 267

two-point boundary value problem with random parameters can be formulated for the computation of a stochastic optimal state- and costate-trajectory. Due to the linear-quadratic structure of the underlying control problem, the state and costate trajectory can be *determined analytically* to a large extent. Inserting then these trajectories into the H-minimal control, stochastic optimal open-loop controls are found on an arbitrary remaining time interval. According to Definition 2, these controls yield then immediately a stochastic optimal open-loop feedback control law. Moreover, the obtained controls can be realized in *real-time*, which is already shown for applications in optimal control of industrial robots, cf. (Schacher 2010).

Summarizing, we get *optimal (open-loop) feedback controls under stochastic uncertainty* minimizing the effects of external influences on system behavior, subject to the constraints of not having a complete representation of the system, cf. (Dullerud and Paganini 2000). Hence, robust or stochastic optimal active controls are obtained by *new* techniques from *Stochastic Optimization*, see (Marti 2008). Of course, the construction can be extended also to *PID–* controllers.

2 Expected Total Cost Function

The performance function F for active structural control systems is defined, cf. (Marti 2001, 2004, 2008), by the conditional expectation of the total costs being the sum of costs L along the trajectory, arising from the displacements $z = z(t, \omega)$ and the control input $u = u(t, \omega)$, and possible terminal costs G arising at the final state z_f. Hence, on the remaining time interval $t_b \leq t \leq t_f$ we have the following conditional expectation of the total cost function with respect to the information $\mathfrak{A}_{t_b}$ available up to time t_b:

$$F := \mathbb{E} \left(\int_{t_b}^{t_f} L \Big(t, \omega, z(t, \omega), u(t, \omega) \Big) dt + G(t_f, \omega, z(t_f, \omega)) \Big| \mathfrak{A}_{t_b} \right). \tag{4a}$$

Supposing quadratic costs along the trajectory, the function L is given by

$$L(t, \omega, z, u) := \frac{1}{2} z^T Q(t, \omega) z + \frac{1}{2} u^T R(t, \omega) u \tag{4b}$$

with positive (semi) definite $2m \times 2m, n \times n$, resp., matrix functions $Q = Q(t, \omega)$, $R = R(t, \omega)$. In the simplest case the weight matrices Q, R are fixed. A special selection for Q reads

$$Q = \begin{pmatrix} Q_q & 0 \\ 0 & Q_{\dot{q}} \end{pmatrix} \tag{4c}$$

with positive (semi) definite weight matrices $Q_q, Q_{\dot{q}}$, resp., for $q, \dot{q}$. Furthermore, $G = G(t_f, \omega, z(t_f, \omega))$ describes possible terminal costs. In case of endpoint control G is defined by

$$G(t_f, \omega, z(t_f, \omega)) := \frac{1}{2}(z(t_f, \omega) - z_f(\omega))^T S(z(t_f, \omega) - z_f(\omega)), \quad (4d)$$

where S is a positive definite (semi) weight matrix, and $z_f = z_f(\omega)$ denotes the (possible probabilistic) final state.

Remark 1. Instead of $\frac{1}{2}u^T Ru$, in the following we also use a more general convex control cost function $C = C(u)$.

3 Open-Loop Control Problem on the Remaining Time Interval $[t_b, t_f]$

Having the differential equation with random coefficients describing the behavior of the open-loop control $u^* = u^*(t), t_b \leq t \leq t_f$, is a solution of the following optimal control problem under stochastic uncertainty:

$$\min \quad \mathbb{E}\left(\int_{t_b}^{t_f} \frac{1}{2}\left(z(t, \omega)^T Qz(t, \omega) + C(u)\right) dt + G(t_f, \omega, z(t_f, \omega))\Big|\mathfrak{A}_{t_b}\right) \quad (5a)$$

$$\text{s.t.} \quad \dot{z}(t, \omega) = Az(t, \omega) + Bu(t) + b(t, \omega), \text{ a.s.}, \quad t_b \leq t \leq t_f \quad (5b)$$

$$z(t_b, \omega) = z_b \text{ (given)} \quad (5c)$$

$$u(t) \in D_t, \quad t_b \leq t \leq t_f. \quad (5d)$$

An important property of (5a-d) is stated next:

Lemma 1. *If the terminal cost function $G = G(t_f, \omega, z)$ is convex in z, and the feasible domain D_t is convex for each time point t, $t_0 \leq t \leq t_f$, then the stochastic optimal control problem (5a-d) is a convex optimization problem.*

4 Stochastic Hamiltonian of (5a–d)

According to (Marti 2008), the stochastic Hamiltonian H related to the stochastic optimal control problem (5a-d) reads:

$$H(t, \omega, z, y, u) := L(t, \omega, z, u) + y^T g(t, \omega, z, u)$$

$$= \frac{1}{2}z^T Qz + C(u) + y^T (Az + Bu + b(t, \omega)) . \quad (6a)$$

Stochastic Optimal Open-Loop Feedback

4.1 Expected Hamiltonian (with Respect to the Time Interval $[t_b, t_f]$ and Information $\mathfrak{A}_{t_b}$)

For the definition of a $H-$minimal control the conditional expectation of the stochastic Hamiltonian is needed:

$$
\begin{aligned}
\overline{H}^{(b)} &:= \ \mathbb{E}\left(H(t,\omega,z,y,u)\big|\mathfrak{A}_{t_b}\right) \\
&= \ \mathbb{E}\left(\frac{1}{2}z^T Qz + y^T\left(Az + b(t,\omega)\right)\big|\mathfrak{A}_{t_b}\right) + C(u) + \mathbb{E}\left(y^T Bu\big|\mathfrak{A}_{t_b}\right) \\
&= \ C(u) + \mathbb{E}\left(B^T y(t,\omega)\big|\mathfrak{A}_{t_b}\right)^T u + \ldots \\
&= \ C(u) + h(t)^T u + \ldots
\end{aligned}
\tag{6b}
$$

with

$$
h(t) := \mathbb{E}\left(B^T(y(t,\omega)\big|\mathfrak{A}_{t_b}\right) = h(t, t_b), \quad t \geq t_b .
\tag{6c}
$$

4.2 H-Minimal Control on $[t_b, t_f]$

In order to formulate the two-point boundary value problem for a stochastic optimal open-loop control $u^* = u^*(t), t_b \leq t \leq t_f$, we need first a H-minimal control

$$
\widetilde{u}^* = \widetilde{u}^*\left(t, z(t,\cdot), y(t,\cdot)\right), t_b \leq t \leq t_f,
$$

defined, cf. (Marti 2008), for $t_b \leq t \leq t_f$ as a solution of the following convex stochastic optimization problem , cf. (Marti 2008):

$$
\min \mathbb{E}\left(H(t,\omega,z(t,\omega),y(t,\omega),u)\big|\mathfrak{A}_{t_b}\right)
\tag{7a}
$$

s.t.

$$
u \in D_t ,
\tag{7b}
$$

where $z = z(t,\omega), y = y(t,\omega)$ are certain trajectories.

According to (7a,b) the H-minimal control

$$
\widetilde{u}^* = \widetilde{u}^*\left(t, z(t,\cdot), y(t,\cdot)\right) = \widetilde{u}^*(t, h)
\tag{8a}
$$

is defined by

$$
\widetilde{u}^*(t, h) := \underset{u \in D_t}{\mathrm{argmin}}\, C(u) + h(t)^T u \qquad \text{for } t \geq t_b .
\tag{8b}
$$

For strictly convex, differentiable cost functions $C = C(u)$, as e.g. $C(u) = \frac{1}{2}u^T Ru$ with positive definite matrix R, the necessary and sufficient condition for $\widetilde{u}^*$ reads in case $D_t = \mathbb{R}^n$

$$\nabla C(u) + h(t) = 0. \tag{9a}$$

If $u \mapsto \nabla C(u)$ is a 1-1-operator, then the solution of (9a) reads

$$u = v(h) := \nabla C^{-1}(-h). \tag{9b}$$

With (6c) and (8b) we then have

$$\widetilde{u}^*(t,h) = v(h(t)) = \nabla C^{-1}\left(-B^T \mathbb{E}\left(y(t,\omega)\big|\mathfrak{A}_{t_b}\right)\right) = \widetilde{u}^*(h(t)). \tag{9c}$$

5 Canonical Hamiltonian System

In the following we suppose that a H-minimal control $\widetilde{u}^* = \widetilde{u}^*\left(t, z(t,\cdot), y(t,\cdot)\right)$, $t_b \leq t \leq t_f$, i.e., a solution $\widetilde{u}^* = \widetilde{u}^*(t,h) = v(h(t)))$ of the stochastic optimization problem (7a,b) is available. Moreover, the conditional expectation $\mathbb{E}\left(\xi\big|\mathfrak{A}_{t_b}\right)$ of a random variable ξ is also denoted by $\overline{\xi}^{(b)}$, cf. (6b). According to (Marti 2008), a stochastic optimal open-loop control $u^* = u^*(t), t_b \leq t \leq t_f$,

$$u^*(t) = \widetilde{u}^*\left(t, z^*(t,\cdot), y^*(t,\cdot)\right), t_b \leq t \leq t_f, \tag{10}$$

of the stochastic optimal control problem (5a-d),can be obtained by solving the following stochastic two-point boundary value problem related to (5a-d):

Theorem 1. *If $z^* = z^*(t,\omega), y^* = y^*(t,\omega), t_0 \leq t \leq t_f$, is a solution of*

$$\dot{z}(t,\omega) = Az(t,\omega) + B\nabla C^{-1}\left(-B^T \overline{y(t,\omega)}^{(b)}\right) + b(t,\omega), \quad t_b \leq t \leq t_f \tag{11a}$$

$$z(t_b,\omega) = z_b \tag{11b}$$

$$\dot{y}(t,\omega) = -A^T y(t,\omega) - Qz(t,\omega) \tag{11c}$$

$$y(t_f,\omega) = \nabla G(t_f,\omega, z(t_f,\omega)), \tag{11d}$$

then the function $u^* = u^*(t), t_b \leq t \leq t_f$, defined by (10) is a stochastic optimal open-loop control for the remaining time interval $t_b \leq t \leq t_f$.

Stochastic Optimal Open-Loop Feedback

271

6 Minimum Energy Control

In this case we have $Q = 0$, i.e., there are no costs for the displacements $z = \begin{pmatrix} q \\ \dot{q} \end{pmatrix}$. In this case the solution of (11c,d) reads

$$y(t, \omega) = e^{A^T (t_f - t)} \nabla G(t_f, \omega, z(t_f, \omega)), \qquad t_b \leq t \leq t_f . \tag{12a}$$

Since the matrix B is deterministic, this yields

$$\widetilde{u}^*(t, h(t)) = v(h(t)) = \nabla C^{-1} \left(-B^T e^{A^T (t_f - t)} \overline{\nabla G(t_f, \omega, z(t_f, \omega))}^{(b)} \right), \tag{12b}$$

$t_b \leq t \leq t_f$.

Having (12a,b), for the state trajectory $z = z(t, \omega)$ we get, see (11a,b), the following system of ordinary differential equations

$$\dot{z}(t, \omega) = Az(t, \omega) + B\nabla C^{-1} \left(-B^T e^{A^T (t_f - t)} \overline{\nabla G(t_f, \omega, z(t_f, \omega))}^{(b)} \right)$$
$$+ b(t, \omega), \qquad t_b \leq t \leq t_f, \tag{13a}$$
$$z(t_b, \omega) = z_b. \tag{13b}$$

The solution of system (13a,b) reads

$$z(t, \omega) = e^{A(t - t_b)} z_b + \int_{t_b}^{t} e^{A(t - s)} \Bigg(b(s, \omega)$$
$$+ B\nabla C^{-1} \left(-B^T e^{A^T (t_f - s)} \overline{\nabla G(t_f, \omega, z(t_f, \omega))}^{(b)} \right) \Bigg) ds,$$

$t_b \leq t \leq t_f . \tag{14}$

For the final state $z = z(t_f, \omega)$ we get the relation:

$$z(t_f, \omega) = e^{A(t_f - t_b)} z_b + \int_{t_b}^{t_f} e^{A(t_f - s)} \Bigg(b(s, \omega)$$
$$+ B\nabla C^{-1} \left(-B^T e^{A^T (t_f - s)} \overline{\nabla G(t_f, \omega, z(t_f, \omega))}^{(b)} \right) \Bigg) ds . \tag{15}$$

6.1 Endpoint Control

In the case of endpoint control, the terminal cost function is given by the following definition (16a), where $z_f = z_f(\omega)$ denotes the desired – possible random – final state:

$$G(t_f, \omega, z(t_f, \omega)) := \frac{1}{2}\|z(t_f, \omega) - z_f(\omega)\|^2 . \tag{16a}$$

Hence,

$$\nabla G(t_f, \omega, z(t_f, \omega)) = z(t_f, \omega) - z_f(\omega) \tag{16b}$$

and therefore

$$\overline{\nabla G(t_f, \omega, z(t_f, \omega))}^{(b)} = \overline{z(t_f, \omega)}^{(b)} - \overline{z_f}^{(b)} \tag{16c}$$

$$= \mathbb{E}\left(z(t_f, \omega)\middle|\mathfrak{A}_{t_b}\right) - \mathbb{E}\left(z_f\middle|\mathfrak{A}_{t_b}\right) .$$

Thus

$$z(t_f, \omega) = e^{A(t_f - t_b)} z_b + \int_{t_b}^{t_f} e^{A(t_f - s)} \Bigg(b(s, \omega)$$

$$+ B \nabla C^{-1}\left(-B^T e^{A^T(t_f - s)} \left(\overline{z(t_f, \omega)}^{(b)} - \overline{z_f}^{(b)}\right)\right) \Bigg) ds . \tag{17a}$$

Taking expectations $\mathbb{E}(\ldots|\mathfrak{A}_{t_b})$ in (17a), we get the following condition for $\overline{z(t_f, \omega)}^{(b)}$:

$$\overline{z(t_f, \omega)}^{(b)} = e^{A(t_f - t_b)} z_b + \int_{t_b}^{t_f} e^{A(t_f - s)} \overline{b(s, \omega)}^{(b)} ds$$

$$+ \int_{t_b}^{t_f} e^{A(t_f - s)} B \nabla C^{-1}\left(-B^T e^{A^T(t_f - s)} \left(\overline{z(t_f, \omega)}^{(b)} - \overline{z_f}^{(b)}\right)\right) ds .$$

$$\tag{17b}$$

6.1.1 Quadratic Control Costs

Here, the control cost function $C = C(u)$ reads

$$C(u) = \frac{1}{2} u^T R u \,, \tag{18a}$$

hence,

$$\tag{18b}$$

$$\nabla C = R u \tag{18c}$$

and therefore

$$\nabla C^{-1}(w) = R^{-1} w \,. \tag{18d}$$

Consequently, (17b) reads

$$\overline{z(t_f, \omega)}^{(b)} = e^{A(t_f - t_b)} z_b + \int_{t_b}^{t_f} e^{A(t_f - s)} \overline{b(s, \omega)}^{(b)} \, ds$$

$$- \int_{t_b}^{t_f} e^{A(t_f - s)} B R^{-1} B^T e^{A^T (t_f - s)} \, ds \overline{z(t_f, \omega)}^{(b)}$$

$$+ \int_{t_b}^{t_f} e^{A(t_f - s)} B R^{-1} B^T e^{A^T (t_f - s)} \, ds \overline{z_f}^{(b)} \,. \tag{19}$$

Define now

$$U := \int_{t_b}^{t_f} e^{A(t_f - s)} B R^{-1} B^T e^{A^T (t_f - s)} \, ds \,. \tag{20}$$

Lemma 2. $I + U$ *is regular.*

Proof. Due to the previous considerations, U is a positive semidefinite $2m \times 2m$ matrix. Hence, U has only nonnegative eigenvalues.

Assuming that the matrix $I + U$ is singular, there is a $2m$-vector $w \neq 0$ such that

$$(I + U) w = 0.$$

However, this yields

$$U w = -I w = -w = (-1) w,$$

which means that $\lambda = -1$ is an eigenvalue of U. Since this contradicts to the above mentioned property of U, the matrix $I + U$ must be regular. $\qquad\square$

From (19) we get

$$(I + U)\overline{z(t_f, \omega)}^{(b)} = e^{A(t_f - t_b)}z_b + \int_{t_b}^{t_b} e^{A(t_f - s)}\overline{b(s, \omega)}^{(b)}\, ds + U\overline{z_f}^{(b)},$$

$$(21a)$$

hence,

$$\overline{z(t_f, \omega)}^{(b)} = (I + U)^{-1} e^{A(t_f - t_b)}z_b + (I + U)^{-1}\int_{t_b}^{t_b} e^{A(t_f - s)}\overline{b(s, \omega)}^{(b)}\, ds$$

$$+ (I + U)^{-1} U\overline{z_f}^{(b)}. \tag{21b}$$

Now, (21b) and (16b) yield

$$\overline{\nabla_z G(t_f, \omega, z(t_f, \omega))} = \overline{z(t_f, \omega)}^{(b)} - \overline{z_f}^{(b)} = \overline{z(t_f, \omega)}^{(b)} - \overline{z_f}^{(b)}$$

$$= (I + U)^{-1} e^{A(t_f - t_b)}z_b$$

$$+ (I + U)^{-1}\int_{t_b}^{t_f} e^{A(t_f - s)}\overline{b(s, \omega)}^{(b)}\, ds$$

$$+ \left((I + U)^{-1} U - I\right)\overline{z_f}^{(b)}. \tag{22}$$

Thus, a stochastic optimal open-loop control $u^*(t)$, $t_b \le t \le t_f$, on $[t_b, t_f]$ is given by, cf. (9b),

$$u^*(t) = -R^{-1} B^T e^{A^T(t_f - t)}\left((I + U)^{-1} e^{A(t_f - t_b)}z_b\right.$$

$$+ (I + U)^{-1}\int_{t_b}^{t_f} e^{A(t_f - s)}\overline{b(s, \omega)}^{(b)}\, ds$$

$$\left. + \left((I + U)^{-1} U - I\right)\overline{z_f}^{(b)}\right), \qquad t_b \le t \le t_f. \tag{23}$$

Finally, the stochastic optimal open-loop feedback control law $\varphi = \varphi(t, z(t))$ is then given by

Stochastic Optimal Open-Loop Feedback

$$\varphi(t_b, z(t_b)) := u^*(t_b)$$

$$= -R^{-1}B^T e^{A^T(t_f-t_b)} (I+U)^{-1} e^{A(t_f-t_b)} z_b$$

$$- R^{-1}B^T e^{A^T(t_f-t_b)} (I+U)^{-1} \int_{t_b}^{t_f} e^{A(t_f-s)} \overline{b(s,\omega)}^{(b)} ds$$

$$- R^{-1}B^T e^{A^T(t_f-t_b)} \left((I+U)^{-1}U - I\right) \overline{z_f}^{(b)} \tag{24}$$

with $z_b := z(t_b)$.

Replacing $t_b \to t$, we find this result:

Theorem 2. *The stochastic optimal open-loop feedback control law $\varphi = \varphi(t, z(t))$ is given by*

$$\varphi(t, z(t)) = \underbrace{-R^{-1}B^T e^{A^T(t_f-t)} (I+U)^{-1} e^{A(t_f-t)}}_{\Psi_0(t)} z(t)$$

$$\underbrace{-R^{-1}B^T e^{A^T(t_f-t)} (I+U)^{-1} \int_t^{t_f} e^{A(t_f-s)} \overline{b(s,\omega)}^{(t)} ds}_{\Psi_1(t, \overline{b(\cdot,\omega)}^{(t)})}$$

$$\underbrace{-R^{-1}B^T e^{A^T(t_f-t)} \left((I+U)^{-1}U - I\right) \overline{z_f}^{(t)}}_{\Psi_2(t)}, \tag{25a}$$

hence,

$$\varphi(t, z) = \Psi_0(t) z + \Psi_1(t, \overline{b(\cdot,\omega)}^{(t)}) + \Psi_2(t) \overline{z_f}^{(t)}. \tag{25b}$$

Remark 2. Note that the stochastic optimal open-loop feedback law $z \mapsto \varphi(t, z)$ is not linear in general, but affine-linear.

6.2 Endpoint Control with Different Cost Functions

In this section we consider more general terminal cost functions G. Hence, suppose

$$G(t_f, \omega, z(t_f, \omega)) := \kappa(z(t_f, \omega) - z_f(\omega)), \tag{26a}$$

$$\nabla G(t_f, \omega, z(t_f, \omega)) = \nabla \kappa(z(t_f, \omega) - z_f(\omega)). \tag{26b}$$

Consequently,

$$\widetilde{u}^*(t, h(t)) = v^*(h(t)) = \nabla C^{-1} \left(B^T e^{A^T(t_f - t)} \overline{\nabla \kappa(z(t_f, \omega) - z_f(\omega))}^{(b)} \right) \quad (27a)$$

and therefore, see (15)

$$z(t_f, \omega) = e^{A(t_f - t_b)} z_b + \int_{t_b}^{t_f} e^{A(t_f - s)} b(s, \omega) \, ds$$

$$+ \int_{t_b}^{t_f} e^{A(t_f - s)} B \nabla C^{-1} \left(-B^T e^{A^T(t_f - s)} \overline{\nabla \kappa(z(t_f, \omega) - z_f(\omega))}^{(b)} \right) \, ds ,$$

$$t_b \leq t \leq t_f. \quad (27b)$$

Special case:
Now a special terminal cost function is considered in more detail:

$$\kappa(z - z_f) := \sum_{i=1}^{2m} (z_i - z_{f_i})^4 \quad (28a)$$

$$\nabla \kappa(z - z_f) = 4 \left((z_1 - z_{f_1})^3, \ldots, (z_{2m} - z_{f_{2m}})^3 \right)^T . \quad (28b)$$

Here,

$$\overline{\nabla \kappa(z - z_f)}^{(b)} = 4 \left(\mathbb{E} \left((z_1 - z_{f_1})^3 | \mathfrak{A}_{t_b} \right), \ldots, \mathbb{E} \left((z_{2m} - z_{f_{2m}})^3 | \mathfrak{A}_{t_b} \right) \right)^T$$

$$= 4 \left(m_3^{(b)}(z_1(t_f, \cdot); z_{f_1}(\cdot)), \ldots, m_3^{(b)}(z_{2m}(t_f, \cdot); z_{f_{2m}}(\cdot)) \right)^T$$

$$=: 4 m_3^{(b)}(z(t_f, \cdot); z_f(\cdot)) . \quad (29)$$

Thus,

$$z(t_f, \omega) = e^{A(t_f - t_b)} z_b + \int_{t_b}^{t_f} e^{A(t_f - s)} b(s, \omega) \, ds$$

$$+ \underbrace{\int_{t_b}^{t_f} e^{A(t_f - s)} B \nabla C^{-1} \left(-B^T e^{A^T(t_f - s)} 4 m_3^{(b)}(z(t_f, \cdot); z_f(\cdot)) \right) \, ds}_{J\left(m_3^{(b)}(z(t_f, \cdot); z_f(\cdot)) \right)} . \quad (30)$$

Stochastic Optimal Open-Loop Feedback

Equation (30) yields then

$$\left(z(t_f,\omega) - z_f(\omega)\right)^3 \Bigg|_{c.-by-c.}$$

$$= \left(e^{A(t_f-t_b)}z_b - z_f + \int_{t_b}^{t_f} e^{A(t_f-s)}b(s,\omega)\,ds + J\left(m_3^{(b)}(z(t_f,\cdot);z_f(\cdot))\right)\right)^3 \Bigg|_{c.-by-c.},$$

(31a)

where "*c.-by-c.*" means "*component-by-component*". Taking expectations in (31a), we get the following relation for the moment vector $m_3^{(b)}$:

$$m_3^{(b)}(z(t_f,\cdot);z_f(\cdot)) = \Psi\left(m_3^{(b)}(z(t_f,\cdot);z_f(\cdot))\right).$$

(31b)

Remark 3.

$$\mathbb{E}\left(\left(z(t_f,\omega) - z_f(\omega)\right)^3 \big| \mathfrak{A}_{t_b}\right)\Bigg|_{c.-by-c.}$$

$$= \mathbb{E}^{(b)}\left(z(t_f,\omega) - \bar{z}^{(b)}(t_f) + \bar{z}^{(b)}(t_f) - z_f(\omega)\right)^3$$

$$= \mathbb{E}^{(b)}\Bigg(\left(z(t_f,\omega) - \bar{z}^{(b)}(t_f)\right)^3 + 3\left(z(t_f,\omega) - \bar{z}^{(b)}(t_f)\right)^2\left(\bar{z}^{(b)}(t_f) - z_f(\omega)\right)$$

$$+ 3\left(z(t_f,\omega) - \bar{z}^{(b)}(t_f)\right)\left(\bar{z}^{(b)}(t_f) - z_f(\omega)\right)^2 + \left(\bar{z}^{(b)}(t_f) - z_f(\omega)\right)^3\Bigg).$$

(31c)

Assuming that $z(t_f,\omega)$ and $z_f(\omega)$ are stochastic independent, then

$$\mathbb{E}\left(\left(z(t_f,\omega) - z_f(\omega)\right)^3 \big| \mathfrak{A}_{t_b}\right)$$

$$= m_3^{(b)}(z(t_f,\cdot)) + 3\sigma^{2(b)}(z(t_f,\cdot))(\bar{z}^{(b)}(t_f) - \overline{z_f}^{(b)}) + \overline{\left(\bar{z}^{(b)}(t_f) - z_f(\omega)\right)^3}^{(b)},$$

(31d)

where $\sigma^{2(b)}(z(t_f,\cdot)$ denotes the conditional variance of the state reached at the final time point t_f, given the information at time t_b.

6.3 Weighted Quadratic Terminal Costs

With a certain (possibly random) weight matrix $\Gamma = \Gamma(\omega)$, we consider the following terminal cost function:

$$G(t_f, \omega, z(t_f, \omega)) := \frac{1}{2} \left\| \Gamma(\omega) \left(z(t_f, \omega) - z_f(\omega) \right) \right\|^2. \tag{32a}$$

This yields

$$\nabla G(t_f, \omega, z(t_f, \omega)) = \Gamma(\omega)^T \Gamma(\omega)(z(t_f, \omega) - z_f(\omega)), \tag{32b}$$

and from (12a) we get

$$y(t, \omega) = e^{A^T(t_f - t)} \nabla_z G(t_f, \omega, z(t_f, \omega))$$

$$= e^{A^T(t_f - t)} \Gamma(\omega)^T \Gamma(\omega)(z(t_f, \omega) - z_f(\omega)), \tag{33a}$$

hence,

$$\overline{y(t)}^{(b)} = e^{A^T(t_f - t)} \overline{\Gamma(\omega)^T \left(\Gamma(\omega)(z(t_f, \omega) - \Gamma(\omega)z_f(\omega)) \right)}^{(b)}$$

$$= e^{A^T(t_f - t)} \left(\overline{\Gamma^T \Gamma z(t_f, \omega)}^{(b)} - \overline{\Gamma^T \Gamma z_f}^{(b)} \right). \tag{33b}$$

Thus, for the $H-$ minimal control we find

$$\widetilde{u}^*(t, h) = v(h(t))$$

$$= \nabla C^{-1} \left(-B^T \overline{y(t)}^{(b)} \right)$$

$$= \nabla C^{-1} \left(-B^T e^{A^T(t_f - t)} \left(\overline{\Gamma^T \Gamma z(t_f, \omega)}^{(b)} - \overline{\Gamma^T \Gamma z_f}^{(b)} \right) \right). \tag{34}$$

We obtain therefore, see (14),

$$z(t, \omega) = e^{A(t - t_b)} z_b + \int_{t_b}^{t} e^{A(t - s)} \Bigg(b(s, \omega)$$

$$+ B \nabla C^{-1} \left(-B^T e^{A^T(t_f - s)} \left(\overline{\Gamma^T \Gamma z(t_f, \omega)}^{(b)} - \overline{\Gamma^T \Gamma z_f}^{(b)} \right) \right) \Bigg) \, ds. \tag{35a}$$

6.3.1 Quadratic Control Costs

Assume that the control costs and its gradient are given by

$$C(u) = \frac{1}{2}u^T R u, \quad \nabla C(u) = R u. \tag{35b}$$

Here, (35a) yields

$$z(t_f, \omega) = e^{A(t_f - t_b)} z_b + \int_{t_b}^{t_f} e^{A(t_f - s)} \left(b(s, \omega) \right.$$

$$\left. -BR^{-1}B^T e^{A^T(t_f - s)} \left(\overline{\Gamma^T \Gamma z(t_f, \omega)}^{(b)} - \overline{\Gamma^T \Gamma z_f}^{(b)} \right) \right) ds. \tag{35c}$$

Multiplying with $\Gamma(\omega)^T \Gamma(\omega)$ and taking expectations, from (35c) we get

$$\overline{\Gamma^T \Gamma z(t_f, \omega)}^{(b)} = \overline{\Gamma^T \Gamma}^{(b)} e^{A(t_f - t_b)} z_b + \int_{t_b}^{t_f} \overline{\Gamma^T \Gamma e^{A(t_f - s)} b(s, \omega)}^{(b)} ds$$

$$- \overline{\Gamma^T \Gamma}^{(b)} \int_{t_b}^{t_f} e^{A(t_f - s)} BR^{-1} B^T e^{A^T(t_f - s)} ds$$

$$\times \left(\overline{\Gamma^T \Gamma z(t_f, \omega)}^{(b)} - \overline{\Gamma^T \Gamma z_f}^{(b)} \right). \tag{36a}$$

According to a former lemma, we define the matrix

$$U = \int_{t_b}^{t_f} e^{A(t_f - s)} BR^{-1} B^T e^{A^T(t_f - s)} ds.$$

From (36a) we get then

$$\left(I + \overline{\Gamma^T \Gamma}^{(b)} U \right) \overline{\Gamma^T \Gamma z(t_f, \omega)}^{(b)}$$

$$= \overline{\Gamma^T \Gamma}^{(b)} e^{A(t_f - t_b)} z_b + \int_{t_b}^{t_f} \overline{\Gamma^T \Gamma e^{A(t_f - s)} b(s, \omega)}^{(b)} ds$$

$$+ \overline{\Gamma^T \Gamma}^{(b)} U \overline{\Gamma^T \Gamma z_f}^{(b)}. \tag{36b}$$

Lemma 3. $I + \overline{\Gamma^T \Gamma}^{(b)} U$ *is regular.*

Proof. First notice that not only U, but also $\overline{\Gamma^T \Gamma}^{(b)}$ is positive semidefinite:

$$v^T \overline{\Gamma^T \Gamma}^{(b)} v = \overline{v^T \Gamma^T \Gamma v} = \overline{(\Gamma v)^T \Gamma v}^{(b)} = \overline{\|\Gamma v\|_2^2}^{(b)} \geq 0.$$

Then their product $\overline{\Gamma^T \Gamma}^{(b)} U$ is positive semidefinite as well. This follows immediately from (Ostrowski 1959) as $\Gamma(\omega)^T \Gamma(\omega)$ is symmetric. $\square$

Since the matrix $I + \overline{\Gamma^T \Gamma}^{(b)} U$ is regular, we get cf. (21a,b),

$$\overline{\Gamma^T \Gamma z(t_f, \omega)}^{(b)} = \left(I + \overline{\Gamma^T \Gamma}^{(b)} U\right)^{-1} \overline{\Gamma^T \Gamma}^{(b)} e^{A(t_f - t_b)} z_b$$

$$+ \left(I + \overline{\Gamma^T \Gamma}^{(b)} U\right)^{-1} \int_{t_b}^{t_f} \overline{\Gamma^T \Gamma e^{A(t_f - s)} b(s, \omega)}^{(b)} ds$$

$$+ \left(I + \overline{\Gamma^T \Gamma}^{(b)} U\right)^{-1} \overline{\Gamma^T \Gamma}^{(b)} U \overline{\Gamma z_f}^{(b)}. \tag{36c}$$

Putting (36c) into (34), corresponding to (23) we get the stochastic optimal open-loop control

$$u^*(t) = - R^{-1} B^T e^{A^T (t_f - t)} \left(\overline{\Gamma^T \Gamma z(t_f, \omega)}^{(b)} - \overline{\Gamma^T \Gamma z_f}^{(b)}\right)$$

$$= \ldots, \quad t_b \leq t \leq t_f, \tag{37}$$

which yields then the related stochastic optimal open-loop feedback control $\varphi = \varphi(t, z(t))$ law corresponding to Theorem 2.

7 Nonzero Costs for Displacements

Suppose here that $Q \neq 0$. According to (11a-d), for the adjoint trajectory $y = y(t, \omega)$ we have the system of differential equations

$$\dot{y}(t, \omega) = - A^T y(t, \omega) - Q z(t, \omega)$$

$$y(t_f, \omega) = \nabla G(t_f, \omega, z(t_f, \omega)),$$

which has the following solution for given $z(t, \omega)$ and $\nabla G(t_f, \omega, z(t_f, \omega))$:

$$y(t, \omega) = \int_t^{t_f} e^{A^T (s-t)} Q z(s, \omega) \, ds + e^{A^T (t_f - t)} \nabla G(t_f, \omega, z(t_f, \omega)). \tag{38}$$

Indeed, we get

$$y(t_f, \omega) = 0 + I \nabla_z G(t_f, \omega, z(t_f, \omega)) = \nabla_z G(t_f, \omega, z(t_f, \omega))$$

$$\dot{y}(t, \omega) = -e^{A^T \cdot 0} Q z(t, \omega)$$

$$- \int_t^{t_f} A^T e^{A^T(s-t)} Q z(s, \omega) \, ds - A^T e^{A^T(t_f-t)} \nabla G(t_f, \omega, z(t_f, \omega))$$

$$= -e^{A^T \cdot 0} Q z(t, \omega)$$

$$- A^T \left(\int_t^{t_f} e^{A^T(s-t)} Q z(s, \omega) \, ds + e^{A^T(t_f-t)} \nabla G(t_f, \omega, z(t_f, \omega)) \right)$$

$$= -A^T y(t, \omega) - Q z(t, \omega).$$

From (38) we then get

$$\overline{y(t)}^{(b)} = \mathbb{E}^{(b)}(y(t, \omega)) = \mathbb{E}\left(y(t, \omega) | \mathfrak{A}_{t_b} \right)$$

$$= \int_t^{t_f} e^{A^T(s-t)} Q \overline{z(s)}^{(b)} \, ds + e^{A^T(t_f-t)} \overline{\nabla G(t_f, \omega, z(t_f, \omega))}^{(b)}. \tag{39}$$

The unknown function $\overline{z(t)}^{(b)}$, and the vector $z(t_f, \omega)$ in this equation are both given, based on $\overline{y(t)}^{(b)}$, by the initial value problem, see (11a,b),

$$\dot{z}(t, \omega) = A z(t, \omega) + B \nabla C^{-1} \left(-B^T \overline{y}^{(b)}(t) \right) + b(t, \omega) \tag{40a}$$

$$z(t_b, \omega) = z_b. \tag{40b}$$

Taking expectations, considering the state vector at the final time point t_f, resp., yields the expressions:

$$\overline{z(t)}^{(b)} = e^{A(t-t_b)} z_b + \int_{t_b}^t e^{A(t-s)} \left(\overline{b(s)}^{(b)} + B \nabla C^{-1} \left(-B^T \overline{y(s)}^{(b)} \right) \right) ds, \tag{41a}$$

$$z(t_f, \omega) = e^{A(t_f-t_b)} z_b + \int_{t_b}^{t_f} e^{A(t_f-s)} \left(b(s, \omega) + B \nabla C^{-1} \left(-B^T \overline{y(s)}^{(b)} \right) \right) ds.$$

$$\tag{41b}$$

7.1 Quadratic Control and Terminal Costs

Corresponding to (16a,b) and (18a,b), suppose

$$\nabla G(t_f, \omega, z(t_f, \omega)) = z(t_f, \omega) - z_f(\omega),$$

$$\nabla C^{-1}(w) = R^{-1}w.$$

According to (10) and (9c), in the present case the stochastic optimal open-loop control is given by

$$u^*(t) = \widetilde{u}^*(t, q(t)) = R^{-1}\left(-\mathbb{E}\left(B^T y(t, \omega)\big|\mathfrak{A}_{t_b}\right)\right) = -R^{-1}B^T \overline{y(t)}^{(b)}. \tag{42a}$$

Hence, we need the function $\overline{y}^{(b)} = \overline{y(t)}^{(b)}$. From (39) and (16a,b) we have

$$\overline{y(t)}^{(b)} = e^{A^T(t_f - t)}\left(\overline{z(t_f)}^{(b)} - \overline{z_f}^{(b)}\right) + \int_t^{t_f} e^{A^T(s-t)} Q\overline{z(s)}^{(b)}\, ds. \tag{42b}$$

Inserting (41a,b) into (42b), we have

$$\overline{y(t)}^{(b)} = e^{A^T(t_f - t)}\left(e^{A(t_f - t_b)} z_b - \overline{z_f}^{(b)}\right.$$

$$+ \int_{t_b}^{t_f} e^{A(t_f - s)}\left(\overline{b(s)}^{(b)} - BR^{-1}B^T \overline{y(s)}^{(b)}\right) ds\bigg)$$

$$+ \int_t^{t_f} e^{A^T(s-t)} Q\left(e^{A(s-t_b)} z_b\right.$$

$$+ \int_{t_b}^{s} e^{A(s-\tau)}\left(\overline{b(\tau)}^{(b)} - BR^{-1}B^T \overline{y(\tau)}^{(b)}\right) d\tau\bigg) ds. \tag{42c}$$

In the following we develop a condition that guarantees the existence and uniqueness of a solution $\overline{y}^b = \overline{y(t)}^{(b)}$ of (42c):

Theorem 3. *In the space of continuous functions, the above (42c) has a unique solution if*

$$c_B < \cfrac{1}{c_A\sqrt{c_{R^{-1}}(t_f - t_0)}\left(1 + \frac{(t_f - t_0)c_Q}{2}\right)}. \tag{43}$$

Here,

$$c_A := \sup_{t_b \le t \le s \le t_f} \|e^{A(t-s)}\|_F \quad c_B := \|B\|_F \quad c_{R^{-1}} := \|R^{-1}\|_F \quad c_Q := \|Q\|_F,$$

and the index F denotes the Frobenius-Norm.

Proof. The proof of the existence and uniqueness of such an solution is based on the Banach fix-point theorem. For applying this theorem, we consider the Banach space

$$\mathscr{X} = \{ f : [t_b; t_f] \to \mathbb{R}^{2m} \ : \ f \text{ continuous} \} \tag{44a}$$

equipped with the supremum norm

$$\|f\|_L := \sup_{t_b \le t \le t_f} \|f(t)\|_2, \tag{44b}$$

where $\|\cdot\|_2$ denotes the Euclidean norm on $\mathbb{R}^{2m}$.

Now we study the operator $\mathscr{T} : \mathscr{X} \to \mathscr{X}$ defined by

$$(\mathscr{T} f)(t) = e^{A^T(t_f-t)} \left(e^{A(t_f-t_b)} z_b - \overline{z_f}^{(b)} \right.$$

$$+ \int_{t_b}^{t_f} e^{A(t_f-s)} \left(\overline{b(s)}^{(b)} - BR^{-1}B^T f(s) \right) ds \Bigg)$$

$$+ \int_{t}^{t_f} e^{A^T(s-t)} Q \left(e^{A(s-t_b)} z_b \right.$$

$$+ \int_{t_b}^{s} e^{A(s-\tau)} \left(\overline{b(\tau)}^{(b)} - BR^{-1}B^T f(\tau) \right) d\tau \Bigg) ds. \tag{45}$$

The norm of the difference $\mathscr{T} f - \mathscr{T} g$ of the images of two different elements $f, g \in \mathscr{X}$ with respect to $\mathscr{T}$ may be estimated as follows:

$$\|\mathscr{T} f - \mathscr{T} g\|$$

$$= \sup_{t_b \le t \le t_f} \left\{ \left\| e^{A^T(t_f-t)} \int_{t_b}^{t_f} e^{A(t_f-s)} BR^{-1}B^T (g(s) - f(s)) ds \right. \right.$$

$$+ \int_{t}^{t_f} e^{A^T(s-t)} Q \int_{t_b}^{s} e^{A(s-\tau)} BR^{-1}B^T (g(\tau) - f(\tau)) d\tau ds \bigg\|_2 \right\}. \tag{46a}$$

Note that the Frobenius norm is sub-multiplicative and compatible with the Euclidian norm. Using these properties, we get

$$\|\mathscr{T} f - \mathscr{T} g\|$$

$$\leq \sup_{t_b \leq t \leq t_f} \left\{ c_A \int_{t_b}^{t_f} c_A c_B c_{R^{-1}} c_B \| f(s) - g(s) \|_2 \, ds \right.$$

$$+ c_A c_Q \int_{t}^{t_f} \int_{t_b}^{s} c_A c_B c_{R^{-1}} c_B \| f(\tau) - g(\tau) \|_2 \, d\tau \, ds \bigg\}$$

$$\leq \sup_{t_b \leq t \leq t_f} \left\{ c_A \int_{t_b}^{t_f} c_A c_B c_{R^{-1}} c_B \sup_{t_b \leq t \leq t_f} \| f(s) - g(s) \|_2 \, ds \right.$$

$$+ c_A c_Q \int_{t}^{t_f} \int_{t_b}^{s} c_A c_B c_{R^{-1}} c_B \sup_{t_b \leq t \leq t_f} \| f(\tau) - g(\tau) \|_2 \, d\tau \, ds \bigg\}$$

$$= \| f - g \| c_A^2 c_B^2 c_{R^{-1}} \sup_{t_b \leq t \leq t_f} \left\{ (t_f - t_b) + \frac{c_Q}{2} \left((t_f - t_b)^2 - (t - t_b)^2 \right) \right\}$$

$$\leq \| f - g \| c_A^2 c_B^2 c_{R^{-1}} (t_f - t_b)(1 + \frac{c_Q}{2}(t_f - t_b)) \,. \tag{46b}$$

Thus, $\mathscr{T}$ is a contraction if

$$c_B^2 < \frac{1}{c_A^2 c_{R^{-1}} (t_f - t_b) \left(1 + \frac{c_Q}{2}(t_f - t_b) \right)} \tag{46c}$$

and therefore

$$c_B < \frac{1}{c_A \sqrt{c_{R^{-1}} (t_f - t_b) \left(1 + \frac{c_Q}{2}(t_f - t_b) \right)}} \,. \tag{46d}$$

In order to get a condition that is independent of t_b, we take the worst case $t_b = t_0$, hence,

$$c_B < \frac{1}{c_A \sqrt{c_{R^{-1}} (t_f - t_0) \left(1 + \frac{(t_f - t_0)c_Q}{2} \right)}} \,. \tag{46e}$$

Stochastic Optimal Open-Loop Feedback

Remark 4. Condition (46e) holds if the matrix Γ_u in (1c) has a sufficiently small Frobenius norm. Indeed, according to (2b) we have

$$B = \begin{pmatrix} 0 \\ M^{-1}\Gamma_u \end{pmatrix}$$

and therefore

$$c_B = \|B\|_F = \|M^{-1}\Gamma_u\|_F \leq \|M^{-1}\|_F \cdot \|\Gamma_u\|_F .$$

$\square$

Having $\overline{y(t)}^{(b)}$, according to (42a) a stochastic optimal open-loop control $u^*(t)$, $t_b \leq t \leq t_f$, reads:

$$u^*(t) = -R^{-1}B^T\overline{y(t)}^{(b)}. \tag{47a}$$

Moreover,

$$\varphi(t_b, z_b) := u^*(t_b), t_0 \leq t_b \leq t_f , \tag{47b}$$

is then a stochastic optimal open-loop feedback control law.

Remark 5. Putting $Q = 0$ in (38), we again obtain the stochastic optimal open-loop feedback control law (24) in Sect. 6.1.1.

8 Example

We consider the structure according to Fig. 2, see (Block 2008), where we want to control the supplementary active system while minimizing the expected total costs for the control and the terminal costs.

The behavior of the vector of displacements $q(t, \omega)$ can be described by a system of differential equations of second order:

$$M\begin{pmatrix} \ddot{q}_0(t, \omega) \\ \ddot{q}_z(t, \omega) \end{pmatrix} + D\begin{pmatrix} \dot{q}_0(t, \omega) \\ \dot{q}_z(t, \omega) \end{pmatrix} + K\begin{pmatrix} q_0(t, \omega) \\ q_z(t, \omega) \end{pmatrix} = f_0(t, \omega) + f_a(t) \tag{48}$$

with

$$M = \begin{pmatrix} m_0 & 0 \\ 0 & m_z \end{pmatrix} \qquad \text{mass matrix} \tag{49a}$$

$$D = \begin{pmatrix} d_0 + d_z & -d_z \\ -d_z & d_z \end{pmatrix} \qquad \text{damping matrix} \tag{49b}$$

$$K = \begin{pmatrix} k_0 + k_z & -k_z \\ -k_z & k_z \end{pmatrix} \qquad \text{stiffness matrix} \tag{49c}$$

Fig. 2 Principle of active structural control

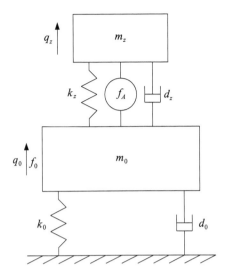

$$f_a(t) = \begin{pmatrix} -1 \\ +1 \end{pmatrix} u(t) \qquad \text{actuator force} \qquad (49d)$$

$$f_0(t,\omega) = \begin{pmatrix} f_{01}(t,\omega) \\ 0 \end{pmatrix} \qquad \text{applied load.} \qquad (49e)$$

Here we have $n = 1$, i.e. $u(\cdot) \in \mathscr{C}(T, \mathbb{R})$, and the weight matrix R becomes a positive real number.

To represent the equation of motion (48) as a first order differential equation we set

$$z(t,\omega) := (q(t,\omega), \dot{q}(t,\omega))^T = \begin{pmatrix} q_0(t,\omega) \\ q_z(t,\omega) \\ \dot{q}_0(t,\omega) \\ \dot{q}_z(t,\omega) \end{pmatrix}.$$

This yields the dynamical equation

$$\dot{z}(t,\omega) = \begin{pmatrix} \mathbf{0} & I_2 \\ -M^{-1}K & -M^{-1}D \end{pmatrix} z(t,\omega) + \begin{pmatrix} \mathbf{0} \\ M^{-1}f_a(s) \end{pmatrix} + \begin{pmatrix} \mathbf{0} \\ M^{-1}f_0(s,\omega) \end{pmatrix}$$

$$= \underbrace{\begin{pmatrix} 0 & 0 & 1 & 0 \\ 0 & 0 & 0 & 1 \\ -\dfrac{k_0+k_z}{m_0} & \dfrac{k_z}{m_0} & -\dfrac{d_0+d_z}{m_0} & \dfrac{d_z}{m_0} \\ \dfrac{k_z}{m_z} & -\dfrac{k_z}{m_z} & \dfrac{d_z}{m_z} & -\dfrac{d_z}{m_z} \end{pmatrix}}_{=:A} z(t,\omega) + \underbrace{\begin{pmatrix} 0 \\ 0 \\ -\dfrac{1}{m_0} \\ \dfrac{1}{m_z} \end{pmatrix}}_{=:B} u(s) + \underbrace{\begin{pmatrix} 0 \\ 0 \\ \dfrac{f_0(s,\omega)}{m_0} \\ 0 \end{pmatrix}}_{=:b(t,\omega)},$$

(50)

where I_p denotes the $p \times p$ identity matrix. Furthermore, we have the optimal control problem under stochastic uncertainty:

$$\min \quad F(u(\cdot)) := \mathbb{E}\frac{1}{2}\left(\int_{t_b}^{t_f} R\left(u(s)\right)^2 \, ds + z(t_f,\omega)^T Gz(t_f,\omega)\Big| \mathfrak{A}_{t_b}\right) \tag{51a}$$

$$s.t. \quad z(t,\omega) = z_b + \int_{t_b}^{t} \left(Az(s,\omega) + Bu(s) + b(s,\omega)\right)ds \tag{51b}$$

$$u(\cdot) \in \mathscr{C}(T,\mathbb{R}). \tag{51c}$$

Note that this problem is of the "Minimum-Energy Control"-type, if we apply no extra costs for the displacements, i.e. $Q \equiv 0$.
The two-point-boundary problem to be solved reads then, cf. (11a-d),

$$\dot{z}(t,\omega) = Az(t,\omega) - \frac{1}{R}BB^T \overline{y(t)}^{(b)} + b(\omega,t) \tag{52a}$$

$$\dot{y}(t,\omega) = -A^T y(t,\omega) \tag{52b}$$

$$z(t_b,\omega) = z_b \tag{52c}$$

$$y(t_f,\omega) = Gz(t_f,\omega). \tag{52d}$$

Hence, the solution of (52a)–(52d), i.e. the optimal trajectories, reads, cf. (12a), (35a),

$$y(t,\omega) = e^{A^T(t_f-t)}Gz(t_f,\omega) \tag{53a}$$

$$z(t,\omega) = e^{A(t-t_b)}z_b + \int_{t_b}^{t} e^{A(t-s)}\Bigg(b(s,\omega)$$

$$- \frac{1}{R}BB^T e^{A^T(t_f-s)}G\overline{z(t_f,\omega)}^{(b)}\Bigg) \, ds. \tag{53b}$$

Finally, we get the optimal control, see (36c) and (37):

$$u^*(t) = -\frac{1}{R}B^T e^{A^T(t_f-t)}\left(I_4 + GU\right)^{-1}Ge^{At_f}\left(e^{-At_b}z_b + \int_{t_b}^{t_f} e^{-As}\overline{b(s,\omega)}^{(b)} \, ds\right) \tag{54}$$

with

$$U = \frac{1}{R}\int_{t_b}^{t_f} e^{A(t_f-s)}BB^T e^{A^T(t_f-s)} \, ds. \tag{55}$$

9 Conclusion

Active regulator strategies are considered for stabilizing dynamic mechanical structures under stochastic applied loadings. The problem has been modeled in the framework of stochastic optimal control for minimizing the expected total costs arising from the displacements of the structure and the regulation costs. Based on the concept of *open-loop feedback control*, in recent years the so-called *Model Predictive Control* became very popular in solving optimal control problems in practice. Hence, due to the great advantages of open-loop feedback controls, *stochastic optimal open-loop feedback controls* have been constructed by taking into account the random parameter variations in the structural control problem under stochastic uncertainty. For finding stochastic optimal open-loop controls, on the remaining time intervals $t_b \leq t \leq t_f$ with $t_0 \leq t_b \leq t_f$, the stochastic Hamilton function of the control problem has been introduced. Then, the class of $H-$ minimal controls can be determined by solving a finite-dimensional stochastic optimization problem for minimizing the conditional expectation of the stochastic Hamiltonian subject to the remaining deterministic control constraints at each time point t. Having a $H-$minimal control, the related two-point boundary value problem with random parameters can be formulated for the computation of stochastic optimal state- and costate-trajectories. Due to the linear-quadratic structure of the underlying control problem, the state and costate trajectory can be determined analytically to a large extent. Inserting then these trajectories into the H-minimal control, stochastic optimal open-loop controls are found on an arbitrary remaining time interval. These controls yield then immediately a stochastic optimal open-loop feedback control law.

References

Allgöwer, F. E. A. (Ed.) (2000). *Nonlinear Model Predictive Control*. Basel: Birkhäuser Verlag.

Aoki, M. (1967). *Optimization of Stochastic Systems: Topics in Discrete-Time Systems*. New York, London: Academic Press.

Block, C. (2008). *Aktive Minderung Personeninduzierter Schwingungen An Weit Gespannten Strukturen im Bauwesen*. No. 336 in Fortschrittberichte VDI, Reihe 11, Schwingungstechnik. Düsseldorf: VDI-Verlag GmbH.

Dullerud, G., & Paganini, F. (2000). *A Course in Robust Control Theory - A Convex Approach*. New York [etc.]: Springer.

Kalman, R., Falb, P., & Arbib, M. (1969). *Topics in Mathematical System Theory*. New York [etc.]: McGraw-Hill Book Company.

Ku, R., & Athans, M. (1973). On the adaptive control of linear systems using the open-loop-feedback-optimal approach. *IEEE Transactions on Automatic Control, AC-18*, 489–493.

Marti, K. (2001). Stochastic optimization methods in robuts adaptive control of robots. In M. Groetschel, et al. (Eds.), *Online Optimization of Large Scale Systems*, pp. 545–577. Berlin-Heidelberg-New York: Springer.

Marti, K. (2004). Adaptive optimal stochastic trajectory planning and control (AOSTPC) for robots. In K. Marti, et al. (Eds.), *Dynamic Stochastic Optimization*, pp. 155–206. Berlin-Heidelberg: Springer.

Marti, K. (2008). Approximate solutions of stochastic control problems by means of convex approximations. In B. H. V. Topping, et al. (Eds.), *Proceedings of the 9th Int. Conference on Computational Structures Technology* (CST08), Paper No. 52. Stirlingshire, UK: Civil-Comp Press.

Marti, K. (2008). Stochastic nonlinear model predictive control (SNMPC). In *79th Annual Meeting of the International Association of Applied Mathematics and Mechanics* (GAMM), Bremen 2008, PAMM, vol. 8, Issue 1, pp. 10,775–10,776. Wiley.

Marti, K. (2008). *Stochastic Optimization Problems*, 2nd edn. Berlin-Heidelberg: Springer.

Marti, K. (2010/11). Continuous-time control under stochastic uncertainty. Accepted for publication in the Wiley Encyclopedia of Operations Research abd Management Science (EORMS) (2010/11).

Marti, K. (2010/11). Optimal control of dynamical systems and structures under stochastic uncertainty: Stochastic optimal feedback control. accepted for publication in Advances in Engineering Software (AES) (2010/11). DOI information (online publication): 10.1016/j.advengsoft.2010.09.008.

Nagarajaiah, S., & Narasimhan, S. (2007). Optimal control of structures. In J.S. Arora, (Ed.), *Optimization of Structural and Mechanical Systems*, pp. 221–244. New Jersey [etc.]: World Scientific.

Ostrowski, A. (1959). Über eigenwerte von produkten hermitescher matrizen. In *Abhandlungen Aus Dem Mathematischen Seminar Der Universität Hamburg*, vol. 23(1), pp. 60–68. Berlin-Heidelberg: Springer.

Richalet, J., Rault, A., Testud, J., & Papon, J. (1978). *Model Predictive Heuristic Control: Applications to Industrial Processes*. Automatica **14**. Elsevier.

Schacher, M. (2010). Stochastisch optimale Regelung von Robotern. Ph.D. thesis, Faculty for Aerospace Engineering and Technology, Federal Armed Forces University Munich (2010).

Soong, T. T. (1988). Active structural control in civil engineering. *Engineering Structures, 10*, 74–84.

Soong, T. T. (1990). *Active Structural Control: Theory and Practice*. New York: Longman Scientific and Technical, J. Wiley.

Soong, T. T., & Costantinou, M. C. (1994). *Passive and Active Structural Vibration Control in Civil Engineering*. CISM Courses and Lectures No. 345. Wien-New York: Springer.

Spencer, B., & Nagarajaiah, S. (2003). State of the art of structural control. *Journal of Structural Engineering, ASCE, 129*(7), 845–856.

Yang, J. N., & Soong, T.T. (1988). Recent advance in active control of civil engineering structures. *Probabilistic Engineering Mechanics, 3*(4), 179–188.

Modeling and Processing of Uncertainty in Civil Engineering by Means of Fuzzy Randomness

Uwe Reuter, Jan-Uwe Sickert, Wolfgang Graf, and Michael Kaliske

Abstract The paper focuses on the adequate quantification of uncertainty which usually influences all numerical simulations of structures in the field of civil engineering. Fuzzy randomness provides adequate modeling of specific uncertainty phenomena, not only in the field of civil engineering. In this paper, approaches for modeling of data and model uncertainty by means of convex fuzzy random variables, including fuzzy variables and random variables as special cases, are presented. Numerical processing of those uncertain variables succeeds with the help of fuzzy stochastic structural analysis. By means of fuzzy stochastic analysis, it is possible to map fuzzy random input variables onto fuzzy random result variables. Thus, safety assessment of structures under precise distinction of the different kinds of uncertainty is feasible. The principal approaches are illustrated by means of two model problems in the field of civil engineering in order to show the significance and the applicability of the methods.

1 Introduction

Numerical simulation of structures in the field of civil engineering is usually characterized by data and model uncertainty. Data uncertainty is uncertainty in the input variables of the numerical simulation, whereas model uncertainty is uncertainty in the underlying model of the simulation induced by uncertain parameters.

U. Reuter (✉)
Department of Civil Engineering, Technische Universität Dresden, Dresden, Germany
e-mail: Uwe.Reuter@tu-dresden.de

J.-U. Sickert · W. Graf · M. Kaliske
Institute for Structural Analysis, Technische Universität Dresden, Dresden, Germany
e-mail: Jan-Uwe.Sickert@tu-dresden.de; Wolfgang.Graf@tu-dresden.de;
Michael.Kaliske@tu-dresden.de

Y. Ermoliev et al. (eds.), *Managing Safety of Heterogeneous Systems*, Lecture Notes
in Economics and Mathematical Systems 658, DOI 10.1007/978-3-642-22884-1_14,
© Springer-Verlag Berlin Heidelberg 2012

Realistic numerical simulation of civil engineering structures requires a precise distinction and an adequate modeling of those uncertain phenomena with respect to their sources. Regarding the source, uncertainty may be classified into epistemic uncertainty and aleatory uncertainty.

Aleatory uncertainty results from variability of materials and structural processes which leads to variability in parameters. The lack of knowledge about processes and material behavior as well as about their variability is linked to epistemic uncertainty.

If sufficient statistically supported information exists for a parameter and the reproduction conditions are constant, the parameter may be described as random variable. Aleatory uncertainty is then described adequately. However, the choice of type of probability distribution function affects the result considerably.

Further advancement of the traditional probabilistic uncertainty model enables the additional consideration of epistemic uncertainty. Thereby, epistemic uncertainty is associated with human cognition, which is not limited to a binary measure. Advanced concepts allow a gradual assessment of intervals. This extension can be realized with the uncertainty characteristic fuzziness, quantified by means of fuzzy set theory.

In order to quantify both aleatory and epistemic uncertainty, imprecise probability concepts have been developed. In view of mathematical modeling, the term imprecise probability is used in a variety of models such as upper and lower probabilities, upper and lower previsions/expectations, possibilities and necessities, belief and plausibility functions, Choquet capacities, sets of probability measures, interval probabilities and further measures (see e.g., (Klir 2006; Möller and Beer 2008; Walley 1991)). Advantageously, the epistemic uncertainty is described by the uncertainty characteristic fuzziness. The uncertainty consisting of both randomness and fuzziness is summarized in the characteristic fuzzy randomness. Fuzzy random data are assessed with the aid of the uncertain measure fuzzy probability. The model fuzzy randomness describes imprecise probabilities as fuzzy sets of probability measures. It includes both randomness and fuzziness as special cases. If data only show a random characteristic, fuzziness is quantified by zero, i.e. a real-valued random variable is used. Fuzzy data without random properties are quantified by fuzzy variables.

2 Modeling Fuzzy Data

Fuzzy data result from the impossibility of an accurate characterization of single observations or information (see also (Viertl 2008)). They are modeled as fuzzy variables. A fuzzy variable $\tilde{x}$ is defined as an uncertain subset of the fundamental set $\mathbf{X}$

$$\tilde{x} = \{x, \mu_{\tilde{x}}(x) \mid x \in \mathbf{X}\}. \tag{1}$$

The uncertainty is assessed by the membership function $\mu_{\tilde{x}}(x)$. A normalized membership function $\mu_{\tilde{x}}(x)$ is defined as

$$0 \leq \mu_{\tilde{x}}(x) \leq 1 \quad \forall\, x \in \mathbb{R} \tag{2}$$

$$\exists\, x_l,\ x_r \ \text{with}\ \mu_{\tilde{x}}(x) = 1 \quad \forall\, x \in [x_l; x_r]. \tag{3}$$

A fuzzy variable $\tilde{x}$ is referred to as convex if its membership function $\mu_{\tilde{x}}(x)$ monotonically decreases on each side of the maximum value, i.e. if

$$\mu_{\tilde{x}}(x_2) \geq \min\left[\mu_{\tilde{x}}(x_1);\ \mu_{\tilde{x}}(x_3)\right] \quad \forall\, x_1, x_2, x_3 \in \mathbb{R} \ \text{with}\ x_1 \leq x_2 \leq x_3 \tag{4}$$

applies. Convex fuzzy variables are presupposed in this paper.

For example, the measurement of the position of an interface between two complementary states 1 and 2 yields only a grey-tone image. Such a measurement result may be characterized by a continuous, monotonic increasing function $B(x)$. The function $B(x)$ maps the measured values onto the interval $[0; 1]$. All measured values assignable to state 1 are rated with $B(x) = 0$ and all measured values assignable to state 2 are rated with $B(x) = 1$. The shape of the function in between depends on the gradient of the grey-tones. By differentiating function $B(x)$, the position of the interface may be modeled as fuzzy variable $\tilde{x}$. Subsequent standardization of the differentiated function yields the membership function $\mu_{\tilde{x}}(x)$ of $\tilde{x}$ (see Fig. 1).

Differing measurements for a single observation may also serve as an example. Table 1 shows an excerpt of joint width measurements. In this example three different measured values were obtained in each case for each measurement date and each measurement location, i. e., the joint width at the respective measurement locations could not be measured unequivocally as real values, but only in uncertain terms.

Such measurements must not to be mistaken for random data, because they result from the impossibility of an accurate characterization of a single observation. Random data result from the variation within a sample of single observations. They are modeled as random variables. For example, test results for evaluation of concrete strengths. Each concrete test cube represents an element of the random sample. Statistical evaluation of the sample supports quantification of the probability distribution function which belongs to the underlying random variable.

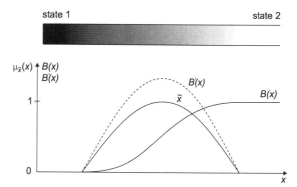

Fig. 1 Modeling of a smooth transition between two complementary states 1 and 2 as fuzzy variable $\tilde{x}$

Table 1 Excerpt of joint width measurements (courtesy of Staedtisches Vermessungsamt Dresden)

Location	Date	1st measurement [mm]	2nd measurement [mm]	3rd measurement [mm]
⋮				
5	10/05/2002	301.45	301.60	301.40
6	10/05/2002	297.00	296.90	296.95
7	10/05/2002	299.00	299.10	298.95
⋮				

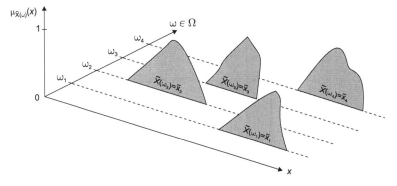

Fig. 2 Convex fuzzy realizations $\tilde{X}(\omega)$ of a fuzzy random variable $\tilde{X}$, e.g. as a result of uncertain measurements

Fuzzy random data occur if a sample of fuzzy data is observed (see also (Viertl 2011)). Modeling of fuzzy random data presented in this paper is based on the definition of fuzzy random variables according to (Möller and Beer 2004). The space of the random elementary events Ω is introduced. Each elementary event $\omega \in \Omega$ generates a fuzzy realization $\tilde{X}(\omega) = \tilde{x}$, in which $\tilde{x}$ is an element of the set $\mathbf{F}(\mathbb{R})$ of all fuzzy variables on $\mathbb{R}$. Each fuzzy variable is defined as a normalized fuzzy set, whose membership function $\mu_{\tilde{x}}(x)$ is at least segmentally continuous. Accordingly, a fuzzy random variable $\tilde{X}$ is the fuzzy result of the mapping given by

$$\tilde{X} : \Omega \to \mathbf{F}(\mathbb{R}). \tag{5}$$

Figure 2 shows four fuzzy realizations $\tilde{X}(\omega)$ of a fuzzy random variable $\tilde{X}$.

Under the assumption of convex fuzzy realizations $\tilde{x}$, a fuzzy random variable $\tilde{X}$ is characterized by a family of random α-level sets X_α

$$\tilde{X} = (X_\alpha = [X_{\alpha l}; X_{\alpha r}] \,|\, \alpha \in [0, 1]). \tag{6}$$

Modeling and Processing of Uncertainty in Civil Engineering

The interval boundaries

$$X_{\alpha l}(\omega) = \min[x \in \mathbb{R} | \mu_{\tilde{X}(\omega)}(x) \geq \alpha], \qquad (7)$$

$$X_{\alpha r}(\omega) = \max[x \in \mathbb{R} | \mu_{\tilde{X}(\omega)}(x) \geq \alpha] \qquad (8)$$

are real-valued random variables and form closed intervals.

An alternative, discrete representation of a fuzzy random variable $\tilde{X}_\tau$ is given by the $l_\alpha r_\alpha$-discretization (Möller and Reuter 2007)

$$\tilde{X} = \left(X_{\alpha_i} = [X_{\alpha_{i+1}l} - \Delta X_{\alpha_i l}; X_{\alpha_{i+1}r} + \Delta X_{\alpha_i r}] \, | \, \alpha_i \in [0, 1); \qquad (9)$$

$$X_{\alpha_n} = [X_{\alpha_n l}; X_{\alpha_n l} + \Delta X_{\alpha_n r}] \qquad | \alpha_n = 1)$$

$$\text{for } i = 1, 2, \ldots, n - 1.$$

In this definition, the terms $\Delta X_{\alpha_i l}$ and $\Delta X_{\alpha_i r}$ are correlated random variables and called random $l_\alpha r_\alpha$-increments. It holds:

$$\Delta X_{\alpha_i l} \geq 0 \text{ for } i = 1, 2, \ldots, n - 1 \qquad (10)$$

$$\Delta X_{\alpha_i r} \geq 0 \text{ for } i = 1, 2, \ldots, n. \qquad (11)$$

Numerical processing of fuzzy, random, and fuzzy random variables requires enhanced methods for structural analysis. Fuzzy stochastic analysis presented in the next section is an appropriate computational approach, which allows the mapping of fuzzy random input variables onto fuzzy random result variables. Two different approaches for mapping of fuzzy random variables are possible. The first variant is based on the bunch parameter representation of a fuzzy random variable by (Möller and Beer 2004) and is equivalent to a real-valued random variable characterized by fuzzy parameters. The second variant utilizes the $l_\alpha r_\alpha$-representation of fuzzy random variables and is equivalent to a fuzzy-valued random variable. Both variants are presented in the following.

3 Processing Fuzzy Random Variables

Numerical processing of fuzzy, random, and fuzzy random variables succeeds with the help of fuzzy stochastic structural analysis. By means of fuzzy stochastic analysis, it is possible to map fuzzy random input variables $\tilde{X}_1, \tilde{X}_2, \ldots, \tilde{X}_l$ onto fuzzy random result variables $\tilde{Z}_1, \ldots, \tilde{Z}_m$

$$M_{FSA} : \underline{\tilde{X}} \mapsto \underline{\tilde{Z}} \qquad (12)$$

$$\underline{\tilde{Z}} = (\tilde{Z}_1, \tilde{Z}_2, \ldots, \tilde{Z}_m) = M_{FSA}(\tilde{X}_1, \tilde{X}_2, \ldots, \tilde{X}_l). \qquad (13)$$

The mapping task may be achieved by means of a hierarchical three-loop computational model. The computational model can be organized in two different variants.

3.1 Variant I

An unified approach for fuzzy random variables is presented in (Krätschmer 2001). This approach is based amongst others on the similar definitions given in (Kwakernaak 1978) and (Puri and Ralescu 1986). On this basis, the bunch parameter representation has been introduced in (Möller and Beer 2004) in order to enable an efficient numerical processing. Thereby, fuzzy probability density functions $\tilde{f}(x)$ are applied, representing a fuzzy set of real-valued probability density functions $f(x)$. Therefore, $\tilde{f}(x)$ is also referred to as assessed bunch of functions $f(x)$. The bunch is described by means of fuzzy bunch parameters $\tilde{s}$. If the bunch depends on more than one $\tilde{s}$, all bunch parameters are joined in the vector $\underline{\tilde{s}}$, which represents a vector of fuzzy variables. The fuzzy probability density function $\tilde{f}(x)$ results therewith in a function $f(\underline{\tilde{s}}, x)$. This leads to the bunch parameter representation

$$\tilde{f}_X(x) = \{(f_X(\underline{s}, x), \mu(f_X(\underline{s}, x))) \mid \underline{s} \in \underline{\tilde{s}}, \mu(f_X(\underline{s}, x)) = \mu(\underline{s})\} \qquad (14)$$

of fuzzy probability density functions.

Typical fuzzy bunch parameters in engineering applications are moments of the fuzzy random variable or parameters of the function $\tilde{f}(x)$. For instance, a GUMBEL distributed fuzzy random variable, that means each $f(x) \in \tilde{f}(x)$ is GUMBEL distributed, may depend on the fuzzy bunch parameters $\tilde{s}_1 = \tilde{a}$ and $\tilde{s}_2 = \tilde{b}$. Then the fuzzy probability density function is

$$f(\underline{\tilde{s}}, x) = \tilde{s}_1 \exp(-\tilde{s}_1(x - \tilde{s}_2) - \exp(-\tilde{s}_1(x - \tilde{s}_2))) . \qquad (15)$$

This approach is different to the approaches introduced e.g. in (Feng 2001; Körner 1997) where only the expected value can be fuzzified. However, regarding the engineering application, the generalization to arbitrary bunch parameters introduced in (Möller and Beer 2004) is more appropriate. Further, the approach conforms to the definitions given in (Krätschmer 2001) as shown in (Möller and Beer 2004).

Data uncertainty (e.g. for geometry and material parameters, loading or boundary conditions) may also be characterized by fuzzy random fluctuations, which depend on external conditions. External conditions include, for example, time τ, the spatial coordinates $\underline{\theta} = \{\theta_1, \theta_2, \theta_3\}$, air pressure or temperature, which are lumped together in the parameter vector $\underline{t} = \{\underline{\theta}, \tau, \varphi\}$. The varying uncertainty of parameters (depending on arbitrary arguments $\underline{t}$) is quantified using fuzzy random functions. Based on the theories of fuzzy random variables (Möller and Beer 2004), fuzzy sets (Zimmermann 1992), and random processes (Thoft-Christensen and Baker 1982), fuzzy random functions are defined as a set of discrete fuzzy random variables

$$\tilde{X}(\underline{t}) = \{\tilde{X}_t = \tilde{X}(\underline{t}) \quad \forall \underline{t} \mid \underline{t} \in \mathbf{T}\} . \qquad (16)$$

For a fuzzy random function, a joint fuzzy probability distribution function $\tilde{F}(\underline{x}) = F(\underline{\tilde{s}}, \underline{x})$ and a joint fuzzy probability density function $\tilde{f}(\underline{x}) = f(\underline{\tilde{s}}, \underline{x})$ may be determined also in dependency of fuzzy bunch parameters (see (Sickert 2005)).

On the basis of the fuzzy bunch parameter representation, the numerical solution of the mapping may be formulated according to (13). Thereby, the fuzzy bunch parameters of all fuzzy random variables are lumped together in the vector $\underline{\tilde{s}}$ with elements $\tilde{s}_k | k = 1, \ldots, n_1$. The bunch parameters of selected fuzzy random result variables $\underline{\tilde{Z}}$ are summarized in the vector $\underline{\tilde{\sigma}}$ with elements $\tilde{\sigma}_j | j = 1, \ldots, m_1$. Typical fuzzy bunch parameters $\underline{\tilde{\sigma}}$ are the fuzzy moments or fuzzy quantiles of $\tilde{Z}_i$. The mapping of (13) is therewith transformed into the mapping

$$m : \underline{\tilde{s}} \mapsto \underline{\tilde{\sigma}} . \tag{17}$$

$$\underline{\tilde{\sigma}} = (\tilde{\sigma}_1, \ldots, \tilde{\sigma}_j, \ldots, \tilde{\sigma}_{m_1}) = m(\tilde{s}_1, \ldots, \tilde{s}_k, \ldots, \tilde{s}_{n_1}) . \tag{18}$$

By means of fuzzy analysis, it is possible to map fuzzy input variables onto fuzzy result variables. The fuzzy result variables of a fuzzy analysis may be found by applying the extension principle. Under the condition that the fuzzy variables are convex, however, α-level optimization is numerically more efficient. Applying α-discretization to the fuzzy bunch parameter, an optimization problem is solved in order to determine the α-level sets of the fuzzy bunch parameters $(\tilde{\sigma}_1, \tilde{\sigma}_2, \ldots, \tilde{\sigma}_{m_1})$. This algorithm is referred to as fuzzy analysis and described, e.g., in (Möller and Beer 2004). Each element of the input α-level sets yields a stochastic analysis. Within the stochastic analysis, a deterministic fundamental solution is processed repeatedly. Therewith, a three-loop computational algorithm is constituted (see Fig. 3). Fuzzy analysis in the form of α-level optimization establishes the outer loop, stochastic analysis the middle loop, and the deterministic fundamental solution the inner loop.

By means of stochastic analysis, random input variables are mapped onto random result variables by applying e.g. Monte Carlo techniques. The deterministic fundamental solution represents an arbitrary computational model, e.g. a finite element model. No special properties of the applied computational model, such as linearity, monotony or convexity, are presupposed.

The mapping algorithm variant I has been extended to the mapping

$$M_{FSA} : \underline{\tilde{X}}(\underline{t}) \mapsto \underline{\tilde{Z}}(\underline{t}) \tag{19}$$

with fuzzy random functions $\underline{\tilde{X}}(\underline{t})$ and $\underline{\tilde{Z}}(\underline{t})$ in the input space as well as in the result space (see (Sickert 2005)). This extension permits the development of the fuzzy stochastic finite element method (FSFEM). Known and extended approaches of the stochastic finite element method are applied within the stochastic analysis.

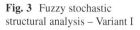
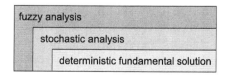

Fig. 3 Fuzzy stochastic structural analysis – Variant I

Fig. 4 Fuzzy stochastic structural analysis – Variant II

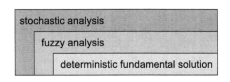

3.2 Variant II

The second variant utilizes the $l_\alpha r_\alpha$-representation of fuzzy random variables. In comparison with the first variant, the sequence of stochastic analysis and fuzzy analysis is changed (see Fig. 4).

A precondition for this variant is that the individual fuzzy random variables of (13) are represented in each case by a multivariate probability distribution function for $l_\alpha r_\alpha$-increments.

This probability distribution function $F_{\tilde{X}}(\tilde{x})$ of a fuzzy random variable $\tilde{X}$ discretized by $2n$ random $l_\alpha r_\alpha$-increments $\Delta X_{\alpha_1 l}, \Delta X_{\alpha_2 l}, \ldots, \Delta X_{\alpha_1 r}$ is defined as the $2n$-dimensional probability distribution function

$$_{lr}F_{\tilde{X}}(\tilde{x}) = P\left(\{\omega \mid \Delta X_{\alpha_1 l}(\omega) \le \Delta x_{\alpha_1 l}, \ldots, \Delta X_{\alpha_1 r}(\omega) \le \Delta x_{\alpha_1 r}\}\right) \quad (20)$$
$$= P\left(\{\Delta X_{\alpha_1 l} \le \Delta x_{\alpha_1 l}, \ldots, \Delta X_{\alpha_1 r} \le \Delta x_{\alpha_1 r}\}\right),$$

whereby $\Delta x_{\alpha_1 l}, \Delta x_{\alpha_2 l}, \ldots, \Delta x_{\alpha_1 r}$ are the $l_\alpha r_\alpha$-increments of the fuzzy realization $\tilde{x}$.

If the probability distribution functions are known for the fuzzy random variables $\tilde{X}_1, \tilde{X}_2, \ldots, \tilde{X}_l$, the stochastic analysis begins with the simulation of s sequences of the fuzzy realizations $\tilde{x}_1, \tilde{x}_2, \ldots, \tilde{x}_l$. This marks a distinction between the second variant and the first one. By means of the latter, it is only possible to simulate real-valued realizations of the individual trajectories. Because fuzzy realizations are immediately available, these may be used as input variables for a fuzzy analysis. By means of fuzzy analysis, the sequence of fuzzy result variables $\tilde{z}_1, \tilde{z}_2, \ldots, \tilde{z}_m$ corresponding to each sequence $\tilde{x}_1, \tilde{x}_2, \ldots, \tilde{x}_l$ may be computed with the aid of α-level optimization. The algorithm given in (Möller and Beer 2004) for solving the α-level optimization may also be applied in this case. This results in s sequences of fuzzy result variables $\tilde{z}_1, \tilde{z}_2, \ldots, \tilde{z}_m$, i.e. a sample comprised of s fuzzy realizations is obtained for each fuzzy random variable $\tilde{Z}_j$. A statistical evaluation of the samples yields an empirical $2n$-dimensional probability distribution function for the $l_\alpha r_\alpha$-increments of each fuzzy random variable $\tilde{Z}_j$, which serve as unbiased estimators for the distributions of $\tilde{Z}_j$. Theoretical probability distribution functions may also be derived from the latter as required.

The preference for a variant of fuzzy stochastic structural analysis depends on the available uncertain data and the engineering problem. Both bunch parameter representation and $l_\alpha r_\alpha$-representation of a fuzzy random variable may be obtained by statistical evaluation of a concrete sample comprised of fuzzy elements. Bunch parameter representation, however, does not permit the precise reproduction of the underlying sample elements (e.g. by Monte Carlo Simulation). $l_\alpha r_\alpha$-representation permits the precise reproduction of samples comprised of fuzzy elements. The interrelation between the two variants is reflected in the associated fuzzy probability distribution functions. The marginal distributions of the multidimensional probability distribution function according to (20) correspond to the left and right boundary functions of the fuzzy probability distribution function according to the bunch parameter representation. The uncoupled treatment of the marginal distributions does not take account of the dependencies between the different α-level sets. The interaction between the α-level sets of the fuzzy realizations of a fuzzy random variable is only taken into account using the $l_\alpha r_\alpha$-representation. For this reason, it is possible to reproduce the underlying fuzzy sample elements. However, in the field of engineering, both variants of fuzzy stochastic structural analysis can be applied for static and dynamic structural analysis and for the assessment of structural safety and durability as well as robustness. Applications from the field of civil engineering, particularly in structural engineering are presented in the following.

4 Examples

4.1 Analysis of a Strengthened Hypar Shell Roof Structure[1]

The capabilities of the mapping according to variant I is demonstrated by means of an example. The example focuses on the assessment of the structural behavior of a strengthened hypar shell roof structure. The shell was built of reinforced concrete in the sixties and recently strengthened by means of a thin textile reinforced concrete layer (Ortlepp et al. 2008). Textile reinforced concrete is a new composite material made of fine-grained concrete and multiaxial warp knitted fabrics (textiles). The textiles consist of filament yarns (rovings) which are connected with the aid of stitching yarn. Each roving is composed by a large number of single filaments. The filaments can consist of different material, e.g. alkali-resistant glass (AR glass) or carbon. Here, a study is presented in which some uncertain parameters are modeled as fuzzy and fuzzy random variables. The study serves to show the applicability of the proposed methods in principle.

An extended layer model with specific kinematics, a so-called multi-reference-plane model (MRM) is used to describe the load-bearing behavior of RC constructions with textile strengthening. The MRM consists of concrete layers and steel reinforcement layers of the old construction, the strengthening layers consist of the inhomogeneous material textile concrete, and the interface layers. The MRM is repeatedly applied as deterministic fundamental solution within the fuzzy stochastic

---

[1] In collaboration with Dipl.-Ing. Stephan Pannier.

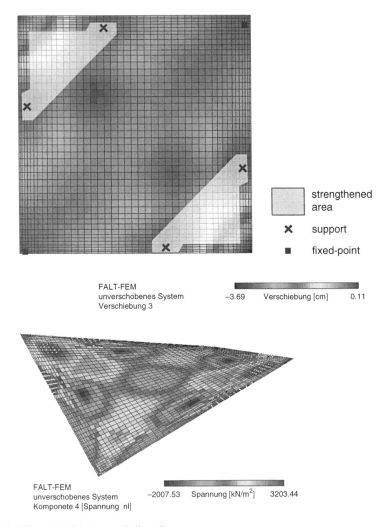

Fig. 5 FE model of the hypar shell roof

analysis. Figure 5 shows the FE model. The structure is discretized by means of 2025 MRM elements. The elements are modeled with 19 layers. The steel reinforcement and textiles are specified as an uniaxial smeared layer in each case.

In order to describe the composite structure comprised of reinforced concrete and textile strengthening, different nonlinear material laws are applied to the individual sub-layers of concrete, steel and textile. Endochronic material laws for concrete and steel are applied for general loading, unloading and cyclic loading processes, and taking into account the accumulated material damage during the load history (Möller et al. 1997). In the case of cyclic loading, the textile-reinforced concrete layers are split into sub-layers of fine-grained concrete and of textile reinforcement.

Modeling and Processing of Uncertainty in Civil Engineering

Table 2 Uncertain input variables

Name	Type	Parameter 1 / 3	Parameter 2
old concrete			
f_c (compressive strength)	log-normal	$\mu = 43$ N/mm^2 $x_0 = 18$ N/mm^2	$\tilde{\sigma} = \langle 4,\ 5,\ 6 \rangle$ N/mm^2
c (factor for tensile strength)	fuzzy	$\tilde{c} = \langle -1.0, 0.0, 1.0 \rangle$	
e_{fak} (tension stiffening)	fuzzy	$\tilde{e}_{fak} = \langle 10,\ 15,\ 20 \rangle$	
fine-grained concrete			
f_c	log-normal	$\mu = 85$ N/mm^2 $x_0 = 24$ N/mm^2	$\tilde{\sigma} = \langle 6,\ 7,\ 8 \rangle$ N/mm^2
c	fuzzy	$\tilde{c} = \langle -2.5, -2.0, -1.5 \rangle$	
e_{fak}	fuzzy	$\tilde{e}_{fak} = \langle 10,\ 15,\ 20 \rangle$	
carbon textiles			
E (Young's modulus)	fuzzy	$\tilde{E} = \langle 2.0, 2.2, 2.4 \rangle 10^5$ N/mm^2	
f_{cr} (tensile strength)	log-normal	$\mu = 1.1 \cdot 10^3$ N/mm^2 $x_0 = 5 \cdot 10^2$ N/mm^2	$\tilde{\sigma} = \langle 1.5, 1.75, 2.0 \rangle 10^2$ N/mm^2
loads			
distributed snow load s	Ex-max- type I	$\mu = 0.87$ kN/m^2	$\tilde{\sigma} = \langle 0.25,\ 0.28,\ 0.31 \rangle$ kN/m^2
distributed wind load w	Ex-max- type I	$\mu = 0.29$ kN/m^2	$\tilde{\sigma} = \langle 0.07,\ 0.09,\ 0.11 \rangle$ kN/m^2

The endochronic material law for concrete is adapted to the fine-grained concrete. A nonlinear elastic-brittle material law is used for the textile reinforcement. Under cyclic loading, damage occurs in the strengthening layer in the fine-grained concrete matrix and the textile structure as well as disruption of the bond between the old concrete and the strengthening layer. These forms of damage and the additional plastic deformations may be described theoretically by means of plasticity and continuum damage theory (Möller et al. 2005). The material laws are dependent on uncertain material parameters which are summarized in Table 2. The listed fuzzy parameters are triangular fuzzy numbers using the common abridged notation.

The shell roof is loaded by uncertain distributed snow and wind loads. The summarized load

$$\tilde{q} = \nu(\tilde{s} + \tilde{w}) \qquad (21)$$

is increased incrementally up to system failure. System failure is defined as the situation, when the applied modified Newton-Raphson iteration cannot find an equilibrium for load factors $\nu < 1$.

The fuzzy failure probability of system failure is computed by means of mapping variant I. Due to the high computational effort of one deterministic fundamental solution, only 370 FE runs were possible. In order to perform the Monte Carlo simulation within the stochastic analysis, a surrogate meta-model is created which fits very well to the results of the FE analysis. The meta-model is based on an artificial neural network (see e.g. (Pannier et al. 2009; Papadrakakis et al. 1996)). Figure 6

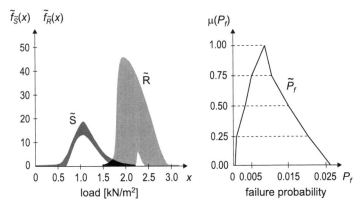

Fig. 6 R-S-plot and fuzzy failure probability $\tilde{P}_f$

displays the computed fuzzy failure probability $\tilde{P}_f$. The figure contains further a R-S-plot comparing the empirical fuzzy density functions of the load sum S and the structural resistance against a unique increase of q according to (21). The overlap of density functions is an evidence for realizations which lead to structural failure.

4.2 Safety Assessment of a Shoring Wall

Variant II of fuzzy stochastic structural analysis is demonstrated in the following by considering the safety assessment of an earth pressure loaded shoring wall. In this example, information on the earth pressure is provided by pressure cell measurements. The measurements started in 1999 and continue. A detailed presentation of the pressure cell measurements can be found in (Franke et al. 2003). Two pressure cells close to each other are installed twice in comparable depth. Nevertheless, different measured values are obtained in each case for each measurement date and each pressure cell location, that is, earth pressure at the respective measurement depth could not be measured precisely as real values, but only in uncertain terms. The different measured values at the same measurement date yield fuzzy data, whereas the variation in the values of a single pressure cell over time signals randomness. These uncertainties follow from uncertain influencing factors like consistency, density, stiffness, angle of internal friction of the soil as well as the characteristics of the interface and the gradient and loads of the surface. Table 3 shows a short excerpt of the measured earth pressure data, computed as the difference of total pressure and water pressure.

The normal conventional approach does not take into account this uncertainty, the uncertain information is reduced to an arithmetic mean. In order to realistically analyze the measured data, however, this uncertainty must be taken into account. The measured values at each measurement data lie in an interval which may

Modeling and Processing of Uncertainty in Civil Engineering 303

Table 3 Excerpt of the measured earth pressure data (Franke et al. 2003) (GL = ground line)

Date	Cell A1 4.94 m under GL [kPa]	Cell B1 4.94 m under GL [kPa]	Cell C2 4.80 m under GL [kPa]	Cell D3 4.83 m under GL [kPa]
11.11.1999	19.3	17.3	33.5	40.0
⋮	⋮	⋮	⋮	⋮
10.05.2000	26.2	23.7	40.9	50.5
⋮	⋮	⋮	⋮	⋮
04.11.2002	28.5	27.5	42.7	48.0
⋮	⋮	⋮	⋮	⋮
10.03.2004	20.9	19.4	31.1	35.0
⋮	⋮	⋮	⋮	⋮
14.07.2006	31.2	36.7	44.4	53.5
⋮	⋮	⋮	⋮	⋮
15.02.2007	22.1	24.1	29.9	36.5
⋮	⋮	⋮	⋮	⋮

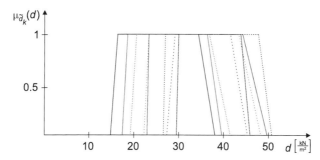

Fig. 7 Fuzzy earth pressures $\tilde{d}_k$ ($k = 1, 2, \ldots, 8$) at eight measurement dates

be considered as a support of a fuzzy variable. It is thus appropriate to model the measured values as fuzzy variables. All values lying between the smallest and largest values measured on each measurement data are possible measurement results. These form the support of the corresponding fuzzy variable. All values lying between the mean values of the two pressure cells installed close to each other are chosen as the 'best possible crisped' measurement results and valuated by membership value one. Figure 7 shows exemplarily the membership functions of the fuzzy earth pressures at eight measurement dates.

Statistical evaluation of all modeled fuzzy values yields a fuzzy random variable for the earth pressure. In order to model the earth pressure as fuzzy random variable, 280 measurement dates are evaluated. For each measurement date, a fuzzy value for the earth pressure is available. Thus, the fuzzy values are regarded as elements of a sample and discretized by $n = 3$ α-levels $\alpha_1 = 0$, $\alpha_2 = 0.5$ and $\alpha_3 = 1$. Statistical evaluation of the sample yields the empirical multivariate probability distribution function for the $l_\alpha r_\alpha$-increments. The obtained empirical probability distribution function is an unbiased estimator for the distribution of the fuzzy random earth pressure $\tilde{D}$.

In the following, the obtained fuzzy random earth pressure $\tilde{D}$ is used exemplarily as a loading of a bottom-fixed shoring wall. The structural analysis is applied for a 1 m wide section of the shoring wall. The earth pressure loading is – following the conventional approach – assumed to be triangular.

Fuzzy stochastic structural analysis variant II begins with the Monte Carlo simulation of s fuzzy realizations $\tilde{d}_1, \tilde{d}_2, \ldots, \tilde{d}_s$ of the fuzzy random earth pressure $\tilde{D}$. Because of the evaluated measurement results, 280 fuzzy realizations $\tilde{d}_1, \tilde{d}_2, \ldots, \tilde{d}_{280}$ are already available. Monte Carlo simulation thus means that a single fuzzy realization is drawn from the 280 fuzzy realizations in each case.

The fuzzy realizations are discretized by $n = 3$ α-levels $\alpha_1 = 0$, $\alpha_2 = 0.5$ and $\alpha_3 = 1$. The α-level set $D^u_{\alpha_i}$ of each fuzzy value $\tilde{d}_u$ ($u = 1, 2, \ldots, 280$) represents the crisp subspace $\underline{X}^u_{\alpha_i}$ in each case. The safety assessment requires the check if a subspace $\underline{X}^u_{\alpha_i}$ lies completely or at least partially in the failure domain $\underline{X}_f$. The interval bounds $P_{f,\alpha_i l}$ and $P_{f,\alpha_i r}$ of the α-level sets P_{f,α_i} of the fuzzy failure probability $\tilde{P}_f$ can thus be estimated according to

$$\hat{P}_{f,\alpha_i l} = \frac{\#\left\{u \mid \underline{X}^u_{\alpha_i} \subseteq \underline{X}_f\right\}}{s}, \tag{22}$$

$$\hat{P}_{f,\alpha_i r} = \frac{\#\left\{u \mid \underline{X}^u_{\alpha_i} \cap \underline{X}_f \neq\right\}}{s}. \tag{23}$$

The symbol $\#\{\cdot\}$ denotes the number of subspaces $\underline{X}^u_{\alpha_i}$ for which the requirements $\underline{X}^u_{\alpha_i} \subseteq \underline{X}_f$ and $\underline{X}^u_{\alpha_i} \cap \underline{X}_f \neq$ are fulfilled respectively.

Structural analysis of the shoring wall requires the analysis of material failure as well as base failure, gliding and tilting. Exemplarily, the load bearing capacity regarding a critical base point moment of $M_{crit} = 196 \frac{\text{kNm}}{\text{m}}$ (material failure) is analyzed. The problem is characterized by monotony. The check if a subspace $\underline{X}^u_{\alpha_i}$ lies completely or at least partially in the failure domain $\underline{X}_f$ is thus reduced to a boundary value problem. The deterministic fundamental solution

$$M = \frac{1}{2} l d \frac{l}{3} = \frac{l^2 d}{6} \quad \text{with} \quad l = 4,9\,\text{m} \tag{24}$$

is given by the equilibrium condition of the 4.9 m high shoring wall. The obtained fuzzy failure probability $\tilde{P}_f$ is given in Fig. 8.

The result shows that no subspace $\underline{X}^u_{\alpha_i}$, i.e., no α-level set $D^u_{\alpha_i}$ lies completely in the failure domain $\underline{X}_f$. 50 subspaces lie at least partially in the failure domain for α-level $\alpha_1 = 0$ and 3 subspaces for α-level $\alpha_3 = 1$, respectively. There is the

Fig. 8 Fuzzy failure probability $\tilde{P}_f$

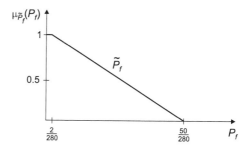

possibility – evaluated with the membership value $\mu_{\tilde{P}_f}(\frac{3}{280}) = 1$ and $\mu_{\tilde{P}_f}(\frac{50}{280}) = 0$, respectively – of a failure probability of $\frac{3}{280}$ and $\frac{50}{280}$, respectively.

The interval bounds of the fuzzy failure probability for α-level $\alpha_3 = 1$ represent the safety level obtained by a pure stochastic safety assessment. That is, a pure stochastic evaluation of the safety level with a reduction of the measurement results to the arithmetic mean value and modeling of the earth pressure as a random variable would lead to an underestimation of the safety level.

5 Conclusion

In this paper, two variants of fuzzy stochastic structural analysis for modeling and processing of uncertain data in civil engineering are presented. Uncertain data are classified into fuzzy, random, and fuzzy random data. Fuzzy data result from the impossibility of an accurate characterization of single observations or information. Random data result from the variation within a sample of single observations. Fuzzy random data occur if a sample of fuzzy data is observed. Fuzzy stochastic structural analysis is performed by the mapping of fuzzy random input variables onto fuzzy random result variables with the aid of a hierarchical three-loop computational model consisting of fuzzy analysis, stochastic analysis, and deterministic fundamental solution, which may be organized in two different variants. The presented methods presuppose convex fuzzy variables. The application of non-convex fuzzy variables is currently under investigation (Reuter 2008).

References

Feng, Y. (2001). The variance and covariance of fuzzy random variables and their applications. *Fuzzy Sets and Systems, 120*, 487–497.

Franke, D., Arnold, M., Bartl, U., & Vogt, L. (2003). Erddruckmessungen an der Kelleraußenwand eines mehrgeschossigen Massivbaus. *Bauingenieur, 78*, 125–130.

Klir, G. J. (2006). *Uncertainty and Information: Foundations of Generalized Information Theory*. Hoboken: Wiley-Interscience.

Körner, R. (1997). *Linear Models with Random Fuzzy Variables*. Bergakademie Freiberg: Dissertation.

Krätschmer, V. (2001). A unified approach to fuzzy random variables. *Fuzzy Sets and Systems, 123*, 1–9.

Kwakernaak, H. (1978). Fuzzy random variables – I. Definitions and theorems. *Information Sciences, 15*, 1–29.

Möller, B., & Beer, M. (2004). *Fuzzy Randomness – Uncertainty in Civil Engineering and Computational Mechanics*. Berlin: Springer.

Möller, B., & Beer, M. (2008). *Engineering computation under uncertainty – Capabilities of nontraditional models. Computers & Structures, 86*, 1024–1041.

Möller, B., Graf, W., & Kluger, J. (1997). Endochronic material modelling in nonlinear FE-analysis of folded plates. In P. Anagnostopoulos, G. M. Carlomagno, & C. A. Brebbia (Eds.), *CMEM VIII, Computational Mechanics Publ.*, 97–106. Southampton.

Möller, B., Graf, W., Hoffmann, A., & Steinigen, F. (2005). Numerical simulation of RC structures with textile reinforcement. *Computers & Structures, 83*, 1659–1688.

Möller, B., & Reuter, U. (2007). *Uncertainty Forecasting in Engineering*. Berlin: Springer.

Ortlepp, R., Weiland, S., & Curbach, M. (2008). Restoration of a hypar concrete shell using carbonfibre textile reinforcement concrete. In M.C. Limbachiya, & H.Y. Kew (Eds.), *Proceedings of the International Conference Excellence in Concrete Construction Through Innovation*, pp. 357–364. London: Taylor & Francis.

Pannier, S., Sickert, J.-U., & Graf, W. (2009). Patchwork approximation scheme for reliability assessment and optimization. In H. Furuta, D. M. Frangopol, M. Shinozuka (Eds.), *Safety, Reliability and Risk of Structures, Infrastructures and Engineering Systems, Proceedings of the 10th Int. Conference on Structural Safety and Reliability*, 482–489. London: Taylor & Francis.

Papadrakakis, M., Papadopoulos, V., & Lagaros, N. D. (1996). Structural reliability analysis of elastic-plastic structures using neural networks and Monte Carlo simulation. *Computer Methods in Applied Mechanics and Engineering, 136*, 145–163.

Puri, M. L., & Ralescu, D. A. (1986). Fuzzy random variables. *Journal of Mathematical Analysis and Applications, 114*, 409–422.

Reuter, U. (2008). Application of non-convex fuzzy variables to fuzzy structural analysis. In D. Dubois, et al. (Eds.), *Soft Methods for Handling Variability and Imprecision, Advances in Soft Computing* vol. 48, pp. 369–375. Berlin: Springer.

Sickert, J.-U. (2005). *Fuzzy Random Functions and their Application in Structural Analysis and Safety Assessment* (in german). TU Dresden, Veröffentlichungen Institut für Statik und Dynamik der Tragwerke, Heft 9, Dresden: Dissertation.

Thoft-Christensen, P., & Baker, M.J. (1982). *Structural Reliability Theory and Its Applications*. Berlin: Springer.

Viertl, R. (2008). Fuzzy models for precision measurements. *Mathematics and Computers in Simulation, 79*, 874–878.

Viertl, R. (2011). *Statistical Methods for Fuzzy Data*. Chichester: Wiley.

Walley, P. (1991). *Statistical Reasoning with Imprecise Probabilities*. New York: Chapman Hall.

Zimmermann, H.-J. (1992). *Fuzzy Set Theory and Its Applications*. Boston: Kluwer.

Optimal Design and Sensitivity of Large Spatial Trusses Under Uncertainty

Simone Zier

Abstract One trend in civil works moves towards higher and higher buildings. But with its height, not only the monumentality and impressiveness rise, but also the sensitivity with respect to applied loads increases, and external influences become more crucial. Because of that, this paper deals with the study of the change of the design and the robustness in dependence of the height of the structures. The basis for our consideration is provided by a spatial n-storey truss which will be increased successively, and which is affected by applied random forces. The recourse problem will be formulated in general and in the standard form of stochastic linear programming (SLP). After the formulation of the stochastic optimization problem, the Recourse Problem based on Discretization (RPD) and the Expected Value Problem (EVP) are introduced as representatives of substitute problems. The resulting (large) linear programs (LP) can be solved efficiently by means of usual LP-solvers. Several numerical results are presented.

1 General Formulation of the Problem

A truss is a structure consisting of a certain number B of rods which are pin-connected among each other and with the foundation at a certain number of nodes (Spillers 1972).

In optimal design the aim is to minimize a certain cost function, here the weight of the structure

$$G_0(A) = \sum_{i=1}^{B} \gamma_i L_i A_i, \tag{1}$$

S. Zier (✉)
Federal Armed Forces University Munich, Aerospace Engineering and Technology, Neubiberg/Munich, Germany
e-mail: simone.zier@unibw.de

Y. Ermoliev et al. (eds.), *Managing Safety of Heterogeneous Systems*, Lecture Notes in Economics and Mathematical Systems 658, DOI 10.1007/978-3-642-22884-1_15,
© Springer-Verlag Berlin Heidelberg 2012

where L_i is the length of rod i, A_i is its cross-sectional area and γ_i is a weight factor. Of course, only non-negative cross-sectional areas, i.e., $A_i \geq 0, i = 1, \ldots, B$, are reasonable, and for practical reasons, we restrict the cross-sectional areas by an upper bound $A_{\max}$.

Using the first collapse theorem, the necessary and sufficient survival conditions of an elasto-plastic structure consist of the yield condition and the equilibrium condition. For the equilibrium condition the constraint

$$CF = P \tag{2}$$

has to hold where C represents the equilibrium matrix, F the interior normal forces and P denotes the external load vector. We suppose that the applied load is not deterministic, but depends on certain stochastic parameters, describing random load variations due to e.g. wave, wind, snow loads, traffic, etc.. If we denote the stochastic elements as ω we get the external loads $P(\omega)$ as functions of ω, and the equilibrium condition (2) is therefore given by

$$CF = P(\omega) \qquad \text{almost surely } (a.s.). \tag{3}$$

While the actual realization of the random element ω is not known, we suppose that its distribution is known.

Since we suppose that the truss is built of elasto-plastic material, the behavior under load is at first elastic and can be described linearly by Hooke's Law and then after exceeding a limit – called yield stress – the material starts yielding. This behavior is described by the yield condition which can be written as

$$F^L \leq F \leq F^U \tag{4}$$

for trusses. These bounds are given by the tension and compression, resp., plastic capacities

$$F^U = N_{ipl}^U = \sigma_{yi}^U \cdot A_i \tag{5a}$$

and

$$F^L = N_{ipl}^L = |\sigma_{yi}^L| \cdot A_i = -\sigma_{yi}^L \cdot A_i, \tag{5b}$$

where σ_{yi}^U and σ_{yi}^L is the upper and lower, resp., yield stress.

For stability reasons, the lower bound of the yield stress σ_y^L is reduced by a factor ϕ, $0 < \phi < 1$, which defines the buckling stress $\sigma_{bL} = \phi \sigma_L$.

Altogether the stochastic problem to obtain the optimal design of trusses can be formulated as

$$\min G_0(A) = \sum_{i=1}^{B} \gamma_i L_i A_i \tag{6a}$$

Optimal Design and Sensitivity of Large Spatial Trusses Under Uncertainty

s.t.

$$CF = P(\omega) \qquad a.s. \tag{6b}$$

$$F^L \le F \le F^U \tag{6c}$$

$$\sigma \ge \sigma_{bL} = \phi \sigma_y^L \tag{6d}$$

$$0 \le A \le A_{\max}. \tag{6e}$$

1.1 Recourse Problem as LP

A basic principle to cope with uncertainty is based on compensation, corrections, hence, recourse (Kall and Wallace 1994). Recourse is the ability to take corrective action – such as repair, strengthening, etc. – after a random event has taken place. We add so called recourse costs $Q_1(y)$ to the initial costs $G_0(A)$ in the case that the yield condition is violated (Marti 2008). Here, y are auxiliary variables introduced to describe whether the yield condition is violated or not. Such an approach, considering the sum of the appearing costs, is a standard one in the reliability-based structural optimization. More precisely, the objective function represents the life cycle costs consisting of the initial construction costs and the expected cost consequences related to partial or total system failure (Gasser and Schuëller 1998, 2002).

By introducing the technology matrix T and the recourse matrix W and by using some further notations of (Zier 2008) the problem can be formulated in the standard form of a stochastic linear programming with recourse (Kall and Wallace 1994)

$$\min c^T x + q^T y(\omega) \tag{7a}$$

s.t.

$$Tx + Wy(\omega) = h(\omega) \qquad a.s. \tag{7b}$$

$$x, y(\omega) \ge 0, \tag{7c}$$

with the initial costs $G_0(x) = c^T x$ and the recourse costs $Q_1(y) = q^T y(\omega)$ in case of violation of the yield condition.

1.2 Substitute Problems

In SLP (7a-c) we have the problem that the vector $h(\omega)$ is random because it includes the stochastic external load vector $P(\omega)$. Thus, the building of appropriate substitute problems is necessary.

1.2.1 Expected Value Problem

A first possibility is to consider the Expected Value Problem (EVP). It is characterized by replacing all stochastic variables by their expectations (Stöckl 2003). In the present case, the random vector $h(\omega)$ is replaced by its expectation $\overline{h}$. The rest remains unchanged. Problem (7a-c) turns then to

$$\min c^T x + q^T y \tag{8a}$$

s.t.

$$Tx + Wy = \overline{h} \tag{8b}$$

$$x, y \geq 0 \tag{8c}$$

with

$$\overline{h} = \left(0^T, EP(\omega)^T\right)^T. \tag{8d}$$

That is a simple linear program with the same size as the original one which can be solved by an ordinary LP-solver. The disadvantage is that the resulting optimal design fulfills the survival condition only for one single load case, the expected load. Therefore it is not very robust, because a marginal change of the load may lead to a failure of the structure.

1.2.2 Recourse Problem with Discretization

Alternatively, it is possible to build the Recourse Problem with Discretization (RPD) of the applied load distribution $P(\omega)$ (Stöckl 2003). Each realization P_r of $P(\omega)$ is assumed to be taken with probability $p_r, r = 1, \ldots R,$ where R is the number of realizations.

The mathematical formulation of the Recourse Problem with Discretization is

$$\min E(c^T x + q^T y) = c^T x + \sum_{r=1}^{R} p_r q^T y^r \tag{9a}$$

s.t.

$$Tx + Wy^1 \qquad\qquad = h^1$$
$$\vdots \qquad\qquad \ddots \qquad\qquad \vdots \tag{9b}$$
$$Tx + \qquad\qquad Wy^R = h^R$$

$$x, \ y^r \geq 0, \ r = 1, \ldots, R \tag{9c}$$

with

$$h^r = \left(0^T, P_r^T\right)^T, \ r = 1, \ldots, R. \tag{9d}$$

The occurrence of R constraints of the form $Tx + Wy = h$, one for each realization, is a certain disadvantage of this method since the problem size increases with the number of realizations.

2 Numerical Example

The basis for our consideration is provided by a spatial n-storey truss. Four deterministic forces are assumed to act on the top and $4n$ stochastic ones from the left and the back, respectively. The 1-storey and general n-storey truss can be seen in Figs. 1 and 2, respectively. The truss consists of $4n + 4$ nodes and $16n$ rods.

The material properties and acting forces can be seen in Table 1. The structure is assumed to have a length of $1m$ and each single storey has a height of $0.5\ m$. Altogether the height of the n-storey is $n \cdot 0.5\ m$. The rods are made of steel with an elasticity modulus of $208\,000 N/mm^2$ and yield stresses $\sigma^U = -\sigma^L = 2\,400 N/mm^2$. In the following, circular cross-sectional areas are considered. Let the stochastic force be normally distributed with an expectation of $100\,000\ N$. For the Recourse Problem with Discretization and for the calculations of the probabilities of failure, we discretize the normal distribution by using 17 realizations as shown in Fig. 3.

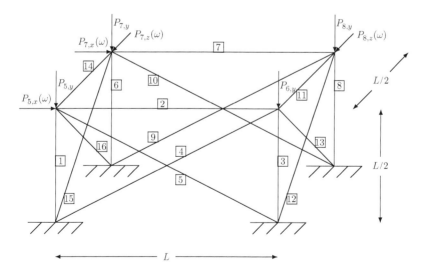

Fig. 1 1-storey spatial truss

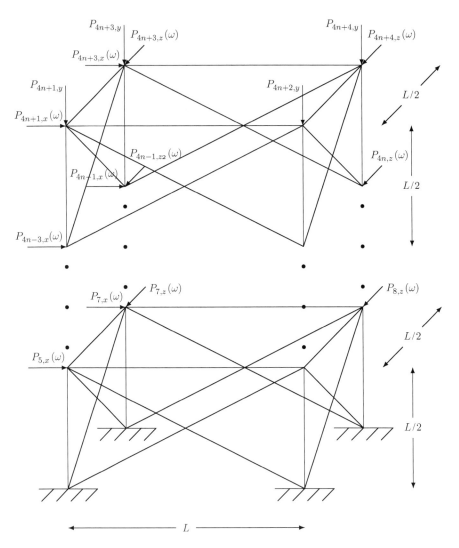

Fig. 2 n-storey spatial truss

For practical reasons, we restrict the diameter $d_i, i = 1, \ldots, B$, of each rod to 10% of its length. Therefore, we get upper limits of the cross-sectional areas of

$$A_{i,\max} = r_{i,\max}^2 \cdot \pi = \left(\frac{1}{2} d_{i,\max}\right)^2 \cdot \pi = \left(\frac{1}{2} \cdot 0.1 \cdot L_i\right)^2 \cdot \pi$$
$$= 2.5 \cdot 10^{-3} \cdot L_i^2 \pi, \ i = 1, \ldots, B. \tag{10}$$

Optimal Design and Sensitivity of Large Spatial Trusses Under Uncertainty   313

Table 1 Input parameters

PARAMETER	VALUE
length of the structure	$1000 mm$
height of the structure	$n \cdot L/2 = n \cdot 500 mm$
material: steel (100Cr6)	
elastic modulus E	$20.8 \cdot 10^4 N/mm^2$
yield stress concerning tension σ_U	$2.4 \cdot 10^3 N/mm^2$
yield stress concerning compression σ_L	$-2.4 \cdot 10^3 N/mm^2$
acting deterministic forces	$P_{4n+1,y} = P_{4n+2,y} = P_{4n+3,y} = P_{4n+4,y} = -10^4 N$
acting stochastic forces	$P_{2k+1,x} = P_{2k+3,x} = P_{2k+3,z} = P_{2k+4,z}$
	$\sim N(\mu, \sigma^2) = N(10^5 N, \sigma^2), k = 1, \ldots, n$

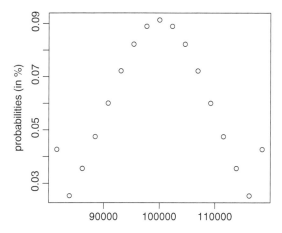

Fig. 3 Discretization of the normal distribution $N \sim (10^5, 10^8)$ with 17 realizations

2.1 Variation of the Number of Storeys

At first, we will study the influence of the height of the structure on the optimal design. Therefore, we will increase the number of storeys successively. The standard deviation of the normally distributed loads is assumed to be 10% of the expectation.

2.1.1 Expected Value Problem

We will first consider the Expected Value Problem (EVP). Since most of the applied load weigh on the lower rods, we will focus our consideration on the rods of the first floor. In the following, a selection of these rods will be studied. Due to clearness, the n-storey will be redisplayed with the labels of the studied rods.

Since the load has to be drained off to the foundation on the bottom, most of the load acts on the vertical rods. The more storeys we have, the more forces act, and

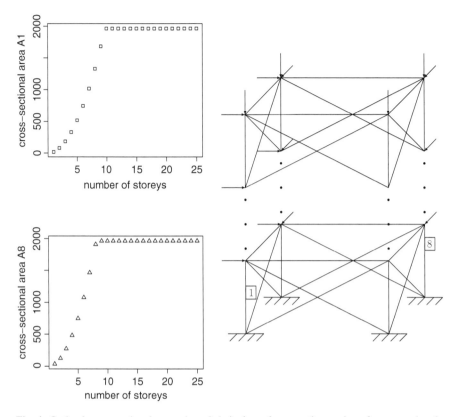

Fig. 4 Optimal cross-sectional areas A_1 and A_8 in dependence on the number of storeys using the EVP

therefore, the cross-sectional areas A_1 and A_8 increase with the number of storeys, as can be seen in Fig. 4. They rise till they reach a certain limit, namely the upper bound given in (10). Then they remain constant.

As soon as the limit is adopted, the vertical rods can carry no additional loads and therefore, it is tried to distribute them to the diagonal rods. Up to now, the diagonal rods on the side surface have hardly been necessary yet. But now, the cross-sectional areas of the diagonal rods of the front and right-hand side in Fig. 5 rise steeper with increasing number of storeys. Taking a look at the cross-sectional areas of the rods 4, 5, 12 and 13, we notice an interesting phenomenon. The 15-storey seems to be kind of an outliers, since it destroys the monotone behavior. Numerical instabilities cannot be excluded, since different solvers lead to slightly different values, however none of it finds a suitable result for the 15-storey. But the data is insofar consistent since the comparison of the rods among each other is identical. While for example the cross-sectional area A_4 is at the beginning smaller than A_5, see Fig. 6b, these relationship turns around when the vertical rods adopt their limit. There is no exception of this behavior, also not at the outlier, the 15-storey.

Optimal Design and Sensitivity of Large Spatial Trusses Under Uncertainty   315

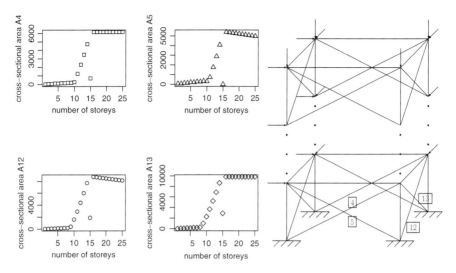

Fig. 5 Optimal cross-sectional areas A_4, A_5, A_{12} and A_{13} in dependence on the number of storeys using the EVP

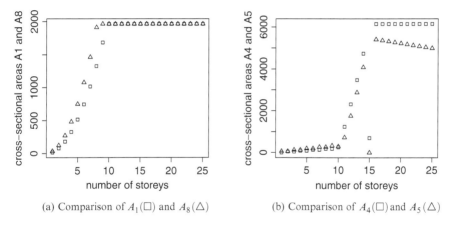

Fig. 6 Comparison of different cross-sectional areas in dependence on the number of storeys using the EVP

While the vertical rods A_1 and A_8 seem to be identical at the first glance, see Fig. 4, the closer look in Fig. 6a shows a difference. The cross-sectional area A_8 rises faster and adopts the upper limit earlier. The reason for that is the right angle between the stochastic forces and these rods. Since trusses can't carry shear forces, there is no possibility to drain off the stochastic forces directly. Only in connection with the diagonal rods, they can take some of the horizontal load. Since the rods which connect the left- with the right-hand side – which are necessary for the usage of rod 8 – are twice as long and therefore anchored in a flatter angle than the rods which connect the back with the front side – necessary for rod 1 – it is easier for them to carry more load. Because of that, the load acting on rod 8 is bigger and so its cross-sectional area has to be bigger, too. The difference gets more considerable with the number of storeys since the applied forces increase. As soon as the upper limit is achieved by rod 8 – that is at the 9-storey – it is tried to divert the force and therefore the horizontal edged rods 7 and 11 to rod 8 get a positive cross-sectional area which can be seen in Fig. 7. Short after that, rod 1 adopts its upper limit and its horizontal neighbor rods 2 and 14 are added to the structure, too. With exception of the 15-storey again, the cross-sectional areas increase with raising storeys till they adopt their upper limit.

Due to raising cross-sectional areas, the volume and therefore also the weight which is presented by the initial costs in Fig. 8a is increasing. Since the number of

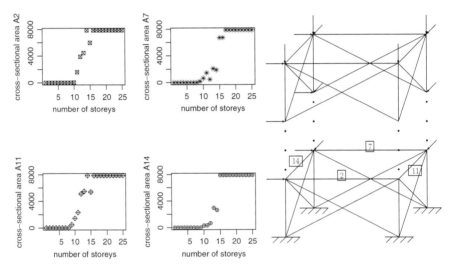

Fig. 7 Optimal cross-sectional areas A_2, A_7, A_{11} and A_{14} in dependence on the number of storeys using the EVP

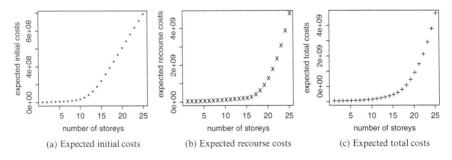

Fig. 8 Expected initial, recourse and total costs using the EVP

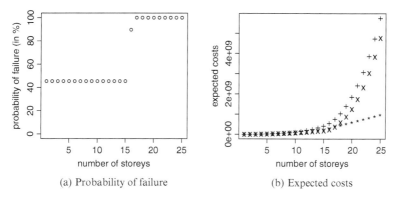

Fig. 9 Probability of failure and expected initial (*), recourse (x) and total (+) costs using the EVP

rods is raising with the number of storeys, the weight of the structure is not limited in contrast to the cross-sectional areas.

Taking into account only the expectation of the load but not the variations, the obtained design using the EVP is not very robust but a small change of the applied load may lead to a failure of the structure which is confirmed taking a look at the probability of failure in Fig. 9a. Since it is always greater than 45%, this structure should not be used. The resulting expected recourse costs increase with the raising number of storeys, as seen in Fig. 8b, and with them also the expected total costs, presented in Fig. 8c.

The relationship between the three cost functions, i.e., the total costs, the initial and the recourse costs, is presented in Fig. 9b where all of them are drawn in one diagram. Here, it is very interesting to see, that there is a point when the slope of the recourse costs gets steeper. This behavior around takes place when most of the cross-sectional areas reach their upper limits and the probability of failure raises sharply. Short after that, the probability of failure increases to 100%, which means that the structure fails in each realization. It can be mentioned that the highest structure until the probability of failure jumps the first time to a higher level is the 15-storey – our outlier.

2.1.2 Recourse Problem with Discretization

If we consider now the Recourse Problem with Discretization based on the 17 realizations, there is no difference at first glance in regarding the progress of the optimal cross-sectional areas as shown in Fig. 10. Using this substitute problem, the rods behave the same way they have done using the EVP. However, comparing the optimal design of the RPD and the EVP in more detail by opposing a selection of cross-sectional areas directly, see Fig. 11, we recognize that in the Recourse Problem with Discretization the cross-sectional areas of most of the rods are the same or a

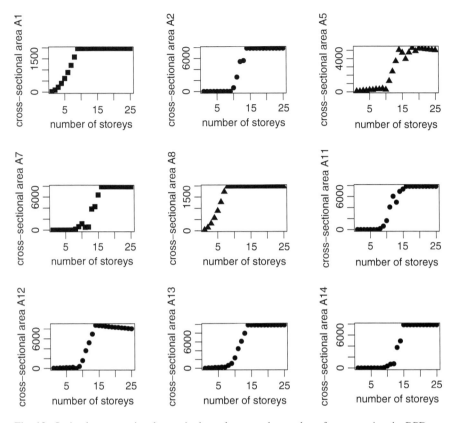

Fig. 10 Optimal cross-sectional areas in dependence on the number of storeys using the RPD

little bit larger. This reinforcement is in fact not very large, but it is sufficient to protect the structure against a failure if the number of storeys is not too large as can be seen in Fig. 12.

As soon as the cross-sectional areas adopt their upper limits, the structure can't again resist all loads and failure appears. But when using the RPD, the probability of failure does not jump so fast to 100%. Instead of that, the structure fails at single realizations. The number of realizations where failure occurs is indeed increasing. And with it, the probability of failure is raising. But it is slow in growth. Corresponding to the zero failure probability at the lower storeys, there are no recourse costs. But with the failure, they also appear. Because of the higher cross-sectional areas in the RPD in contrast to the EVP, the expected initial costs are higher, too, of course. But in some sense the increase of the initial costs is compensated by the low recourse costs, the expected total costs of the RPD are smaller.

Altogether we found out, that using the Recourse Problem with Discretization of the normally distributed load leads to a more robust structure with lower costs. Therefore, the RPD is obviously preferred to the EVP.

Optimal Design and Sensitivity of Large Spatial Trusses Under Uncertainty 319

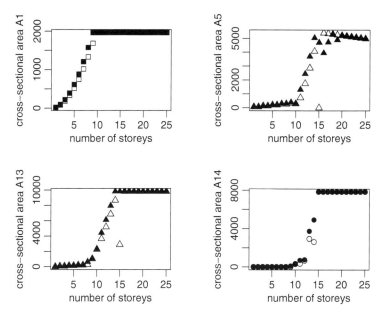

Fig. 11 Comparison of the optimal cross-sectional areas in dependence on the number of storeys using the EVP ($\Box$, $\triangle$, $\circ$) and the RPD ($\blacksquare$, $\blacktriangle$, $\bullet$)

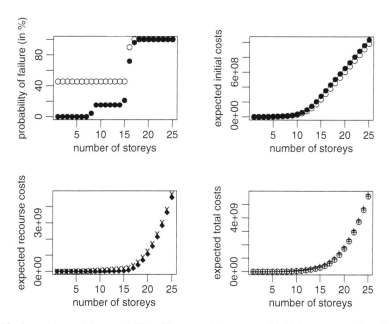

Fig. 12 Comparison of the probability of failure and the expected initial, recourse and total costs using the EVP ($\circ$, x, +) and the RPD ($\bullet$, $\blacklozenge$, $\oplus$)

2.2 Variation of the Standard Deviation

Next, we will study the sensitivity of spatial trusses with respect to increasing stochastic uncertainty. Therefore, we vary the standard deviation in case of the 3-, 5-, 7- and 10-storey.

Increasing the variability in the 3-storey truss, the cross-sectional areas of the vertical rods 1 and 8 and all the diagonal ones (rods 4, 5, 9, 10, 12, 13, 15 and 16) raise linearly which is shown exemplary for rod 8 in Fig. 13a, whereas the vertical rods 3 and 6 are constant and the horizontal rods (rods 2, 7, 11 and 14) are not necessary at all. The resulting structure is shown in Fig. 14. For clearness reasons, the rods which are not visible – i.e. the rods on the left-hand and back side – are drawn dashed. With the cross-sectional areas also the expected initial costs raise. Since the truss can resist each load realization, the probability of failure and therefore also the expected recourse costs are zero, and the total costs comply with the initial costs which is presented in Fig. 13b.

Considering now the 5-storey, the increase of the standard deviation leads at first to a linear ascent of the cross-sectional areas again. But due to the high variability, the rods once adopt the upper limit, which can be seen for rod 8 in Fig. 15a. That is the moment, when failure appears for the first time in one realization which tells us the probability of failure of 4.27%, see Fig. 15b. Increasing the standard deviation, more and more realizations can't be resisted, and the probability of failure grows continuously till 10.35%.

Considering the expected costs in Fig. 15c we find again the linear dependence of the initial costs when the standard deviation is low. The slope gets smaller as soon as the cross-sectional areas reach the upper limit. Due to the probability of failure of zero, at the beginning, no recourse costs appear. The occurring failure while growing standard deviation is reflected by increasing recourse costs.

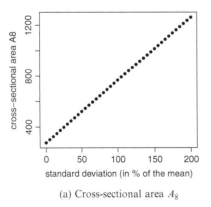
(a) Cross-sectional area A_8

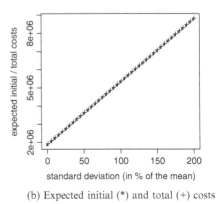
(b) Expected initial (*) and total (+) costs

Fig. 13 Optimal cross-sectional area A_8 and expected costs in dependence on the standard deviation considering the 3-storey

Optimal Design and Sensitivity of Large Spatial Trusses Under Uncertainty   321

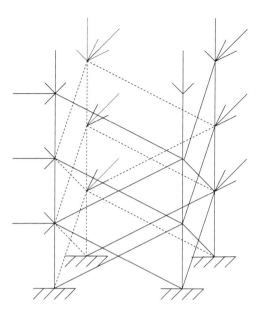

Fig. 14 Optimal 3-storey spatial truss

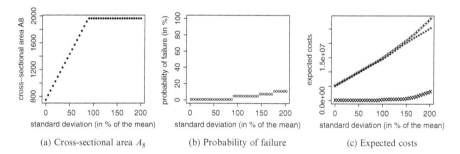

(a) Cross-sectional area A_8 (b) Probability of failure   (c) Expected costs

Fig. 15 Optimal cross-sectional area A_8, probability of failure and expected initial (∗), recourse (x) and total (+) costs in dependence on the standard deviation considering the 5-storey

This behavior is continued, if we add several storeys. We can notice in Fig. 16a, that the more storeys our structure has, the faster the upper limit is adopted by rod 8. The 10-storey needs the full cross-sectional area even at a standard deviation of zero, i.e. when there is no stochastic influence at all.

The observation of the appearance of failure when the first rod adopts its upper limit can be confirmed by taking a look at the probability of failure in Fig. 16b. The higher the structure is, the earlier failure appears and the higher is the probability of failure.

This example shows very well, that higher structures are much more sensitive with respect to variations in the applied loads and therefore, it is much more likely to fail.

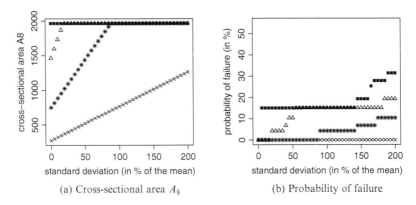

Fig. 16 Optimal cross-sectional area A_8 and probability of failure in dependence on the standard deviation considering the 3- (x), 5- (∗), 7- (△) and 10- (■) storey

2.3 Further Numerical Aspects

The numerical calculations have been made on an ordinary Intel(R) Pentium(R) 4 personal computer with 2.66 GHz CPU and 2.096 Mb RAM. The problems have been formulated in GAMS (General Algebraic Modeling System, see http://www.gams.com). From the different solvers embedded in this software, XPress (see http://www.dashoptimization.com), CPLEX (see http://www.ilog.com) and Minos (see http://www.sbsi-sol-optimize.com) have been used to solve the occurring linear programs.

The size of the resulting LPs of the structures is dependent on the number of storeys n and the number of realizations R and will be calculated in the following.

Since we have $4n + 4$ nodes which are pin-connected among each other and the whole structure, we have $3 \cdot (4n + 4)$ degrees of freedom, i.e., for each node the displacement in $x-, y-$ and $z-$direction. Due to the anchoring of the four nodes on the ground, $4 \cdot 3$ degrees of freedom are deleted which results in $12n$ degrees of freedom. With a number of $B = 16n$ rods, the equilibrium matrix C is therefore a $12n \times 16n$ matrix (Spillers 1972).

Besides the $16n$ design variables $A_1, \ldots, A_B$, for each realization we need four auxiliary variables $y_i^{L+}, y_i^{L-}, y_i^{U+}, y_i^{U-}$ for each rod i, $i = 1, \ldots, B$, which tell us, whether the lower or upper bound of the yield condition is fulfilled or not. This results in $64n$ additional variables for each realization (Zier 2008). Since the technology matrix T and the recourse matrix W have the dimension $28n \times 16n$ and $28n \times 64n$, resp., we get $28n$ constraints for each realization.

Due to the structure of the Expected Value Problem – which can be interpreted as problem with one realization only – and the Recourse Problem with Discretization, given in (8a-c) and (9a-c), resp., the resulting problems consist of one objective function and $R \cdot 28n$ constraints with $16n + R \cdot 64n$ non-negative variables.

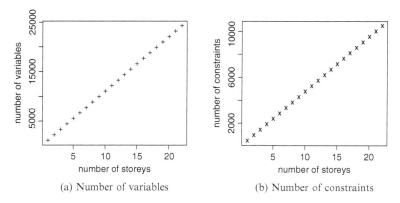

Fig. 17 Number of variables and constraints in dependence of the number of storeys using 17 realizations

Figure 17a shows the number of variables in dependence of the storeys and Fig. 17b visualizes the number of constraints. In both cases, we took $R = 17$ for the number of realizations.

For the optimal design of the 15-storey truss with the RPD with 17 realizations, we get therefore 7140 constraints with 16 560 variables for example.

3 Conclusion

We will conclude with emphasizing that the numerical results confirm the advantage of the incorporation of the stochastic aspects. Due to lack of robustness and therefore high probability of failure, we come to the conclusion that the EVP should not be used. In contrast, the RPD achieves very good results.

Another important result of the numerical example is the confirmation, that the sensitivity with respect to random forces is increasing with the number of storeys of the building. That means, the higher the structure the less robust it is.

References

Gasser, M., & Schuëller, G. I. (1998). Some basic principles of reliability-based optimization (RBO) of structures and mechanical components, *Stochastic Programming Methods and Technical Applications*. In K. Marti, & P. Kall (Eds.), *Lecture Notes in Economics and Mathematical Systems* (LNEMS), vol. 458, pp. 80–103. Berlin/Heidelberg/New York: Springer.

Gasser, M., & Schuëller, G. I. (2002). Reliability-based optimization using approximate methods, *Stochastic Optimization Techniques: Numerical Methods and Technical Applications*. In K. Marti (Ed.), *Lecture Notes in Economics and Mathematical Systems* (LNEMS), vol. 513, pp. 223–293. Berlin/Heidelberg/New York: Springer.

Kall, P., & Wallace, S. W. (1994). *Stochastic Programming*. Chichester: Wiley.

Marti, K. (2008). *Stochastic Optimization Methods*, Second Edition. Berlin/Heidelberg/New York: Springer.

Spillers, W. R. (1972). *Automated Structural Analysis*. New York: Pergamon Press.

Stöckl, G. (2003). *Optimaler Entwurf Elastoplastischer Mechanischer Strukturen Unter Stochastischer Unsicherheit, Fortschritt-Berichte VDI*, vol. 18(278). Düsseldorf: VDI-Verlag GmbH.

Zier, S. (2008). Optimal design of trusses considering uncertainty: a comparison of two approaches, In B. H. V. Topping, & M. Papadrakakis (Eds.), *Proceedings of the Ninth International Conference on Computational Structures Technology* (CST), paper 53. Kippen, Stirlingshire, United Kingdom: Civil-Comp Press.

Part V
Analysis and Optimization of Economic Systems Under Uncertainty

Sustainable Agriculture in China: Estimation and Reduction of Nitrogen Impacts

Günther Fischer, Wilfried Winiwarter, Tatiana Ermolieva, Gui-Ying Cao, Harrij van Velthuizen, Zbigniew Klimont, Wolfgang Schoepp, Wim van Veen, David Wiberg, and Fabian Wagner

Abstract In this chapter we present an integrated model for long term and geographically explicit planning of agricultural activities to meet demands under resource constraints and ambient targets. Environmental, resource and production feasibility indicators permit estimating impacts of agricultural practices on environment to guide agricultural policies regarding production allocation, intensification, and fertilizer application while accounting for local constraints. Physical production potentials of land are incorporated in the model, together with demographic and socio-economic variables and behavioral drivers to reflect spatial distribution of demands and production intensification levels. The application of the model is demonstrated with a case study of nitrogen accounting at the level of China counties. We discuss current intensification trends and estimate the ranges of agricultural impacts on China's environment under plausible pollution mitigation scenarios with a particular focus on nitrogen sources and losses.

G. Fischer (✉) · T. Ermolieva · G.-Y. Cao · sH. van Velthuizen · Z. Klimont · W. Schoepp · D. Wiberg · F. Wagner
International Institute for Applied Systems Analysis, Schlossplatz 1, A-2361, Laxenburg, Austria
e-mail: fisher@iiasa.ac.at; ermol@iiasa.ac.at; cao@iiasa.ac.at; velt@iiasa.ac.at;
klimont@iiasa.ac.at; schoepp@iiasa.ac.at; wiberg@iiasa.ac.at; fwagner@iiasa.ac.at

W. van Veen
Centre for World Food Studies, Vrije Universiteit, Amsterdam, De Boelelaan 1105 1081 HV Amsterdam, The Netherlands
e-mail: W.C.M.vanVeen@sow.vu.nl

W. Winiwarter
International Institute for Applied Systems Analysis and AIT Austrian Institute of Technology, Schlossplatz 1, A-2361, Laxenburg, Austria and Donau-City Str. 1, A-1220 Vienna, Austria
e-mail: winiwarter@iiasa.ac.at

Y. Ermoliev et al. (eds.), *Managing Safety of Heterogeneous Systems*, Lecture Notes in Economics and Mathematical Systems 658, DOI 10.1007/978-3-642-22884-1_16,
© Springer-Verlag Berlin Heidelberg 2012

1 Introduction

Economic growth, increasing demand for food, feed, fiber and biofuels speed up industrialization of agricultural activities characterized by new technologies, specialization and concentration, higher mechanization, increased chemical and fertilization. Production intensification is primarily guided by profit maximization principles and has a number of comparative advantages. However, there are risks and costs which are often not factored in the production planning process, e.g., loss of food and producers diversity, GHG emissions, environmental and water pollution, problems related to human health and livestock diseases, degradation and decrease of socio-economic conditions in rural areas, poverty, rural-urban migration, loss of cultural heritage, etc.

Adverse implications of production intensification, in particular, environmental impacts and health risks establish the need to identify pathways towards sustainable agriculture planning. Estimation of impacts and mitigation measures to reduce agricultural pollution over large territories is a challenging task. It requires a careful choice of models which is often driven by availability and quality of data on the one hand and on the other, by the reliability and robustness of conclusions. Pollution mitigation measures have to realistically account for location-specific demographic and economic indicators, demand and production, pollution and health risks. They should fulfill various goals and constraints, e.g., environmental norms, ambient targets, required levels of food supply, limits regarding population exposed to environmental risks, etc.

Models for planning agriculture development and assessing impacts are traditionally classified along two main lines. One line involves process-based modeling, which combines resource and production potentials of land with data-intensive biophysical processes and models of agricultural (point and disperse) pollution. The models estimate crop growth, soil carbon dynamics, soil temperature and moisture regimes, nitrogen leaching, and emissions of gases on very fine spatio-temporal resolutions under alternative local agricultural practices (Leonard et al. 1987; Li et al. 1992). Availability of spatio-temporal data, its harmonization and further calibration of the underlying biophysical processes even at local scales is a complex task and the results are essentially subject to underlying uncertainties, data quality and model structure. Cross-comparison of process-based models often shows substantial variability and discrepancies both among the modeled outputs and in comparison to field measurements (Frolking et al. 1998). As pointed out by (Bellocchi et al. 2010), the calibration and validation may require using interdependent multiple criteria, interpolation and statistics for tailoring the validation requirements to the specific objectives of the application.

The second line of models focuses on the socio-economic and behavioral aspects of agricultural producers and consumers, aggregate demand and supply. Models such as IMPACT (Rosegrant et al. 1999) perform on the level of major world regions. Resources like land, water or climatic conditions are described by scenarios in rather aggregate terms. With the focus at global or regional problems, the location-specific trends and heterogeneities can often be overlooked. Within the limits of the natural resources, sustainable land exploitation is largely determined by location-specific anthropogenic factors, i.e. demand concentrations, availability of infrastructure, market access and the complex interaction of behavioral, socio-economic, cultural and technological factors.

In this paper we discuss an integrated agriculture planning model that explicitly combines the two lines. The model employs up- and down-scaling probabilistic robust procedures (Fischer et al. 2006) that permit to match the spatio-temporal resolutions of the biophysical (process-based) models with the resolutions of the socio-economic, behavioral and optimization models, scenarios, and data to produce decisions at scales suitable for policy analysis and implementation. The model simulates different scenarios of demand increases inducing respective location-specific production adjustments. In some locations, the indicators characterizing status of environment, socio-economic conditions, and humans' exposure to adverse impacts may already exceed admissible thresholds, signaling that further production growth in these locations should not take place. The question then becomes how to plan expansion of production facilities to meet demand without exacerbating the problems. For this, the model uses indicators defined by various interdependent factors including the spatial distribution of people and incomes, the current levels of crop and livestock production and intensification, and the conditions and current use of land resources. These indicators are used to discount production locations by the degree of their diverse risks and production suitability. The risk-based preference structure is then used in production allocation algorithms to derive recommendations regarding sustainable and robust production expansion, allocation and intensification. Appendix 1 summarizes production allocation algorithm when inherent risks and contingencies are characterized by ambient constraints. In more general cases of risks and contingencies, Appendix 2 describes an algorithm for production allocation in multi-producer environment under environmental safety and food security constraints in the form of multidimensional risk measures having direct connections with Value-at-Risk (VaR) and Conditional Value-at-Risk (CVaR or expected shortfalls) type indicators. Similar algorithm has been elaborated in the case study of rural developments in Ukraine (Borodina 2011).

The proposed integrated agriculture production planning model is applied to the analysis of plausible agricultural pollution projections in China to 2030 (Ermolieva et al. 2005) under alternative scenarios of population, economic growth, and technological innovations. The objective of the study is to address the following questions:

1. What will be the demand for agricultural products, particularly for meat, under plausible economic, demographic and urbanization development paths to 2030?
2. How will increased demand for feed and food translate into the livestock numbers and crop production?
3. How much nitrogen will become available from livestock manure as a consequence of livestock production intensification? How much mineral fertilizers will be needed in addition to local manure supply?

4. What environmental loads, GHG emissions, and water pollution are expected as a result of agricultural production intensification?
5. What improvements can be achieved by production planning based on risk indicators that jointly reduce environmental pollution through water and air contamination in different stages of agricultural production chain, i.e., from nutrients losses in livestock houses to emissions and nutrients losses on crop fields?

Because of its major role in food production and environmental sustainability, we identify nitrogen as the key nutrient in this study. On the basis of FAO projections ('World Agriculture: Towards 2015/203), it has been concluded (Eickhout et al. 2006) that "...despite improvements in the nitrogen use efficiency, total reactive nitrogen loss will grow strongly in the world's increasingly intensive agricultural systems. In the 1995–2030 period emissions of reactive nitrogen from intensive agricultural systems will continue to rise, particularly in developing countries. Therefore, the increase of nitrogen use efficiency and further improvement of agronomic management must remain high on the priority list of policy makers".

The study benefits from a socio-economic and agricultural data base (ADB) at county level consisting of about 3000 administrative units in China for the years 1997 to 2005. The work has been conducted within EU FP6&7 projects on "Policy Decision Support for Sustainable Adaptation of China's Agriculture to Globalization" (CHINAGRO), "Chinese Agricultural Transition: Trade, Social and Environmental Impacts" (CATSEI), "Atmospheric Composition Change, the European Network of Excellence" (ACCENT), and "Integrated Nitrogen Management in China" (INMIC, an activity of IIASA's Greenhouse Gas Initiative). The ADB is prepared and used in CHINAGRO (Fischer et al. 2007, 2006; Keyzer and van Veen 2005), CATSEI (Fischer et al. 2006, 2008), and INMIC (Ermolieva et al. 2009; Fischer et al. 2010). In our work we only partially borrow data from international assessments, while the majority of the data come from the ADB. The ADB makes it possible to distinguish the heterogeneities of agricultural practices, for example, by crop and livestock types, management systems, level of production intensification, location-specific livestock housing and manure facilities, location-specific emission levels, climatic conditions, time of fertilizer application, etc.

The paper is organized as follows. Section 2 briefly outlines the main characteristics of the integrated model with further references to related publications. Nitrogen fluxes from agriculture to environment are described in more detail. Section 3 presents the summary of the recent results for China case study in which scenarios of uncertain agricultural nutrients/nitrogen impacts are analyzed and mitigation scenarios are formulated. The model permits generating infinitely many scenarios. In this paper we restrict attention to some basic cases which allow to identify potential ranges of uncertain outcomes. General discussions and conclusions are presented in Sect. 4. Appendices 1 and 2 include the structure of basic production allocation algorithms.

2 The Model

Below we provide a brief overview of the model, which is described in more detail in (Ermolieva et al. 2009) and (Fischer et al. 2009, 2006). The model is temporally and geographically explicit. It operates at different spatial scales, e.g., national, subnational, county-level, depending on the objectives of the research. This spatial flexibility allows for fine-tuning the associated policy advice to appropriately capture location-specific heterogeneities. Harmonized integration of socio-economic and demographic modeling components with proper scales of biophysical modeling (Fischer et al. 2002) simplifies production planning with limited resources and provides possibilities to improve production potentials in the presence of inherent uncertainties and risks, e.g., weather, contamination of environment, livestock diseases (Ermolieva et al. 2005).

To estimate levels of demand and agricultural production in China case study, agricultural activities are represented at 31 provinces and about 3000 counties. Relying on economic and demographic scenarios till 2030, developed in (Huang et al. 2003) and (Toth et al. 2003), demand increases and consumption of agricultural products are projected for cereals, four types of meat, milk, and eggs. The projections distinguishing between geographical regions, urban and rural areas, and vary with income.

Modeling of livestock sector dynamics is addressed with special attention. The role of livestock in global agriculture should not be underestimated. It has been recognized as an important part of a multi-faceted, integrated approach to agriculture production planning, rural community development, and environmental sustainability (Borodina 2011; Steinfeld et al. 2006). In many countries, livestock is among the main sources of income. Livestock is growing faster than any other agriculture sector. In contrast to developed countries where livestock consumption has stabilized, in the developing countries, annual per capita consumption of meat has doubled since 1980, from 14 to 28 kg in 2002. Development of the livestock sector has been most dynamic in East Asia and China. Currently, China accounts for 57% of the increase in total meat production in developing countries (Steinfeld et al. 2006). Further developments of the sector are of high priority for Chinese government.

To reflect the importance of livestock sector in Chinese agriculture development, the numbers of animals are projected from the base year consistently with economic and demographic scenarios and demand increase at the county level in such a way that production is assumed to meet the demand. Accounting for essential spatial, cultural, and climatic heterogeneities of China, the main livestock categories include poultry, pigs, dairy, cattle, buffaloes, yaks, sheep and goats, and other large animals (combing horses, donkeys, camels). Essential for nitrogen pollution, the model differentiates three main animals' management systems: traditional, specialized/industrial, and grazing. Information on the systems' shares in total livestock production is available at county level for the base year. The following assumptions estimate the mix of management system beyond the base year:

1. Projections of the livestock distribution for confined traditional systems are linked to the projected decrease in rural population.
2. Industrial livestock systems are modeled to meet the provinces' projected demands for livestock products. These systems compensate for the decreases in traditional systems and evolve consistently with the demand growth at provincial level.
3. The geographical distribution of pastoral livestock is projected in accordance with the availability and productivity of grasslands.

The rapid growth of industrial livestock farming is a major contributor to the worsening environmental quality in China. As the steadily growing population and rising incomes in China speed up demand for livestock products, growth in the animal sector will keep pace. Increasing demands can be met only by further intensification of production operations. Intensification has comparative advantages, but it also creates a number of problems which require proper regulations.

Damages are created through a number of socio-economic, environmental, and health pathways which are taken into account in the model. Intensification shifts production from rural to urban and peri-urban areas with higher demand closer to feed sources, separates the source of nutrient intake from the cycle of direct nutrient replenishment, and produces high volumes of concentrated animal waste that over-burdens urban water and waste management. Large amounts of water are being consumed in industrial animal husbandry, leading to overuse of a possibly scarce resource. Most water pollution from agriculture results from the storage and disposal of animal manure and waste. Manure, often stored in tanks or in pools known as "lagoons", may contain pathogenic bacteria and/or antibiotic residues. Leaking lagoons, but also manure application on fields, may lead to the spread of harmful compounds.

The model estimates the pollution level from livestock operations and crop fertilization with the help of a few agricultural, environmental, and biophysical indicators characterizing production intensity, water, soil, and air quality. Human health risks are measured in terms of population exposure to different levels of environmental pollution. The feasible domains of the indicator variables are subdivided into sub-domains of different degrees of impact, severity or suitability. The variables may be combined in risk functions to reflect the levels of different risks in areas associated with agricultural production (Ermolieva et al. 2009; Fischer et al. 2006, 2010). Production is increased primarily through the establishment of new facilities and/or the expansion of existing facilities. In some areas, especially in the vicinity of urban areas, the indicators may signal that a further allocation of production is impossible. For not exacerbating environmental and health problems, production facilities are then adjusted according to a production allocation algorithm summarizes in Appendix 1 (for details see (Fischer et al. 2006, 2008)).

In this paper, we discuss how the model is applied to account for agricultural nitrogen only. Plants require nitrogen for growth as well as for providing protein in food or feed, which leads to the application of nitrogen fertilizer in agricultural practice (Smil 2001). In China, nitrogen fertilizers are being widely used to increase yields on scarce land resources. A large part of the fertilizer applied is not taken up by plants, but released into the environment cascading through diverse environmental pools (Erisman et al. 2007; Galloway et al. 2004). In the model, the environmental effects of soil nitrogen are measured in terms of atmospheric emissions of NH_3, N_2O and NO, and its leaching to ground-water or surface water (Velthof et al. 2009). The model estimates nutrient losses associated with agricultural nitrogen (manure as well as mineral fertilizers) at the level of counties for:

1. *point-source losses* in the form of emissions to the atmosphere and leaching to ground and surface water from specific release points such as livestock housing or manure storage facilities.
2. *non-point losses* resulting from the application of fertilizer and manure to cultivated land or from grazing livestock in pasture areas. Non-point nutrient losses have two components. The first part comprises non-effective nutrients, i.e., nutrients not reaching the crop (including losses due to emissions, runoff and leaching), which depend on the environmental setting and nutrient application practices. These losses occur independently of crop uptake capacity. The second part consists of potentially effective nutrients that reach the crop root zone. Released quantities depend on the crop's uptake capacity.

Based on the spatially explicit distribution of animals and crop production projected by the model, we apply existing schemes of nitrogen releases to assess the loss of agriculturally derived nitrogen compounds along different pathways. The scheme of this model is presented in Fig. 1. As a consequence of agricultural activities, emissions of N_2O and NH_3 into the atmosphere, and of nitrate leached to ground-water are assessed.

For estimating nitrogen leaching to ground-water, we adopt the *MITERRA model* (Velthof et al. 2009), which applies a combined water and nitrogen balance to derive indicators of leaching for a broad range of soil types aggregated into seven classes (sandy, clay, gleyic, stagno-gleyic, peat, loam, and paddy soils) with different leaching characteristics (FAO/IIASA/ISRIC/ISSCAS/JRC 2009; Shi et al. 2004). Soils are also distinguished by the type of crop water management, i.e., separately for irrigated and rain-fed land. For each soil class, climate condition (e.g., precipitation, temperature), and land use, the approach estimates the fraction of nitrogen surplus that moves to the ground water, i.e., the leaching fraction. Soils used for rice paddies are considered impermeable to water and assumed to have no leaching.

In a simplified way, the accounting of nitrogen applied to the field may be described as follows:

$$N\_Surpl = N_{mnr} + N_{fert} + N_{fix} - N_{upt},$$

where N_Surpl denotes surplus of nitrogen applied to the field, N_{mnr} is nitrogen in manure available for field fertilization (net of losses during housing); N_{fert} is nitrogen in chemical fertilizers; N_{fix} is atmospheric nitrogen fixed by N-fixing crops; N_{upt} is nitrogen uptake by all crops (net of nitrogen left in recycled crop residues). The nitrogen surplus, together with the leaching fraction derived according to soil

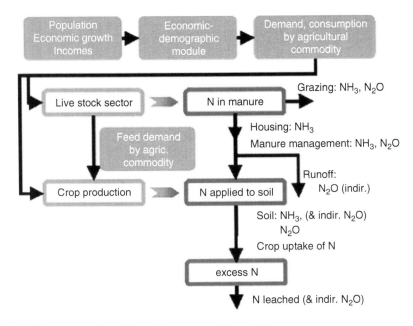

Fig. 1 Nitrogen cascading: Schematic structure of the model

and climate parameters, allows for the estimation of nitrogen leaching, in kilograms N per hectare cultivated land.

For estimating the N_2O emissions to the atmosphere we employ the methodology developed for IIASA's Greenhouse gas–Air pollution INteractions and Synergies (GAINS) model (Winiwarter 2005). GAINS applies IPCC default emission factors (IPCC 2000) recommended for national inventory submissions to the UNFCCC. Thus the derived results can be consistently compared among countries. The N_2O emissions are computed as the product of an emission factor times the respective activity data for manure management, grazing and soils. Indirect emissions (both from leaching and redeposition of gaseous releases) are implicitly covered in the emission factor applied, and therefore the model results generally are very close to those assessed with the IPCC method.

Ammonia emissions (NH_3) from livestock production are estimated at four major stages: in animal houses; during storage of manure; when applying manure; and from livestock grazing. These stages are explicitly distinguished in the livestock model (Fischer et al. 2006, 2009, 2007) and in the GAINS methodology (Klimont 2001; Klimont and Brink 2004). The ammonia accounting is based on experience and parameters developed primarily for Europe. Here we are considering the impact of local manure management practices and reflect in the estimates different levels of livestock productivity (e.g., (Ermolieva et al. 2005; Menzi 2001; NuFlux 2001)).

Emissions of ammonia from mineral fertilizer application depend on multiple factors including type of fertilizer applied, soil properties, meteorological conditions, time of application in relation to a crop canopy, and method of application. The nitrogen losses from fertilizer application are region specific. It must be stressed that the uncertainty range of emission factors is large. Typically, nitrogen losses from synthetic fertilizers vary between 1% and 4%, with the exception of ammonium sulfate (8%), urea (15%–25%) and ammonium bicarbonate (ABC) (20%–30%). Application practice is very different between countries, and may range from mostly non-volatizing fertilizers (typical for most of Europe) to predominantly urea and ABC (China). Figures 3, 4, 6, 7, 9, 10 demonstrate uncertainty ranges only for some basic scenarios. The outcomes of such a scenario-based uncertainty analysis can be further used for designing robust development paths and corresponding nitrogen outputs, which is beyond the scopes of this paper.

3 Numerical Application: A Case Study of China

In the studies, several scenarios of pollution mitigation options in China were analyzed and compared with regard to excess nitrogen:

1. a "business-as-usual" scenario, in which the increase of production is allocated proportionally to the demand increase, which is concentrated in the vicinity of densely populated urban areas;
2. a reallocation scenario that combines the demand driven preference structure of the business-as-usual scenario with information on population densities and urban agglomerations to reduce risks caused by livestock production;
3. an "optimizing fertilizer use" scenario (first apply manure; only then supply nutrients to crops with mineral fertilizer); and
4. a scenario consisting of optimized fertilizer use combined with technological options focused on ammonia abatement ("minimized ammonia" scenario).

The business-as-usual scenario (1) implicitly minimizes the transportation costs as production concentrates in the vicinity of urban areas with high demand. In the alternative scenario (2), the production is shifted to more distant locations characterized by availability of cultivated land, lower livestock and population density, but at the expense of additional transportation. In addition to (2), scenario (3) focuses on reducing fertilizer application, while scenario (4) represents a scenario of drastic ammonia emission reductions. Formally, the scenarios correspond to different priors in the allocation procedure summarized in Appendix 1.

The effects of the four scenarios are compared in terms of nitrogen leaching fraction, (Fig. 2), the quantities of leached nitrogen (Figs. 3, 4, 5), and nitrogen emitted to the atmosphere as N_2O (Fig. 6, 7, 8) or NH_3 (Figs. 9 and 10). These indicators have been computed for each spatial administrative unit, i.e., county. Figures 2 through 10 illustrate spatial heterogeneity of the indicators for the base line scenario only.

It is remarkable how heterogeneous are the leaching fractions among different locations following patters similar to patters of precipitation, i.e. decreasing from

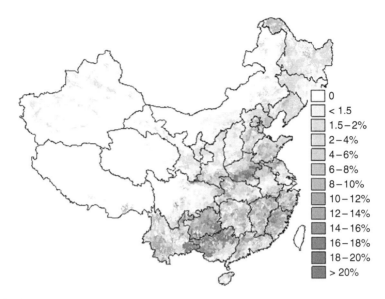

Fig. 2 Nitrogen leaching fraction, in percent terms

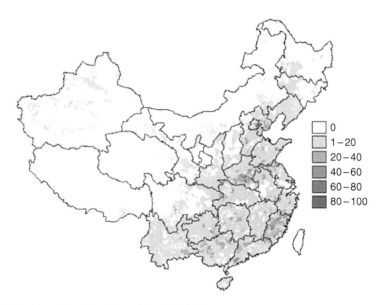

Fig. 3 Leaching in kg / ha cultivated land, in 2000

Sustainable Agriculture in China: Estimation and Reduction of Nitrogen Impacts

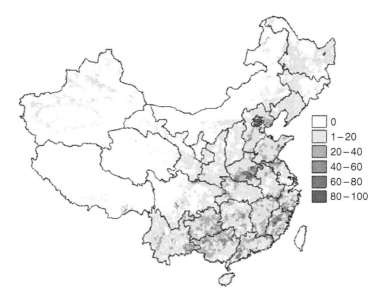

Fig. 4 Leaching in kg / ha cultivated land, in 2030

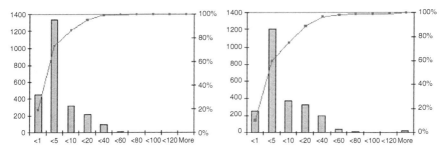

Fig. 5 Leaching in severity classes by number of affected counties, in kg / ha cultivated land, in 2000 (**a**) and 2030 (**b**)

the southeast to the northwest. Higher fractions in the south can be explained by a combination of other factors such as soil type, temperature, land use type. For example, lighter arable soils retain only a small fraction of water in comparison to clay soils. Also, losses of nutrients may be stimulated by agricultural practices such as timing of fertilizers application. Nitrogen leached from fields or livestock facilities is then estimated as a function of nitrogen fraction and the activity. The higher is the intensity of the activity, the more nitrogen is expected to leach. In Figs. 4 intensive color indicating higher leaching in the vicinity of Beijing can be explained by high density of livestock production.

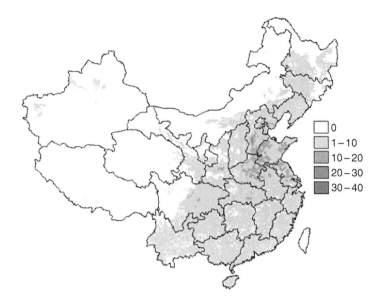

Fig. 6 N2O in kg / ha cultivated land, in 2000

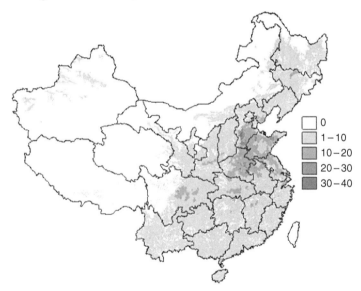

Fig. 7 N2O in kg / ha cultivated land, in 2030

The rate of nitrogen leaching is depicted in Figs. 5. As figure shows, the number of counties in higher nitrogen leaching (severity) classes increases over time. Figures 6 and 7 identify geographical patters of N2O emission rates. and Fig. 8 displays how the distribution of counties in nitrogen emissions classes changes in the period from 2000 to 2030.

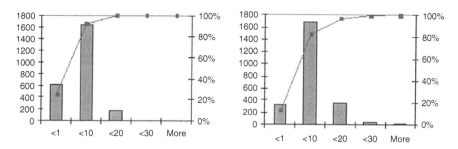

Fig. 8 N2O emissions by size classes and number of affected counties, in kg per ha of cultivated land, for 2000 (**a**) and 2030 (**b**)

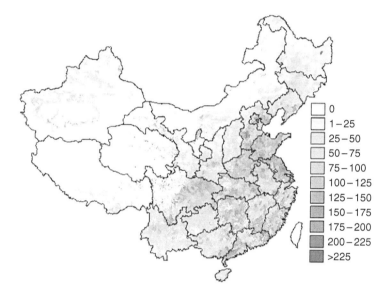

Fig. 9 Ammonia emissions from agriculture (kg ammonia/ha cultivated land) in 2000

Figures 9 and 10 identify geographical patters of ammonia emissions. The rate of ammonia emissions depends primarily on the intensity and the type of agricultural activities, farming and crop production. Central role in determining the level of emissions plays the type of fertilizers. For example, application of urea may cause 15% to 25% and ammonium bicarbonate (ABC) – 20% to 30% emissions, which are predominant fertilizers in China.

At the level of China, in BAU scenario indicators of leaching, N2O-N and NH3-N change from around 701, 855, 7469 (kt N) in the base year to about 1101, 1282, and 10878 (kt N) in 2030, respectively. In sustainable reallocation scenarios, the values towards 2030 become 1077, 1278, 10848 (kt N) for leaching, N2O-N and NH3-N, respectively. Optimization scenario results in 321, 956, 7487 (kt N) and

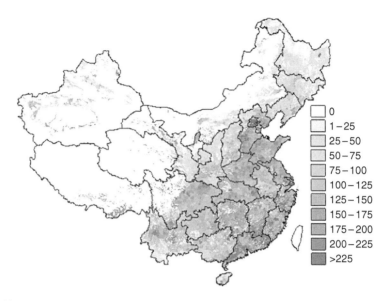

Fig. 10 Ammonia emissions from agriculture, in kg ammonia/ha cultivated land, in 2030

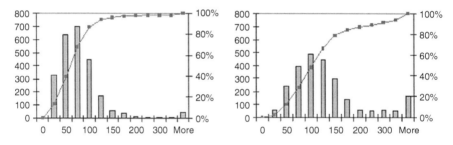

Fig. 11 Ammonia emissions by size classes and number of affected counties, in kg per ha of cultivated land, for 2000 (**a**) and 2030 (**b**)

minimized ammonia in 328, 963, and 6884 (kt N) leaching, N2O-N and NH3-N in 2030, respectively (Fig. 11).

In all scenarios and for all years considered, the greatest part of nitrogen loss is via the ammonia pathway. NH_3 nitrogen emissions are between 40% and 50% of total nitrogen application. Loss in the form of N_2O follows next, and it may seem somewhat surprising that nitrate leaching is even less important. This can be explained by the huge losses in the gas phase, while the leaching fractions (relatively small) apply only to the quantity of soil nitrogen not lost or used elsewhere, and also do not include nitrate runoff to surface water.

The scenarios vary, but it is clear in all cases that nitrogen pollution is likely to increase over time. Even in the scenarios that provide considerable improvements,

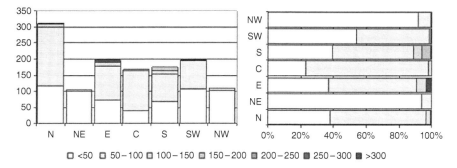

Fig. 12 Absolute (million people) and relative (share of total population) distribution of population according to classes of severity of environmental pressure (measured in terms of kg nitrogen in ammonia emitted per ha cultivated land), 2000. The label on the horizontal axis indicates the regions in China: N, NE, E, C, S, SW, NW stand for North, North-East, East, Center, South, South-West, North-West, respectively, business-as-usual scenario

aggregate environmental pressures associated with livestock manure and use of mineral fertilizer increase between 2000 and 2030 by roughly a third to almost half.

Environmental risk indicators (here, risk indicators are introduced in terms of ambient targets similar to norms on water or air pollution. Fischer et al. (2009) provide further discussion on VaR and CVaR risk indicators.) are determined for each administrative region. However, it is not so much the area that is affected by an adverse environmental situation, but the health of the population which is at stake. In order to estimate this risk, we need to assess the exposure of people. Figure 12 presents population exposure in terms of different classes of ammonia emissions in 2000, and Fig. 13 compares the same indicator for the four alternative scenarios by China regions and aggregated for the whole country. Ammonia is selected not because it is dangerous to humans, although ammonia exposure may be harmful. As mentioned, ammonia is a major nitrogen flux, and may therefore serve as a robust indicator of the overall nitrogen load both on the environment and on humans. While we are not able to provide information on an absolute risk measured in monetary terms, we may compare between the different scenarios. The scenarios (2), (3) and (4) reallocate agricultural production to minimize overall risk (see the procedure in the Appendix 1 and (Ermolieva et al. 2009; Fischer et al. 2010). This function, considering population density implicitly on the demand side, effectively performs reallocation away from population. The altered distribution pattern of production alone will then reduce the exposure of the population. With mitigation measures in place, the gradual improvement of the situation across scenarios (Fig. 13), alleviating the exposure of people to environmental and health risk, becomes visible when comparing the incremental additional measures leading from the *business-as-usual* to the *minimized ammonia* scenario.

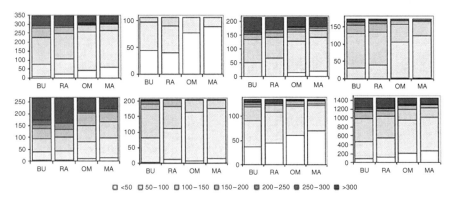

Fig. 13 Number of people by classes of severity of ammonia losses (kg nitrogen in ammonia emitted per ha total area), by economic regions and scenarios (BU = business-as-usual, RA = reallocation, OM = optimized manure; MA = minimized ammonia), in 2030

4 Concluding Remarks

In this paper we discussed model-based estimation of nitrogen fluxes from agriculture in China. Proposed integrated agriculture planning model combines the advantages of the two main types of traditional modeling: process-based and aggregate socio-economic demand-supply modeling. The model derives indicators quantifying environment pollution and health exposure from agricultural activities spatially. This permits to assign a risk preference structure to specific locations and to guide recommendations regarding robust production allocation and intensification. The model has been applied to the analysis of agricultural developments in China to 2030 focusing on pollution abatement options. In this paper, agricultural pollution is measured in terms of nitrogen excess. Nitrogen deserves special attention for a number of reasons. While being the main component in the atmosphere (78%), molecular nitrogen as such is not accessible to plants. And first needs to be "fixed" to form reactive nitrogen, i.e. nitrogen in form of a range of different chemical compounds. Reactive nitrogen is indispensable in agriculture to stimulate plant growth, however recent nitrogen pollution studies show the incredible growth of nitrogen in the environment. While in 1860, humanity produced 15 million metric tons of reactive nitrogen, by 1995, that number was at 156 million tons, and increased to 187 million tons by 2005 (Galloway et al. 2004). In comparison to global CO_2 emissions of 27 billion tons annually, these numbers may seem small, but nitrogen impacts are magnified by the so called nitrogen cascade, i.e., propagation of nitrogen fluxes through the atmosphere, into the soil, into the water, into the coastal systems and back into the atmosphere, which considerably magnifies the impacts.

Some of the reactive nitrogen comes from industry, but the majority comes from agricultural activities – livestock production and crops fertilization. The most important pathways of nitrogen loss to the environmental are atmospheric emissions of

Sustainable Agriculture in China: Estimation and Reduction of Nitrogen Impacts   343

NH_3, N_2O and NO, and leaching to ground-water or surface water. Nitrogen losses are expected to further increase in the future, as demand for agricultural products will continue to rise. Much of the accelerated N cycle is expected in China where remarkable increase in the production of agricultural products was observed over the last years. Four alternative agricultural scenarios of nitrogen pollution reduction in China are developed and discussed. We differentiate between abatement strategies that provide a spatial production reallocation – very effective in terms of costs and reduction of population exposure, and scenarios that provide actual reductions in nutrient (nitrogen) application and release. Environmental sustainability aspects of these scenarios are compared with respect to area and human exposure to different severity classes of risks. The derived geographically explicit results of the four scenarios indicate essential location-specific heterogeneities regarding the level of environmental risks, health exposure, and economic capacity to cope with the risks. However, it is clear that in all locations, the level of pollution is likely to increase. Even in scenarios (3) and (4), which carry considerable improvements, China's environmental deterioration increases by roughly one third to nearly half. Nitrogen losses in the form of ammonia pose the biggest problem by far in all four scenarios. Losses in the form of N_2O follow, while nitrate leaching is of less importance. While we do not intend to provide a full nitrogen balance for China, the results may be useful to look at the respective fluxes in perspective to gain an overall understanding of the system.

A comparison of results obtained here with literature values indicates general agreement on the magnitude of nitrogen fluxes. Because of uncertainties and variability, instead of using absolute estimates for planning pollution mitigation options we propose to employ approaches based on robust indicators. These proved to have a success in evaluating the preference structure of feasible decisions. The results relying on the indicators may then be considered valuable for policy advice. In this application our indicators guide production allocation and intensification in such a way that both the demands targets (food security) and security constraints on the environment and health exposure are met.

Acknowledgements The authors are greatfull to the participants of the IIASA/GAMM workshop on Coping with Uncertainty held at IIASA in 2009, and to the anonymous referees for critical suggestions that led to improvements of this paper.

Appendix 1: Production Allocation Algorithm

Let us briefly summarize the production allocation algorithm (for more details, see (Fischer et al. 2006)). The objective is to allocate new supply facilities in the best possible way to meet the projected increase in national demand d_i for agricultural products i among the production activities/locations k, $k = \overline{1, K}$ while considering various risk indicators. In the following model, the risks are treated as constraints on production expansion (similar to ambient targets in pollution control

344	G. Fischer et al.

models). Therefore, the problem is to determine suitable activity levels y_{ik} given the constraints:

$$\sum_k a_{ik} y_{ik} \geq d_i, \tag{1}$$

$$y_{ik} \geq 0, \tag{2}$$

$$\sum_i y_{ik} \leq b_k, i = \overline{1,m}, k = \overline{1,K}, \tag{3}$$

where b_k denotes thresholds for environmental and health risks and imposes limitations to an increase in production of system or location k, $k = \overline{1,K}$. Apart from b_k, there may be additional limits on y_{ik}, $y_{ik} \leq r_{ik}$, which may be associated with legislation, for example, to restrict production i to a production "belt" or to exclude production i from urban or protected areas, etc. Thresholds b_k and r_{ik} may indicate strictly prohibited levels. The procedure may also allow for the thresholds to be exceeded while imposing taxes or requiring a premium to be paid for the mitigation of certain risks. Coefficients $a_{ik} \geq 0$ describe heterogeneities among products in different locations. Equations (1)–(3) are well established in the literature and belong to the type of generalized transportation problems. They may describe also dynamic problems of allocations by using i to enumerate demands for different products at different time intervals, e.g., indices $i = 1, \ldots, m$ may indicate demand for products at time $t = 1$; $i = m + 1, \ldots, 2m$ indicate demand for products at time $t = 2$, and so on. In general problems, (3) has the structure $\sum_i \beta_{ik} y_{ik} \leq b_k$, $i = \overline{1,m}$, $k = \overline{1,K}$, where β_{ik} may reflect different pollution outputs from i in location k. With deterministic β_{ik} this type of equations can be reduced to (3) by introducing new decision variables $x_{ik} = \beta_{ik} y_{ik}$.

Apart from (1)–(3) there always exist additional information regarding uncertain activity levels y_{ik}, i.e., behavioral uncertainties. These information is used to derive a prior probability q_{ik} reflecting the assumption that a unit of demand d_i for product i should be supplied by activity/location k. For instance, it is reasonable to allocate more livestock to areas with a higher demand increase, higher productivity or better access to animal feeds, transportation services. This preference structure is expressed in prior q_{ik}, $\sum_k q_{ik} = 1$ for all i. The use of priors is consistent with spatial economic theory (see, e.g., in Fujita 1999; Karlqvist et al. 1978). The likelihood q_{ik} can be modeled as inversely proportional to production costs, distances, risks and ambient targets b_k and r_{ik}. In the case a_{1k}, i.e. standard transportation model, an initial estimate of production i allocated to k can be derived as $q_{ik} d_i$. This could, however, result in a violation of applicable restrictions (3). Sequential rebalancing (Fischer et al. 2006) proceeds as follows. For the simplicity of illustration, we assume that $a_{ik} = 1$. The *expected* initial allocation of d_i to k is $y_{ik}^0 = q_{ik} d_i, i = \overline{1,m}$. As this allocation may not comply with constraint $\sum_i y_{ik}^0 \leq b_k, j = \overline{1,n}$, the relative imbalances $\beta_k^0 = b_k / \sum_i y_{ik}^0$ are derived and updated $z_{ik}^0 = y_{ik}^0 \beta_k^0, i = \overline{1,m}$. Now constraint $\sum_i y_{ik} \leq b_k$ is met, $k = 1, 2, \ldots,$

but the estimate z_{ik}^0 may cause an imbalance for relation (1), i.e., $\sum_k z_{ik}^0 \neq d_i$. Continue calculating $\alpha_i^0 = d_i / \sum_k z_{ik}^0$, $i = \overline{1, m}$ and updating the imbalances $y_{ik}^1 = z_{ik}^0 \alpha_i^0$, etc. The estimate y_{ik}^s can be represented as

$$ y_{ik}^s = q_{ik}^k d_i, q_{ik}^s = \left(q_{ik} \beta_k^{s-1} \right) / \left(\sum_j q_{ik} \beta_k^{s-1} \right), \ i = \overline{1, m}, k = 1, 2, \ldots . $$

Assume $y^s = \{ y_{ik}^s \}$ has been calculated. Find

$$ \beta_k^s = \overline{b}_k / \sum_i y_{ik}^s $$

and

$$ q_{ik}^{s+1} = \left(q_{ik} \beta_j^s / \sum_i q_{ik} \beta_j^s \right), \ i = \overline{1, m}, k = 1, 2, \ldots , $$

and so on.

In this form the procedure can be considered as a redistribution of required supply d_i among producers $k = 1, 2, \ldots$ by applying the sequential adjustment q_{ik}^{s+1}, i.e., by using a Bayesian type of rule to update the prior distribution:

$$ q_{ik}^{s+1} = q_{ik} \beta_k^s / \sum_i q_{ik} \beta_k^s, q_{ik}^0 = q_{ik} . $$

The iterative update of q_{ik} is based on an 'observation' of imbalances of the basic constraints rather than calculated for observations of random variables. A simple rebalancing procedure, similar to the one mentioned above for standard transportation constraints (1)–(3) was proposed by G.V. Sheleikovskii (for more details and references, see (Bregman 1967)) for the estimation of passenger flows between old and projected new regions. Similar procedure for general problem (1)–(3) may be used for analyzing of interregional migration, agricultural export-import flows, etc. Verification of its convergence to the optimal solution maximizing the cross-entropy function

$$ \sum_{i,k} y_{ik} \ln \frac{y_{ik}}{q_{ik}} \tag{4} $$

is provided in (Fischer et al. 2006) for general forms of constraints. The alternative scenarios introduced in Sect. 4 correspond to different production allocation priors q_{ik}, $i = \overline{1, m}, k = \overline{1, K}$. The behavioral uncertainty can also be treated in a stochastic manner as a random allocation of demand d_i among points $k = \overline{1 : K}$ with respect to the prior probability q_{ik}, which is a topic of a separate paper.

Appendix 2: Stochastic Model for Production Allocation

The approach presented in Appendix 1 guides production expansion relying on individual behavioral principles set by priors. The risks are characterized by imposing certain standards as additional ambient or "safety" constraints. In general, these constraints may depend on some scenarios of potential future shocks to the system. Let us consider now a more general multi-producer (location) model in a stochastic environment. We may assume that there is a coordinating agency (a principle agent). The agency acts as a social planner and is responsible for maximizing the overall performance of the production chain to stabilize the aggregate production under minimal costs. Suppose that the agency has to determine levels y_{ik} of product i in locations k in order to meet stochastic demand $d_i(\omega)$, where $\omega = (\omega_1, \omega_2, \dots)$ is a vector of all contingencies affecting demand and production. It is naturally to assume that the decision on production expansion has to be made before the information on contingencies arrives. In this case, the total ex-ante production may not exactly correspond to the real demand, i.e., we may face both over-supplies and shortfalls. In other words, the amount of production y_{ik}, $k = 1, \dots, K$, which is planned ex-ante to satisfy the demand $d_i(\omega)$, $y_i(\omega) = \sum a_{ik}(\omega) y_{ik}$ may underestimate ($y_i(\omega) < d_i(\omega)$) or overestimate ($y_i(\omega) > d_i(\omega)$) the real demand $d_i(\omega)$ under revealed contingencies ω and the safety constraints imposed by strict thresholds b_k in (3). The constraint (3) necessitates, in general, additional supply of ex-ante production $z_i \geq 0$ from external sources (say, through international trade). It may also require the ex-post redistribution of the production from internal producers, $k = \overline{1, K}$, to eliminate arising shortfalls and over-supplies in locations. For now, let us ignore these ex-post redistributional aspects assuming that the most significant impacts are associated with ex-ante decisions y_{ik} and z_i. In fact, the presented further model can be easily extended to represent the ex-post adjustments of decisions y_{ik}, z_i, as well as more detailed temporal aspects of production planning.

Let c_{ik} be the unit production cost. In more general model formulation, c_{ik} also include the unit transportation cost for satisfying location-specific demand. Then the model of production planning among the facilities under ambient and other constraints can be formulated as the minimization of the total cost function:

$$f(y, z) = \sum_{i,k} c_{ik} y_{ik} + \sum_{i=1}^{m} e_i z_i,$$

subject to constraints (2), (3), and the following additional safety constraints

$$P\left[\sum_{k=1}^{K} a_{ik}(\omega) y_{ik} + z_i \geq d_i(\omega) \right] \geq p_i, z_i \geq 0, i = \overline{1, m}, \tag{5}$$

Sustainable Agriculture in China: Estimation and Reduction of Nitrogen Impacts 347

where $e_i > 0$, $i = \overline{1, m}$, denotes the unit import cost. A safety level p_i, $0 < p_i < 1$, defines the food security constraint (regulates the supply-demand relations) for all possible scenarios (contingencies) ω. The introduction of constraints (5) is a standard approach for characterizing stability in case of insurance business, security of nuclear power plants and other risky activities. Safety constraints of type (5) are usually used in cases where impacts of random interruptions can not be easily evaluated. In this case, the value p_i is selected such that an expected shortfall occurs only, say, once in 100 month, i.e., $1 - p_i = 1/100$.

The main methodological challenge is concerned with the lack of convexity of constraints (5). Yet, the remarkable fact is that the model defined by (2)–(3), (5) can be effectively solved by linear programming methods due to the following convex reformulation of this model. Let us consider the minimization of the expectation function

$$F(y, z) = f(y, z) + \sum_{i=1}^{m} \alpha_i E \max \left\{ 0, d_i(\omega) - \sum_{k=1}^{K} a_{ik}(\omega) y_{ik} - z_i \right\}, \quad (6)$$

subject to constraints (2), (3), and $z_i \geq 0$, $i = \overline{1, .m}$. The minimization of function $F(y, z)$ is a rather specific case of stochastic minimax models analyzed (both optimality conditions and solution procedures) in Ermoliev and Wets, 1988. In particular, if $F(y, z)$ has continuous derivatives with respect to z_i, e.g., the probability distribution function of ω has continuous density function, then

$$\frac{\partial F}{\partial z_i} = e_i - \alpha_i E I(d_i(\omega) - \sum_{k=1}^{K} a_{ik}(\omega) y_{ik} - z_i \geq 0),$$

where $I(\xi \geq 0)$ is the indicator function: $I(\xi \geq 0) = 1$, if $\xi \geq 0$, and $I(\xi \geq 0) = 0$ otherwise. Therefore, we can rewrite $\frac{\partial F}{\partial z_i}$ as

$$\frac{\partial F}{\partial z_i} = e_i - \alpha_i P \left[d_i(\omega) - \sum_{k=1}^{K} a_{ik}(\omega) y_{ik} - z_i \geq 0 \right], \quad (7)$$

which allows to establish connections between the original model defined by (2), (3), (5) and the minimization of convex function $F(y, z)$ defined by (6).

Assume $(y*, z*)$ minimizes $F(y, z)$ subject to constraints (2), (3), and $z_i \geq 0$, $i = \overline{1, .m}$. Assume also that $e_i < \alpha_i$, $i = \overline{1, .m}$. Then from (7) it follows that for all i with positive components $z_i^* > 0$, i.e., when $\frac{\partial F}{\partial z_i} = 0$, the optimal solution $(y*, z*)$ satisfies the following safety constraints

$$P \left[d_i(\omega) - \sum_{k=1}^{K} a_{ik}(\omega) y_{ik} - z_i \geq 0 \right] = e_i / \alpha_i. \quad (8)$$

Moreover, for all i with $z_i^* = 0$, i.e., when $\dfrac{\partial F(y*, z*)}{\partial z_i} \geq 0$, the optimal $(y*, z*)$ satisfies the following safety constraint

$$
P\left[d_i(\omega) - \sum_{k=1}^{K} a_{ik}(\omega) y_{ik} \geq 0 \right] \leq e_i / \alpha_i. \tag{9}
$$

If we choose α_i as $e_i / \alpha_i = 1 - p_i$, i.e., $\alpha_i = e_i/(1 - p_i)$, then (8)–(9) become equivalent to the safety constraint (5) of the original model (2), (3), (5). In other words, the minimization of convex function $F(y, z)$ defined by (6) subject to (2), (3), and $z_i \geq 0$, $i = \overline{1, m}$, yields the desirable solution of the original model (2), (3), (5). Efficient computational procedures for solving stochastic minimax problems with objective functions defined as in (6) can be found in (Ermoliev and Wets 1988; Rockafellar and Uryasev 2000). In particular, (Rockafellar and Uryasev 2000) discuss the applicability of linear programming methods in cases where the original model defined by a general probability distributions of ω can be sufficiently approximated by models with discrete probability distributions. This paper establishes also important connections between the minimization of (6)-type functions and Conditional-Value-at-Risk risk measure.

The minimization of function (6) can also be solved by a stochastic quasi-gradient method (Ermoliev and Wets 1988). In applying this method to minimization of (6), the differentiability of $F(y)$ and any assumption on probability distribution of ω is not required. Also, the probability distribution of ω may only be given implicitly. For instance, only observations of random $d_i(\omega)$ and $a_{ik}(\omega)$ may be available or only a Monte Carlo procedure ("pseudo-sampling" simulation model as described in Sect. 2) is used to simulate supply and demand. In Sects. 2 and 3 we illustrate application of the rebalancing algorithm described in Appendix 1 while the outlined stochastic allocation algorithm has been elaborated, e.g., in (Borodina 2011; Fischer et al. 2009). Impressive application of stochastic quasi-gradient methods in a form of adaptive Monte Carlo optimization can be found in (Wang 2010).

References

Bellocchi, G., Rivington. M., Donatelli., M., & Matthews, K. (2010). Validation of biophysical models: issues and methodologies: A review. *Agronomy for Sustainable Development, 30*, 1.

Borodina, A., Borodina, E., Ermolieva, T., Ermoliev, Y., Fischer, G., Makowski, M., van Velthuizen. Food security and socio-economic risks of agricultural production intensification in Ukraine: a model-based policy decision support. Y. Ermoliev, K. Marti, M. Makowski (eds.), Managing Safety of Heterogeneous Systems, Lecture Notes in Economics and Mathematical Systems 658, DOI 10.1007/978-3-642-22884-1_16, Springer-Verlag Berlin Heidelberg 2011.

Bregman, L. (1967). Proof of the convergence of Sheleikhovskii's method for a problem with transportation constraints. *Journal of Computational Mathematics and Mathematical Physics, 7*(1), 191–204. *Zhournal Vychislitel'noi Matematiki,* USSR, Leningrad.

Sustainable Agriculture in China: Estimation and Reduction of Nitrogen Impacts 349

Eickhout, B., Bouwman., A. F., van Zeijts, H. (2006). The role of nitrogen in world food production and environmental sustainability. *Agriculture, Ecosystems and Environment, 116*, 4–14.

Erisman, J. W., Bleeker, A., Galloway, J. N. & Sutton, M. A. (2007). Reduced nitrogen in ecology and the environment. *Environmental Pollution, 150*, 140–149.

Ermoliev, Y., & Wets, R. (Eds.) (1988). *Numerical Techniques for Stochastic Optimization. Computational Mathematics.* Berlin: Springer.

Ermolieva T., Fischer, G., & van Velthuizen, H. (2005). Livestock production and environmental risks in China: Scenarios to 2030. FAO/IIASA Research Report. IIASA, Laxenburg, Austria.

Ermolieva, T., Winiwarter, W., Fischer, G., Cao, G. Y., Klimont, Z., Schöpp, W., Li, Y., & Asman, W. A. H. (2009). Integrated nitrogen management in China. IIASA Interim Report, IR-09-005. IIASA, Laxenburg, Austria.

FAO/IIASA/ISRIC/ISSCAS/JRC (2009). *Harmonized World Soil Database (version 1.1).* FAO, Rome, Italy and IIASA, Laxenburg, Austria.

Fischer, G., Ermolieva, T., Ermoliev, Y., & van Velthuizen, H. (2006). Livestock production planning under environmental risks and uncertainties. *Journal of Systems Science and System Engineering, 15*(4), 385–399.

Fischer, G., Cao, G.-Y., Ermolieva, T., & Sun, L.-X. (2008). Urbanization and livestock Production: An approach to health and environmental risks. *Journal of Population and Development, 14*(6), 2–10.

Fischer, G., Ermolieva, T., Ermoliev, Y., & Sun, L. (2009). Risk-adjusted approaches for planning sustainable agricultural development. *Stochastic Environmental Research and Risk Assessment, 23*(4), 441–450.

Fischer, G., Ermolieva, T., Ermoliev, Y., & Sun, L. (2007). Integrated risk management approaches for planning sustainable agriculture. In C. Huang, C. Frey, & J. Feng (Eds.), *Advances in Studies on Risk Analysis and Crisis Response*. Paris, France: Atlantis Press.

Fischer, G., Ermolieva, T., Ermoliev, Y., & van Velthuizen, H. (2006). Sequential downscaling methods for estimation from aggregate data. In K. Marti, Y. Ermoliev, M. Makowski, G. Pflug (Eds.), *Coping with Uncertainty: Modeling and Policy Issue*, pp. 155–169. Berlin, New York: Springer.

Fischer, G., van Velthuizen, H., Mahendra, S., & Nachtergaele, F. (2002). Global agro-ecological assessment for agriculture in the 21st century: methodology and results. IIASA Research report, RR-02-02. IIASA, Laxenburg, Austria.

Fischer, G., Winiwarter, W., Ermolieva, T., Cao, GY., Qui, H., Klimont, Z., Wiberg, D., & Wagner, F. (2010). Integrated modeling framework for assessment and mitigation of nitrogen pollution from agriculture: Concept and case study for China. *Agriculture, Ecosystems and Environment, 136*(1-2), 116–124.

Frolking, S. E., Mosier, A. R., Ojima, D. S., Li, C., Parton, W. J., Potter, C. S., Priesack, E., Stenger, R., Haberbosch, C., Dörsch, P., Flessa, H., & Smith, K. A. (1998). Comparison of N2O emissions from soils at three temperate agricultural sites: simulations of year-round measurements by four models. *Nutrient Cycling in Agroecosystems, 52*, 77–105.

Galloway, J. N., Dentener, F. J., Capone, D. G., Boyer, E. W., Howarth, R. W., Seitzinger, S. P., Asner, G. P., Cleveland, C. C., Green, P.A., Holland, E.A., Karl, D.M., Michaels, A.F., Porter, J.H., Townsend, A. R., & Vörösmarty, C. J. (2004). Nitrogen cycles: past, present and future. *Biogeochemistry, 70*(2), 153–226.

Huang, J., Zhang, L., Li, Q., & Qiu, H. (2003). CHINAGRO project: National and regional economic development scenarios for China's food economy projections in the early 21st Century. Report to Center for Chinese Agricultural Policy. Chinese Academy of Sciences.

IPCC (2000). Good Practice Guidance and Uncertainty Management in National Greenhouse Gas Inventories. J. Penman, D. Kruger, I. Galbally, T. Hiraishi, B. Nyenzi, S. Emmanul, L. Buendia, R. Hoppaus, T. Martinsen, J. Meijer, K. Miwa and K. Tanabe, eds., IPCC National Greenhouse Gas Inventories Programme, Institute for Global Environmental Strategies, Kanagawa, Japan.

Keyzer, M. A., & van Veen, W. (2005). A summary description of the CHINAGRO-welfare model. CHINAGRO report. SOW-VU, Free University, Amsterdam, The Netherlands.

Klimont, Z. (2001). Current and Future Emissions of Ammonia in China. Paper presented at the 10th Annual Emission Inventory Conference: One Atmosphere, One Inventory, Many Challenges. May 1-3, Denver, CO, USA.

Klimont, Z., & Brink, C. (2004). Modeling of emissions of air pollutants and greenhouse gases from agricultural sources in Europe. IIASA Interim Report, IR-04-048. IIASA, Laxenburg, Austria.

Leonard, R. A., Knisel, W. G., & Still, D. A. (1987). GLEAMS: Groundwater loadingeffects of agricultural management systems. *Transactions of the American Society of Agricultural Engineers, 30*, 1403–1418.

Li, C., Frolking, S., & Frolking, T. A. (1992). Model of nitrous oxide evolution from soil driven by rainfall events: 1. Model structure and sensitivity. *Journal of Geophysical Research, 97*, 9759–9776.

Menzi, H. (2001). Area-wide Integration (AWI) of Specialized Crop and Livestock Activities: Assessment of Nutrient Management and Environmental Impacts. Final Report of SCA contribution in China. Swiss College of Agriculture (SCA), Switzerland. Internal Document.

NuFlux (2001). *NuFlux-AWI, User manual for the NuFlux-AWI Nutrient Balance Calculation Program. Main Manual*. Swiss College of Agriculture (SCA), CH-3052, Zollikofen, Switzerland.

Rockafellar, T., & Uryasev, S. (2000). Optimization of conditional value-at-risk. *The Journal of Risk, 2*, 21–41.

Rosegrant, M. W., Ringler, C. & Gerpacio, R. (1999). Water and Land Resources and Global Supply. In G. H. Peters, & J. von Braun (Eds.), *Food Security, Diversification and Resource Management: Refocusing the Role of Agriculture*, Proceedings of the 23rd International conference of Agricultural Economics held at Sacramento, California 10-16 August 1997. England: University of Oxford.

Shi, X. Z., Yu, D. S., Warner, E. D., Pan, X. Z., Petersen, G. W., Gong, Z. G., Weindorf, D. C. (2004). Soil database of 1:1,000,000 digital soil survey and reference system of the Chinese genetic soil classification system. *Soil Survey Horizons, 45*, 129–136.

Smil, V. (2001). *Enriching the Earth: Fritz Haber, Carl Bosch and the Transformation of World Food Production*. Cambridge, Massachusetts: MIT Press.

Steinfeld, H., Gerber, P., Wassenaar, T., Castel, V., Rosales, M., & de Haan, C. (2006). Livestock's long shadow: environmental issues and options. FAO-LEAD, 2006. (Available online at http://books.google.com/books?id).

Toth, F., Cao, G.-Y., & Hizsnyik, E. (2003). Regional population projections for China. IIASA Interim Report, IR-03-042. IIASA, Laxenburg, Austria.

Velthof, G. L., Oudendag, D., Witzke, H. P., Asman, W. A. H., Klimont, Z., & Oenema, O. (2009). Integrated assessment of nitrogen emissions from agriculture in EU-27 using MITERRA EUROPE. *Journal of Environmental Quality, 38*, 402–417. doi:10.2134/jeq2008.0108.

Winiwarter, W. (2005). The GAINS model for greenhouse gases: Version 1.0: nitrous oxide. IIASA Interim Report, IR-05-55. IIASA, Laxenburg, Austria.

Wang, C. (2010). Allocation of resources for protecting public goods against uncertain threats generated by agents. IIASA Interim Report, IR-10-012, IIASA, Laxenburg, Austria.

Evaluation of Portfolio of Financial and Insurance Instruments: Simulation of Uncertainty

Piotr Nowak, Maciej Romaniuk, and Tatiana Ermolieva

Abstract The increasing number of natural catastrophes leads to severe losses for production, in infrastructure and individual property. Classical insurance mechanisms may not be sufficient in dealing with such losses because of dependencies among sources of losses, huge values of damages, problems with adverse selection and moral hazard. To cope with dramatic consequences of such extreme events integrated policy is required. In this paper we discuss the model of portfolio which consists of a few layers of insurance and financial instruments, like catastrophe fund, catastrophe bonds, governmental help, etc. We use approach based on neutral martingale method and simulations. We price the catastrophe bond applying Vasicek model used for zero-coupon bond under assumption of independence between catastrophe occurrence and behavior of financial market. We discuss the effects of uncertainties which arise from estimation of rare events with serious, catastrophic consequences like natural catastrophes.

1 Introduction

The insurance industry faces overwhelming risks caused by natural catastrophes, e.g. losses from Hurricane Andrew hit 30 billion \$ in 1992, the losses from Hurricane Katrina in 2005 are estimated on $40 - 60$ billion \$ (see (Muermann 2008)). To cope with dramatic consequences of such extreme events integrated policy that combines mitigation measures with diversified ex-ante and ex-post

---

P. Nowak · M. Romaniuk
Systems Research Institute Polish Academy of Sciences, Warszawa, Poland
e-mail: pnowak@ibspan.waw.pl; mroman@ibspan.waw.pl

T. Ermolieva
International Institute for Applied Systems Analysis, Laxenburg, Austria
e-mail: ermol@iiasa.ac.at

Y. Ermoliev et al. (eds.), *Managing Safety of Heterogeneous Systems*, Lecture Notes
in Economics and Mathematical Systems 658, DOI 10.1007/978-3-642-22884-1_17,
© Springer-Verlag Berlin Heidelberg 2012

financial instruments is required. Without proper policies the natural catastrophes will increase long-term consequences for societies and economy of many countries, especially poor ones (see e.g. (MacKellar et al. 1999)).

The classical insurance mechanisms are not prepared for such extreme losses caused by natural catastrophes. Even one, single catastrophe could cause problems with reserves for many insurers or even bankruptcy of these enterprises. For example, after Hurricane Andrew more than 60 insurance companies became insolvent (see (Muermann 2008)). The traditional insurance models (see (Borch 1974)) deal with independent, rather small risks like car accidents. In such case the law of large numbers and the central limit theorem justify the ruin probability calculus and simple strategy of selecting an insurance contract portfolio: the greater the number of risks, the better (see (Borch 1974; Ermoliev et al. 2001)). Catastrophic risks require new approaches to the formation of a portfolio of an insurance company. The sources of losses from natural catastrophes are strongly dependent in terms of time and localization, e.g. single hurricane could start fire in many houses. The law of large numbers cannot be applied for such risks, and the traditional strategy of portfolio construction can only increase the probability of bankruptcy of insurer (see (Ermoliev et al. 2001)).

Additionally, classical insurance mechanisms are often criticized because of serious problems with adverse selection and moral hazard – e.g., hope for governmental help or possession of insurance policy may change people's attitude and draw them to growing crops in high risk regions, building houses in threatened area, not preventing additional losses, etc. Moreover, the primary insurers rely on classical reinsurance markets which are affected by price cycles connected with occurrence of natural catastrophes, terrorist attacks, etc.

The single event, e.g. earthquake or hurricane, could result in damages of $50–$100 billion. Keeping in mind that daily fluctuations on worldwide financial markets reach tens of billion $, securitization of losses (e.g. in the form of so called catastrophe bonds – see e.g. (Ermolieva et al. 2007; Nowak and Romaniuk 2009; Romaniuk and Ermolieva 2005)) may be helpful for dealing with results of extreme natural catastrophes (see e.g. (Cummins et al. 2002; Freeman and Kunreuther 1997; Froot 2001; Harrington and Niehaus 2003)).

For example, in agricultural regions natural disasters may lead to severe losses of agricultural production and, thus, to decrease of farmers' income (see e.g. (MacKellar et al. 1999)). The effects of such catastrophes are commonly known and various production planning practices were developed. However, these traditional risk management mechanisms are not always sufficient for dealing with extreme events (see e.g. (Nowak et al. 2008; Skees et al. 2002)). Additionally, due to lower level of income for rural areas, management of risks is especially price-sensitive. The demand for financial instruments depends on willingness of farmers to sacrifice a portion of their often uncertain income. The supply of financial instruments depends on feasibility of financial or insurance instruments and the ability of the insurance industry to manage losses at these prices. Therefore we should take into account the "fairness" requirement for both insurers and insureds (see e.g. (Ermolieva et al. 2007)).

Evaluation of Portfolio of Financial and Insurance 353

In this paper we discuss the model of portfolio which consists of a few layers of insurance and financial instruments, like catastrophe fund, catastrophe bonds, governmental help, etc. We use approach based on neutral martingale method and simulations. We price the catastrophe bond applying Vasicek model used for zero-coupon bond under assumption of independence between catastrophe occurrence and behavior of financial market. Obtained pricing formula is then applied in simulations. The mentioned portfolio should be optimal in some way, i.e. it should fulfill needs of both insureds and insurers. In order to achieve "fairness" for both insurer and insureds, we limit the probability of insurer bankruptcy and the probability of overpayment for insureds. We discuss the effects of uncertainties which arise from estimation of rare events with serious, catastrophic consequences like natural catastrophes. Therefore there is a need to take into account possible errors in estimation, e.g. applying approach based on confidence intervals. These intervals may also incorporate expertise knowledge to overcome lack of precise data.

This paper is organized as follows. In Sect. 2 we describe properties and examples of catastrophe bonds. In Sect. 3 we present the model of portfolio which consists of a few layers of financial and insurance instruments. In Sect. 4 the pricing formula for some examples of cat bonds is discussed. Then in Sect. 5 simulation results for some portfolios are described. In Sect. 6 we discuss possible sources of uncertainties which should be taken into account during portfolio modeling. Then in Sect. 7 the final remarks and some conclusions are provided.

2   Catastrophe Bonds

As it was mentioned before, the single catastrophe event could result in damages of $50 – $100 billion. This could cause the bankruptcy of the insurer or serious problems with coverage of losses (see (Cummins et al. 2002; Froot 2001; Harrington and Niehaus 2003)). Additionally, the classical insurance mechanisms are not adequate in face of catastrophic event because of dependencies among sources of risks, potentially unlimited losses, problems with adverse selection, moral hazard and reinsurance pricing cycles.

Therefore applying alternative financial or insurance instruments may be profitable. The problem is to "package" natural disasters risk / losses into classical forms of tradeable financial assets, like bonds or options. The most popular catastrophe-linked security is the catastrophe bond (in abbreviation *cat bond* or *Act-of-God* bond, see (Cox et al. 2000; D'Arcy and France 1992; Ermolieva et al. 2007; George 1999; Nowak and Romaniuk 2009; O'Brien 1997; Romaniuk and Ermolieva 2005)).

In 1993 catastrophe derivatives were introduced at the Chicago Board of Trade (CBoT). These financial derivatives were based on underlying indexes reflecting insured property losses due to natural catastrophes that were reported by insurance and reinsurance companies. The first type of contracts traded at the CBoT were insurance futures and options on insurance futures. Later they were replaced by catastrophe spread options based on underlying loss indexes provided by an independent statistical agency.

New instruments include over-the-counter (OTC) products, primarily engineered by investment bankers. In 1997 new weather derivatives were introduced in the USA, later also in Europe. In 2000 European bank and an insurance company launched "Meteo transformer" to issue weather derivatives and insurance contracts. The stock exchange Euronext introduced electricity derivatives in 2001.

Cat bonds become wider known in April 1997, when USAA, an insurer from Texas, initiated two new classes of cat bonds: A-1 and A-2. Next successful catastrophe bond was issued in 1997 by Swiss Re to cover earthquake losses. The first cat-bond prepared by a non-financial firm was issued in 1999 in order to cover earthquake losses in the Tokyo region for Oriental Land Company, Ltd., the owner of Tokyo Disneyland (see (Vaugirard 2003)). The cat bond market in year 2003 hit a total issuance of \$1.73 billion, a 42% increase from 2002s record of \$1.22 billion (see (McGhee 2004)). The report also shows that since 1997, 54 cat bond issues have been completed with total risk limits of almost 8 billion (see (Vaugirard 2003)). To the end of 2004 there were about 65 emissions of cat bonds. Insurance and reinsurance companies have issued almost all of the cat bonds, and except for a few from commercial companies, reinsurers have accounted for over 50% of the issuances (see (Vaugirard 2003)). The market of cat bonds is expected to emerge in future.

There is one important difference between cat bonds and other financial instruments. The premiums from cat bond are always connected with additional random variable, i.e. occurrence of some natural catastrophe in specified region and fixed time interval. Such event is called *triggering point* (see (George 1999)). For example, the A-1 USAA bond was connected with hurricane on the east coast of the USA between July 15, 1997 and December 31, 1997. If there had been a hurricane in mentioned above region with more than \$1 billion loses against USAA, the coupon of the bond would have been lost. As usually, the structure of payments for cat bonds depends also on some primary underlying asset. In case of A-1 USAA bond, the payment equaled LIBOR plus 282 basis points. As we can see from this example, the triggering point changes the structure of payments for the cat bond. The other types of cat bonds may be related to various kinds of triggering points — e.g. to magnitude of earthquake, the losses from flood, etc. (see e.g. (Niedzielski 1997; Walker 1997)).

Another example is the cat bond known as Atlas Re II, issued for SCOR Group, intended to cover claims linked to natural catastrophe events from January 1, 2002 during the period of three years. These events were earthquakes in California and Japan, and wind-storms in Northern Europe. The mentioned cat bond complements the USD 100 million per event cover of Atlas Re, the previous cat bond issued for SCOR, which was connected with occurrence of a first event. Atlas Re II provides coverage for a second or third event during a given year, with a USD 100 million per event limit and a USD 150 million limit over three years. The triggers were based on reported earthquake magnitude or an index calculated from wind-speeds in case of wind-storms.

Evaluation of Portfolio of Financial and Insurance

The main aim of cat bonds is to transfer *risk* from insurance markets or governmental budgets to financial markets. Apart from transferring capital, a liquid catastrophe derivatives market allow insurance and reinsurance companies to adjust their exposure to natural catastrophic risk dynamically through hedging with those contracts at lower transaction costs. If the triggering point is connected with industry loss indices or parametric triggers, the moral hazard exposure of bond investors is greatly reduced or eliminated. Cat bonds are often rated by an agency such as Standard & Poor's, Moody's, or Fitch Ratings.

The cash flows for catastrophe bond are managed by special tailor-made fund, called a special-purpose vehicle (SPV) (see (Vaugirard 2003)). The hedger (e.g. insurer) pays an insurance premium in exchange for coverage in case if catastrophic event occurs. The investors purchase an insurance-linked security for cash. The mentioned premium and cash flows are directed to SPV, which issues the catastrophe bonds. Usually, SPV purchases safe securities in order to satisfy future possible demands. Investors hold the issued assets whose coupons and/or principal depend on occurrence of the triggering point. If the pre-specified event occurs during the fixed period, the SPV compensates the insurer and the cash flows for investors are changed, i.e. there is full or partial forgiveness of the repayment of principal and/or interest. If the triggering point does not occur, the investors usually receive the full payment. Triggering point may be connected with the issuer's actual losses, losses modeled by special software based on real parameters of catastrophe, insurance industry index, real parameters of catastrophe or hybrid index related to modeled losses.

3 Portfolio Construction

The portfolio could consist of a whole set of various financial and insurance instruments. Model of such portfolio may be used by insurer or other organization (like government) in order to evaluate parameters and possible scenarios, e.g. to calculate the probability of ruin, the value of maximum losses for given probability level, the necessity of using the considered instrument in the portfolio, etc.

Let $(\Omega, \mathscr{F}, P)$ be a probability space. We consider trading horizon $[0, T']$, $T' > 0$. We fix some $T \in [0, T']$. The structure of catastrophe model is based on subdividing the considered region into m nodes (or cells). Based on this structure, the scenario of catastrophe and losses arising from this scenario should be simulated.

For example, floods involve both meteorological and hydrological processes. Additionally these phenomena are influenced by human facilities and activities (like dams, land use, etc.). To find an estimate of the largest size flood for the given period of time and region, simulations based on historical records or probabilistic distributions are used. The most commonly used distributions are normal family (normal distribution, log-normal, log-normal type 3), the general extreme-value family (GEV, Gumbel, log-Gumbel, Weibull), the Pearson type 3 family (Pearson type 3, log-Pearson type 3) and the generalized Pareto distribution (see (Malamud and Turcotte 2006; Rao and Hamed 2000)). For example, the power-law distribution for flood frequency has the form $Q = Ct^\alpha$, where α and C are regression coefficients, and Q is the maximum discharge associated with a recurrence interval of t years. Apart from modeling the catastrophe itself, the losses caused by such event should be also estimated. Appropriate calculations are based on vulnerability curves or matrices for the flood level, earthquake magnitude, wind speed for hurricane, etc. versus the types of buildings, types of crops, etc. (see e.g. (Ermolieva et al. 2005, 2007)). Using special software (see e.g. (Ermolieva et al. 2007; Pinelli et al. 2008)) the whole set of scenarios, estimators or histograms of losses may be found.

Because of mutual dependencies between cells caused by catastrophe itself (e.g. flood) and by insurance or financial instruments, which may spread the losses, the considerable amount of nodes may be affected by the same, single event. Therefore for each node v we have set V_v of all the neighboring nodes which may be affected if there is destruction in node v. These sets should be taken into account in simulations of losses (see (Ermoliev et al. 2001)).

By $C(j,t)$ we denote claims caused by losses for cell j on the time interval $[0,t]$ and by $C(t)$ the aggregated demands for all cells on the time interval $[0,t]$, i.e. $C(t) = \sum_{i=1}^{m} C(j,t)$. These claims depend on losses $L(j,t)$. For example, for proportional insurance contract we have $C(j,t) = q_j L(j,t)$, where q_j is the proportion parameter and for proportional contract with lower bound we have

$$C(j,t) = q_j L(j,t) I (L(j,t) > k_0), \tag{1}$$

where $I(.)$ is the characteristic function and k_0 is the lower bound. By $L(t)$ we denote the aggregated losses, i.e. $L(t) = \sum_{i=1}^{m} L(j,t)$, by $\Pi(j,t)$ – the insurance premiums for the cell j on the time interval $[0,t]$, and by $\Pi(t)$ – the aggregated premiums for all cells, i.e. $\Pi(t) = \sum_{i=1}^{m} \Pi(j,t)$. Then the classical insurance model (see (Borch 1974)) describing the evolution of the profit process for the insurer has the form

$$R(t) = R(0) + \Pi(t) - C(t), \tag{2}$$

where $R(0)$ is the initial capital, if we assume that the risk-free yield is equal to zero.

The main goal for insurer is the maximization of the profit given by random variable

$$\Theta = R(0) + \Pi(T) - C(T). \tag{3}$$

In order to achieve "fairness" for both insurer and insureds, we limit the probability of insurer bankruptcy by value p_1, i.e.

$$P (R(T) \leq 0) \leq p_1 \tag{4}$$

and the probability of overpayment for insureds by p_2 (see (Nowak and Romaniuk 2009; Stone 1973)), i.e.

$$P (C(j,T) \leq \Pi(j,T)) \leq p_2. \tag{5}$$

Evaluation of Portfolio of Financial and Insurance 357

Therefore the insurer should maximize the function (3) under the constraints (4) and (5).

As it was mentioned before, because of dependency between the losses and the nature of risk itself (small probability of huge catastrophe) standard insurance mechanism could not be adequate. Therefore we add some extra layers in order to create the whole portfolio of various insurance and financial instruments. The second layer in such portfolio is the catastrophe bond issued by government, insurer or other enterprise. If insurer issues such bond, then at the maturity time of the bond T the process (3) has the form

$$\Theta = R(0) + PV(\Pi(T)) - PV(C(T)) + \Pi^{cb} - PV\left[f^{cb}(L(T))\right], \qquad (6)$$

where PV denotes the present value of cash flow, Π^{cb} – the aggregated premiums from cat bond issuing (or with minus sign – premiums directed to SPV), $f^{cb}(.)$ – the payment function for the considered kind of cat bond. The triggering point for such cat bond may be connected e.g. with surpassing the limit k_1 by aggregated losses $L(T)$.

On the next level other layers may be also added to portfolio. For example if the aggregated losses will be above some level k_2, i.e. $L(T) > k_2$, then there is possibility that special governmental fund may be used. We assume that the probability space $(\Omega', \mathscr{F}', P')$ describes the external help. Then the probability of using governmental fund may be modeled as independent random event A_2. We may assume that the value of such fund is denoted by independent random variable X_2 or this value is proportional to the aggregated losses, i.e. $\varphi_2 L(T)$, where φ_2 is the proportion parameter.

In the same way we may add the next layer, e.g. foreign help. Such help may be used if the aggregated losses will be above level k_3, i.e. $L(T) > k_3$ and the probability of using this fund may be modeled as independent random event A_3. The value of such fund may be denoted by independent random variable X_3 or it may be proportional to the aggregated losses, i.e. $\varphi_3 L(T)$.

Taking into account all the layers of the portfolio, the process (3) has now the form

$$\Theta = R(0) + PV(\Pi(T)) - PV(C(T)) + \Pi^{cb} - PV\left[f^{cb}(L(T))\right]$$
$$+ PV(X_2) I(A_2) I(L(T) > k_2) + PV(X_3) I(A_3) I(L(T) > k_3) \qquad (7)$$

or

$$\Theta = R(0) + PV(\Pi(T)) - PV(C(T)) + \Pi^{cb} - PV\left[f^{cb}(L(T))\right]$$
$$+ PV(\varphi_2 L(T)) I(A_2) I(L(T) > k_2) + PV(\varphi_3 L(T)) I(A_3) I(L(T) > k_3)$$
$$(8)$$

depending on the way we model values of the funds, under mentioned previously constraints (4) and (5).

For the formula (7) or (8) we need some additional parameters, which may be also calculated, like price of the considered type of catastrophe bond (see Sect. 4). Because problems similar to the above equations are extremely sensitive for the constraints, special maximization procedure may be used (see e.g. (Ermoliev et al. 2001)). Apart from insurer, the other important entity is the government. It is possible to optimize the function similar to (7) or (8) which takes into account also the government needs (see e.g. (Nowak et al. 2008)).

4 Cat-Bond Pricing

In this section we assume that the number of cells is equal to one ($m = 1$). Let $(W_t)_{t \in [0,T']}$ be a Brownian motion.

Let $(U_i)_{i=1}^{\infty}$ be independent, identically distributed random variables with bounded second moment. In further considerations we treat U_i as value of losses during i-th catastrophic event.

We define compound Poisson process by formula

$$\tilde{N}_t = \sum_{i=1}^{N_t} U_i , \ t \in [0, T'] ,$$

where N_t is Poisson process with intensity κ.

We will assume that $L_t = \tilde{N}_t , \ t \in [0, T']$.

Let

$$k_0 < k_1^1 < ... < k_1^n, \ n > 1$$

be a sequence of constants.

Let $\tau_i : \Omega \to [0, T'] , \ 1 \le i \le n$ be a sequence of stopping times defined as follows

$$\tau_i (\omega) = \inf_{t \in [0,T']} \left\{ \tilde{N} (t) (\omega) > k_1^i \right\} \wedge T', \ 1 \le i \le n.$$

The filtration $(F_t)_{t \in [0,T']}$ is given by formula

$$F_t = \sigma \left(F_t^0 \cup F_t^1 \right) , \ F_t^0 = \sigma (W_s, s \le t) ,$$
$$F_t^1 = \sigma \left(\tilde{N}_s, s \le t \right) , t \in [0, T'] .$$

We assume that

$$F_0 = \{A \in F : P (A) = 0\}$$

and that $(W_t)_{t \in [0,T']}$, $(N_t)_{t \in [0,T']}$ and $(U_i)_{i=1}^{\infty}$ are independent. Then the filtered probability space $\left(\Omega, F, (F_t)_{t \in [0,T']}, P \right)$ satisfies the standard assumptions,

Evaluation of Portfolio of Financial and Insurance 359

i.e. σ-algebra F is P-complete, the filtration $(F_t)_{t \in [0, T']}$ is right continuous, what means that for each $t \in [0, T')$

$$F_{t+} = \bigcap_{s > t} F_s = F_t$$

and F_0 contains all the sets in F of P-probability zero.

We denote by $(B_t)_{t \in [0, T']}$ banking account satisfying the following equation:

$$d B_t = r(t) B_t dt, \ B_0 = 1,$$

where r is a risk-free spot interest rate. The solution of the above equation has the form:

$$B_t = \exp \left(\int_0^t r(u) du \right), \ t \in [0, T'].$$

We denote by $B(t, T)$ the price at the time t zero-coupon bond with maturity date $T \leq T'$ and the face value equal to 1. Let

$$w_1 < w_2 < \ldots < w_n$$

be a sequence of nonnegative constants, for which $\sum_{i=1}^n w_i \leq 1$.

Definition 1. We denote by $IB(T, Fv)$ a catastrophe bond satisfying the following assumptions:

a) If the catastrophe does not occur in the period $[0, T]$, i.e. $\tau_1 > T$, the bond-holder is paid the face value Fv;
b) If $\tau_n \leq T$, the bond-holder receives the face value minus the sum of write-down coefficients in percentage $\sum_{i=1}^n w_i$.
c) If $\tau_{k-1} \leq T < \tau_k$, $1 < k \leq n$, the bond-holder receives the face value minus the sum of write-down coefficients in percentage $\sum_{i=1}^{k-1} w_i$.
d) A cash payments are done at date of maturity T.

Definition 2. $B(t, T)$, $t \leq T \leq T'$ is called the arbitrage-free family of zero-coupon bond prices with respect to r, if the following conditions are satisfied:

a) $B(T, T) = 1$ for each $T \in [0, T']$.
b) There exists a probability Q, equivalent to P, such that for each $T \in [0, T']$ the process of discounted zero-coupon bond price

$$B(t, T) / B_t, \ t \in [0, T],$$

is a martingale with respect to Q. Then we have the following pricing formula

$$B(t, T) = E^Q \left(e^{-\int_t^T r(u) du} | F_t^Q \right), \ t \in [0, T].$$

Let $\lambda_u = -\lambda$ denote the risk premium for risk-free bonds. The following Radon-Nikodym derivative defines a probability measure Q, equivalent to P:

$$\frac{dQ}{dP} = \exp\left(\int_0^T \lambda_u dW_u - \frac{1}{2}\int_0^T \lambda_u^2 du\right) P\text{-a.s.},$$

such that $B(t, T)/B_t$, $t \in [0, T]$, is a martingale with respect to Q.

We assume the Vasicek model of the risk-free spot interest rate r. The interest rate satisfies the following equation

$$dr(t) = a(b - r(t))dt + \sigma dW_t$$

for positive constants a, b and σ.

We also assume that financial market is independent from the catastrophe risk and investors are neutral toward nature jump risk.

We will apply the methodology from (Vaugirard 2003) to price the catastrophe bond.

Theorem 1. *Let $IB(0)$ be the price of a $IB(T, Fv)$ at time 0. Let*

$$\Phi = \sum_{i=1}^{n} w_i \Phi_i,$$

where Φ_i are cumulative distribution function of τ_i. Then

$$IB(0) = Fve^{-TR(T, r(0))}\{1 - \Phi(T)\}, \tag{9}$$

where

$$R(\theta, r) = R_\infty - \frac{1}{a\theta}\left\{(R_\infty - r)(1 - e^{-a\theta}) - \frac{\sigma^2}{4a^2}(1 - e^{-a\theta})^2\right\}$$

and

$$R_\infty = b - \frac{\lambda\sigma}{a} - \frac{\sigma^2}{2a^2}.$$

Proof. We show basic steps of the proof of Theorem 1. From (Vaugirard 2003) it follows that

$$E^Q\left(\exp\left(-\int_0^T r(u)du\right)\right) = Fve^{-TR(T, r(0))}.$$

Since $\exp\left(-\int_0^T r(u)du\right)$ and $1 - \sum_{i=1}^{n} w_i I_{\tau_i \leq T}$ are independent under Q,

$$IB(0) = E^Q\left(\exp\left(-\int_0^T r(u)du\right)\right)\left\{1 - \sum_{i=1}^{n} w_i E^Q(I_{\tau_i \leq T})\right\},$$

Evaluation of Portfolio of Financial and Insurance

Since τ and W are independent,

$$
1 - \sum_{i=1}^{n} w_i E^Q \left(I_{\tau_i \leq T} \right) = 1 - \sum_{i=1}^{n} w_i E^P \left(I_{\tau_i \leq T} \frac{dQ}{dP} \right)
$$

$$
= 1 - \sum_{i=1}^{n} w_i E^P \left(I_{\tau_i \leq T} \right) E^P \left(\frac{dQ}{dP} \right)
$$

$$
= 1 - \sum_{i=1}^{n} w_i E^P \left(I_{\tau_i \leq T} \right)
$$

$$
= 1 - \Phi (T).
$$

Finally, the pricing formula at time $t = 0$ has the form (9). □

The following proposition gives the form of the cumulative distribution functions of τ_i. Its proof follows exactly from the definition of the stopping times.

Proposition 1. *The value of the cumulative distribution function Φ_i, $1 \leq i \leq n$, at the moment T has the form*

$$
\Phi_i (T) = 1 - \sum_{j=0}^{\infty} \frac{(\kappa T)^j}{j!} e^{-\kappa T} \Phi_{\tilde{U}_j} \left(k_1^i \right),
$$

where $\Phi_{\tilde{U}_j}$ is the cumulative distribution function of the sum $\tilde{U}_j = \sum_{p=0}^{j} U_p$. In the above formula we assume that $U_0 \equiv 0$.

5 Numerical Experiments

In order to analyze the features of the portfolio proposed in Sect. 3, the appropriate simulations were conducted. We assume that quantity of losses is modeled by Poisson process with expected value $\mu = 0.05$ and the value of each loss is given by random variable from Gamma distribution with scale parameter $\alpha = 10$ and shape parameter $\beta = 10$. Therefore the generated losses have catastrophic nature, i.e. they are rare, but with high value. Other types of distributions for modeling the value of losses are also possible, e.g. Weibull distribution.

The trading horizon is set on 5 years and constant continuous risk-free yield r is equal to 0.05. For each portfolio we generate $n = 100000$ simulations.

Only insurance contract is taken into account in the *Portfolio I*. It is assumed that insurance premium is equal to 0.02, it is paid by 100 insureds and the insurance contract is proportional with lower bound. For simplicity the lower bond is set for the whole portfolio of insurance contacts to $k_0 = 5$ and proportion parameter is set to $q = 0.95$. For initial capital $R(0) = 45$ the probability of insurer bankruptcy is equal to 2.318%.

Then the price of cat bond is calculated. According to Sect. 4, the Vasicek model with parameters $a = 0.025, b = 0.05, \sigma = 0.01, r(0) = 0.05$ is assumed. The face value of the bond is set to 1 and triggering point is connected with surpassing the limit $k_1 = 40$ by aggregated losses. If the triggering point occurs, then the bond holder receives only 50% of face value, i.e. $w = 0.5$. In such case, the price of the cat bond calculated according to formula (9) via simulations is equal to 0.769732.

All the instruments discussed in Sect. 3 are taken into account in *Portfolio II*. It is assumed that $R(0) = 45$, the features of cat bond are the same as mentioned above, the cat bond is sold with small discount for price 0.76 and the external help is proportional to the losses with parameters $k_2 = 40, k_3 = 60, \varphi_2 = 0.01, \varphi_3 = 0.01$. The occurrence of governmental help is independent event with probability 0.05 and the occurrence of foreign help is independent event with probability 0.02. Then we may calculate measures for our portfolio based on simulations (see Table 1). The probability of ruin is substantially lower with relatively low average value of external help. Because of selling the cat bond with small discount, the average value of flows for cat bond is slightly negative. Both average and median of value of portfolio are similar to initial capital. Also the quantiles for VaR (Value-at-Risk) are calculated. From these quantiles it is seen that value of portfolio is rather stable.

For the *Portfolio III*, the cat bond has the greater impact. The cat bond is sold with greater discount for price 0.75 and $k_3 = 100$. Therefore, the foreign help is more limited. The appropriate measures for such portfolio are given in Table 2. It is seen that average value of flows for cat bond is more negative because of the greater discount for the cat bond price. Also average value and median for the whole portfolio is slightly less than initial capital.

In case of *Portfolio IV* the price for the cat bond is the same as calculated according to formula (9), i.e. 0.769732. The rest of parameters are the same as in *Portfolio III*. The appropriate measures for such portfolio are given in Table 3. In this case the average value of flows for cat bonds are slightly positive. Selling the cat

Table 1 Numerical features of Portfolio II

	Value
Average value of portfolio	44.1992
Median of value of portfolio	45.1199
Standard deviation of value of portfolio	6.90794
5% quantile of value of portfolio	45.1199
9% quantile of value of portfolio	45.1199
1% quantile of value of portfolio	4.4582
99% quantile of value of portfolio	45.1199
Maximal loss for portfolio with probability 5%	−0.119922
Maximal loss for portfolio with probability 1%	40.5472
Probability of ruin	0.82%
Average value of flows for cat bond	−0.902683
Average value of flows for external help	0.00142813

Evaluation of Portfolio of Financial and Insurance

Table 2 Numerical features of Portfolio III

	Value
Average value of portfolio	43.2407
Median of value of portfolio	44.1199
Standard deviation of value of portfolio	6.76293
5% quantile of value of portfolio	44.1199
9% quantile of value of portfolio	44.1199
1% quantile of value of portfolio	6.00394
99% quantile of value of portfolio	44.1199
Maximal loss for portfolio with probability 5%	0.880078
Maximal loss for portfolio with probability 1%	38.9961
Probability of ruin	0.78%
Average value of flows for cat bond	−1.96109
Average value of flows for external help	0.00163002

Table 3 Numerical features of Portfolio IV

	Value
Average value of portfolio	45.2098
Median of value of portfolio	46.0931
Standard deviation of value of portfolio	6.67132
5% quantile of value of portfolio	46.0931
9% quantile of value of portfolio	46.0931
1% quantile of value of portfolio	7.69566
99% quantile of value of portfolio	46.0931
Maximal loss for portfolio with probability 5%	−1.09312
Maximal loss for portfolio with probability 1%	37.3043
Probability of ruin	0.73%
Average value of flows for cat bond	0.0744107
Average value of flows for external help	0.00100092

bond for undiscounted price improves the value of portfolio measured with average value and median.

In case of *Portfolio V* we assume that special premiums from the insurer are directed to SPV which directly sells the cat bonds. This premium is equal to 0.01 for one cat bond and SPV issues 100 cat bonds. The other parameters are the same as in *Portfolio IV*. Then the appropriate measures for such portfolio are given in Table 4. As we could see, because premiums are directed to SPV even if there is no triggering point, the average value of flows for cat bonds for insurer is slightly negative.

6 Uncertainties Problem

There are many sources of uncertainties which affected the described model.

First of them is the model of catastrophe event itself. Natural catastrophes are rare, therefore there may be problems with sparse historical data or with fitting parameters for probability models, e.g. estimation of expected value for Poisson

	Value
Table 4 Numerical features of Portfolio V	
Average value of portfolio	45.1062
Median of value of portfolio	46
Standard deviation of value of portfolio	6.59314
5% quantile of value of portfolio	46
9% quantile of value of portfolio	46
1% quantile of value of portfolio	5.97093
99% quantile of value of portfolio	46
Maximal loss for portfolio with probability 5%	-1
Maximal loss for portfolio with probability 1%	39.0291
Probability of ruin	0.71%
Average value of flows for cat bond	-0.022605
Average value of flows for external help	0.000719589

process. Therefore the interval estimation may be seen as a way to solve such problems – instead of using only one crisp value for each parameter, the appropriate intervals may be used. However, in such case also the simulated results will be achieved as confidence intervals with some degree of uncertainty. But taking into account central limit theorem and the nature of Monte Carlo simulations, these intervals are strongly supported by probability theory.

Other way to solve the mentioned problems is to apply the fuzzy set theory. Then the appropriate parameters may be modeled as fuzzy numbers to incorporate expert knowledge or some historical data. Obtaining results via simulations is more complicated in this setting unless we restrict ourselves to α-level intervals for fuzzy numbers (see e.g. (Nowak and Romaniuk 2009)). There are also some approaches emphasizing similarities between statistical confidence intervals and fuzzy α-level intervals (see e.g. (Buckley 2004)).

Apart from problems with fitting parameters, historical data may not be adequate to properly determine the type of probability distribution. In our simulations we use Gamma distribution for modeling the value of each loss. There are other types of distributions also known in literature as appropriate modeling tool, e.g. Weibull distribution or extreme value distribution. Applying irrelevant distribution may change the simulations output in a serious way.

Because of low probability of the catastrophic event, simulations should be done in specific way. Otherwise, the error of order $\sqrt{n}$ for Monte Carlo method may be too high to properly determine estimators of output values e.g. the expected value of cash flows.

Other source of uncertainty is arising from modeling of financial market, i.e. problems of fitting parameters for stochastic process of interest risk rate. Also in this case instead of crisp, point estimation, other approaches like interval estimation or fuzzy estimation may be useful (see e.g. (Nowak and Romaniuk 2010)).

Also the decision takers may use other set of financial and insurance instruments to fulfill their needs. We restrict our model presented in Sect. 3 to standard insurance

Evaluation of Portfolio of Financial and Insurance 365

policy, catastrophe bond, governmental help and foreign help. Other instruments are also possible, e.g. governmental bonds or contingency credit.

7 Conclusions

The insurance industry face overwhelming risks caused by natural catastrophes. But the classical insurance mechanisms are not prepared for such extreme losses. Even one, single catastrophe could cause problems with reserves for many insurers or even bankruptcy of these enterprises. Catastrophic risks require new approaches to the formation of a portfolio of an insurance company. Keeping in mind that daily fluctuations on worldwide financial markets reach tens of billion \$, securization of losses may be helpful for dealing with results of extreme natural catastrophes.

In this paper we discuss the model of portfolio which consists of a few layers of insurance and financial instruments, like catastrophe fund, catastrophe bonds, governmental help, etc. We use approach based on neutral martingale method and simulations. We price the catastrophe bond applying Vasicek model used for zero-coupon bond under assumption of independence between catastrophe occurrence and behavior of financial market. Obtained pricing formula is then applied in simulations. We discuss the effects of uncertainties which arise from estimation of rare events with serious, catastrophic consequences like natural catastrophes.

References

Borch, K. (1974). *The Mathematical Theory of Insurance*. Lexington: Lexington Books.

Buckley, J. J. (2004). *Fuzzy Statistics*. New York: Springer.

Cox, S. H., Fairchild, J. R., & Pedersen, H. W. (2000). Economic aspects of securitization of risk. *ASTIN Bulletin, 30*(1), 157–193.

Cummins, J. D., Doherty, N., & Lo, A. (2002). Can insurers pay for the "big one"? Measuring the capacity of insurance market to respond to catastrophic losses. *Journal of Banking and Finance, 26*.

D'Arcy, S. P., & France, V. G. (1992). Catastrophe futures: A better hedge for insurers. *Journal of Risk and Insurance, 59*(4), 575–600.

Ermoliev, Yu. M., Ermolyeva, T.Yu., McDonald, G., Norkin, V. I. (2001). Problems on insurance of catastrophic risks. *Cybernetics and Systems Analysis, 37*(2).

Ermolieva, T., & Ermoliev, Y. (2005). Catastrophic risk management: flood and seismic risks case studies. In S. W. Wallace, & W. T. Ziemba (Eds.), *Applications of Stochastic Programming*, MPS-SIAM Series on Optimization. Philadelphia, PA, USA.

Ermolieva, T., Romaniuk, M., Fischer, G., & Makowski, M. (2007). Integrated model-based decision support for management of weather-related agricultural losses. In O. Hryniewicz, J. Studzinski, & Shaker M. Romaniuk (Eds.), *Enviromental informatics and systems research. Vol. 1: Plenary and session papers - EnviroInfo 2007*. Verlag.

Freeman, P. K., & Kunreuther, H. (1997). *Managing Environmental Risk Through Insurance*. Boston: Kluwer.

Froot, K. A. (2001). The market for catastrophe risk: A clinical examination. *Journal of Financial Economics*, *60*(2).

George, J. B. (1999). Alternative reinsurance: Using catastrophe bonds and insurance derivatives as a mechanism for increasing capacity in the insurance markets. *CPCU Journal*, *52*(1).

Harrington, S. E., & Niehaus, G. (2003). Capital, corporate income taxes, and catastrophe insurance. *Journal of Financial Intermediation*, *12*(4).

MacKellar, L., Freeman, P., & Ermolieva, T. (1999). Estimating Natural Catastrophic Risk Exposure and the Benefits of Risk Transfer in Developing Countries. http://www.iiasa.ac.at/Research/CAT/paris.pdf.

Malamud, B. D., & Turcotte, D. L. (2006). The applicability of power-law frequency statistics to floods. *Journal of Hydrology*, *322*.

McGhee, C. (2004). *Market Update: The Catastrophe Bond Market at Year-End 2003*. Guy Carpenter & Company, Inc. and MMC Security Corporation.

Muermann, A. (2008). Market price of insurance risk implied by catastrophe derivatives. *North American Actuarial Journal*, **12**(3), 221–227.

Niedzielski, J. (1997). USAA places catastrophe bonds. *National Underwriter*, *16*

Nowak, P., Romaniuk, M., & Ermolieva, T. (2008). Integrated management of weather – related agricultural losses – computational approach. In E. Wilimowska, L. Borzemski, A. Grzech, & Wroclaw J. Swiatek (Eds.), *Information Systems Architecture and Technology*.

Nowak, P., & Romaniuk M. (2009). Portfolio of financial and insurance instruments for losses caused by natural catastrophes. In Z. Wilimowska, L. Borzemski, A. Grzech, Wroclaw J. Swiatek (Eds.), *Information Systems Architecture and Technology. IT Technologies in Knowledge Oriented Management Process*.

Nowak, P., & Romaniuk, M. (2009). Fuzzy Approach to Evaluation of Portfolio of Financial and Insurance Instruments. In K. T. Atanassov, O. Hryniewicz, J. Kacprzyk, M. Krawczak, Z. Nahorski, E. Szmidt, S. Zadrożny (Eds.) *Advances in Fuzzy Sets, Intuitionistics Fuzzy Sets*.

Nowak, P., & Romaniuk, M. (2010). Computing option price for Levy process with fuzzy parameters. *European Journal of Operational Research*, *201*(1).

O'Brien, T. (1997). Hedging strategies using catastrophe insurance options. *Insurance: Mathematics and Economics*, *21*(2), 153–162.

Pinelli, J.-P., & Gurley, K. R., et al. (2008). Validation of probabilistic model for hurricane insurance loss projections in Florida. *Reliability Enigeeniring and System Safety, 93*.

Rao, A. R., & Hamed, K. H. (2000). *Flood Frequency Analysis*. CRC Press.

Romaniuk, M., & Ermolieva, T. (2005). Application EDGE software and simulations for integrated catastrophe management. *International Journal of Knowledge and Systems Sciences*, *2*(2), 1–9.

Skees, J., Varangis, P., Larson, D., & Siegel, P. (2002). Can Financial Markets be Tapped to Help Poor People Cope with Weather Risks? Policy Research Working Paper *2812*, The World Bank Development Research Group, Rural Development (March 2002).

Stone, J. M. (1973). A theory of capacity and the insurance of catastrophe risks. *The Journal of Risk and Insurance*, *40*.

Vaugirard, V. E. (2003). Pricing catastrophe bonds by an arbitrage approach. *The Quarterly Review of Economics and Finance*, *43*, 119–132.

Walker, G. (1997). Current developments in catastrophe modelling. In N.R. Britton, & J. Olliver (Eds.), *Financial Risks Management for Natural Catastrophes*. Brisbane, Australia: Griffith University.

Pricing Catastrophe Bonds under Safety Constraints

Shuo Liu and Liyan Han

Abstract This chapter proposes an approach for catastrophe bonds (cat-bonds) pricing using stochastic balances of cash flows. Monte Carlo simulation model permits to overcome cat-bonds trading data shortage and sheds the light on the relations between cat-bond coupon rates, their issue volumes, and supply curves. This model controls the moral hazard risk and other stochastic imbalances of the cash flows through probabilistic safety constraints. The model is applied to the analysis of typhoon risk in China.

1 Introduction

The number of natural catastrophes has raised dramatically in the late 20th century. Increased catastrophe losses affect insurance and reinsurance industries significantly, leading a number of insurance companies to bankruptcy. The shortage of capital due to high claims on catastrophe losses became a serious problem for insurers and reinsurers. Insurers began to seek ways to transfer natural catastrophe risks to capital markets with large capital reserves. This has been accomplished using both traditional reinsurance and recent capital market mechanisms. In the 1990s, a series of such mechanisms appeared, which are called insurance-linked securities or catastrophe securities (Sheehan 2003; Swiss 2007).

According to Meyers and Kollar (1999), a catastrophe bond is a "corporate bond with special language that requires investors to forgive some or all principal or interest in the event that catastrophe losses surpass the trigger specified in the bond". We call the catastrophe bond as cat-bond for short. The "trigger specified in the bond" is linked either to the magnitude of a catastrophe (e.g., earthquake of magnitude 7

S. Liu (✉) · L. Han
School of Economics and Management, Beihang University, Beijing, China
e-mail: liushuo.buaa@gmail.com; hanly1@163.com

Y. Ermoliev et al. (eds.), *Managing Safety of Heterogeneous Systems*, Lecture Notes
in Economics and Mathematical Systems 658, DOI 10.1007/978-3-642-22884-1_18,
© Springer-Verlag Berlin Heidelberg 2012

according to Richter scale) or to catastrophe losses if they exceed a specified threshold (e.g. catastrophe losses over 25 million US dollars). When trigger event does not occur before the cat-bond's paid coupon to bond holders, get investment income from trust fund and get payment from insurer for the difference of coupon rate and investment income. Once the trigger event occurs, the bond holder will lose part or all of their coupon income or principals, this money will be paid from SPV to insurer to cover their catastrophe loss claims (Jaffee and Russell 1997). The structure of bond cash flows induced by trigger event is displayed in Fig. 1.

Thus, the main purpose of cat-bond is to transfer risk from insurance markets and governmental budgets to financial markets (Cox and Pedersen 2000). Compared to reinsurance with traditionally high reinsurance premiums, cat-bonds have relatively low risk transfer costs (Geman and Yor 1993; Lane 2008), therefore they have become very popular in recent years (Fig. 2). Now, cat-bond is the most widely used catastrophe security in the world. The issuance of cat-bonds kept stable even in the economic recession in 2008 and rebounded much higher in 2009.

Apart from the increased attention to cat-bonds in developed countries, more and more scholars and practitioners focus on the application of cat-bonds in developing

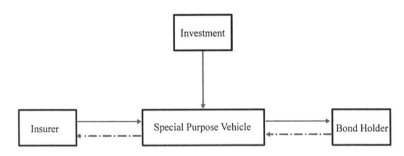

Fig. 1 Cash flow of catastrophe bonds

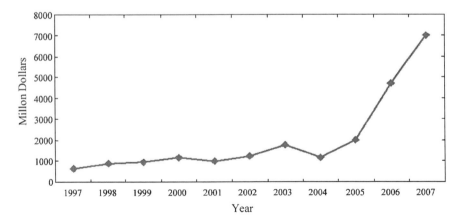

Fig. 2 Total (for the world) number of catastrophe bond issued in 1997-2007. Source: (Guy Carpenter 2008)

countries, especially in countries without systemic casual insurance programs. In the absence of catastrophe insurance, catastrophe securities, including cat-bond, are becoming a powerful tool for catastrophe risk transfer in developing countries (e.g., Han and Liu (2007)).

Existing cat-bonds offer a higher coupon rate compared to corporate bonds of the same credit level (Christensen 1999). Respectively, the criteria for pricing cat-bonds are different from those of corporate bonds. Existing approaches to pricing catastrophe derivatives may, in general, be classified into three major categories. Lane (1998, 2000, 2008) and Lane and Mahul (2008) apply actuarial regression pricing methodology. Embrechts (1996) and Embrechts and Meister (1995) use a generic utility maximization approaches. Wang (2004) and Christensen (1999, 2000) valuate cat-bond by the underlying catastrophe loss measurement. Many scholars, e.g., (Cummins 1993; Cummins and Geman 1995), for pricing of cat-bonds and other derivatives propose arbitrage-based frameworks.

An integrated model for pricing catastrophe risks, in particular, cat-bonds related to flood, earthquake, environmental risk, has been developed at International Institute for Applies Systems Analysis (IIASA) by Ermolieva et al. (2000, 2003); Romaniuk and Ermolieva (2005). The model combines fast Monte Carlo generators of catastrophes and stochastic optimization procedures to address pricing of multivariate catastrophe risks and spatially distributed endogenous catastrophe losses.

Following general ideas of later research, Liu et al. (2009) proposes a pricing method based on so-called behavioral principles and applies it to bond pricing of typhoon risks in China. This is a Monte Carlo simulation model tracking stochastic cash flows and different performance indicators of cat-bonds, i.e., their behavior under different parameters and scenarios of uncertainties capturing in a proper way the relation between the issue volume and the coupon rate. This Monte Carlo-based behavioral model permits to overcome the trading data shortage typical for actuarial regression based pricing. The developed model is applicable in developing countries that do not have mature insurance system and capital markets. Yet, the moral hazard has not been treated in (Liu et al. 2009). The main goal of this paper is to modify the model in (Liu et al. 2009) for dealing with moral hazard risk to insurers.

The paper is organized as follows. Section 2 outlines the model from (Liu et al. 2009). It also discusses the moral hazard problem. Section 3 analyses the improvements of the model and summarizes numerical application to typhoon risk in China. Conclusions are presented in Sect. 4.

2 Catastrophe Bond Pricing Model and Moral Hazard

2.1 Assumptions and Model Structure

The basic assumptions of the model in (Liu et al. 2009) are the following:

- All individuals are rational and committed to survival and maximization of their profit.

- Catastrophe insurance is mandatory.
- SPV(Special Purpose Vehicle) is a non-profit unit with restricted investment activities.
- No friction costs are involved (including the trading cost, management fees and taxes).

This model is a sequential decision selection model from a given finite set of alternatives. Initial inputs are combinations of coupon rate and issue volume of the cat-bond. Coupon rate may vary from 0 to 30% and issue volume – from 0 to 10 billion RMB.[1]

The *two constraints* of the model are:

- Insurers' survival constraint: Insurers want to control the probability of bankruptcy due to a catastrophe under a certain level α.
- Insured survival constraint: Government wants to control the probability that insured losses exceed the accepted level by less than β.

Formally, the safety constraints are formulated as:

$$P(R(\tau) \leq 0 | \tau \in [0, T]) \leq \alpha$$
$$P(IL(\tau) \geq L' | \tau \in [0, T]) \leq \beta \tag{1}$$

where $R(\tau)$ is the risk reserve of an insurer at time τ when catastrophe occurs, $IL(\tau)$ defines losses to insured, L' sets the accepted level of losses. The constraints rely on expert judgments about acceptable safety levels α and β.

The aim of the model is to find acceptable combinations of cat-bond's coupon rate and its issue volume satisfying the safety constraints. The results of the model are useful for regulation commission or government to make decisions regarding cat-bond issuance. The structure of the model is shown in Fig. 3.

Insurer's bankruptcy occurs when the risk reserve $R(\tau)$ turns negative. In a general situation, the risk reserve is defined by the capital reserve, premiums income, losses claims, cat-bond coupon rate payment, and cat-bond gain as follows:

$$R(\tau) = C \cdot v^{-\tau} + \pi \cdot \sum_{i=1}^{\tau} v^{-i} - n \cdot r_c \cdot B_0 \cdot \sum_{i=1}^{\tau} v^{1-i} + B(\tau) - L(\tau) \cdot \varphi \tag{2}$$

where τ defines the first time when trigger event occurs, v is the discount factor $(v = (1 + r_f)^{-1})$ and r_f is risk free rate, C is the initial capital or so-called risk reserve of an insurer, π is the catastrophe insurance premium, n is the issue volume of cat-bond, r_c is the coupon rate of cat-bond, B_0 is cat-bond face value, $B(\tau)$ is the payment the insurer receives from the cat-bond once catastrophe occur at time τ, $L(\tau)$ denotes losses catastrophe causes at time τ, φ is the insurance coverage rate.

---

[1]Chinese currency; in 2010 1 US Dollar $\approx$ 6.7 RMB.

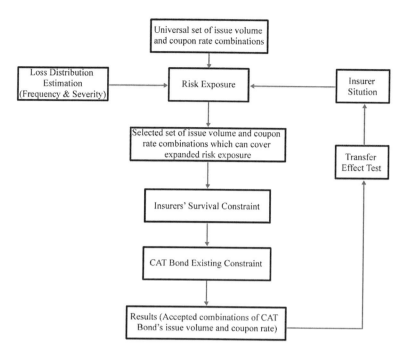

Fig. 3 The sequential decision selection model structure

The second constraint in (1) defines the main purpose of cat-bond: risk transfer. The issuer of cat-bond (in our case – insurer) should guarantee that the probability of insured losses exceeding acceptable level is not greater than β. There may be two situations. If insurer survives after a catastrophe (or catastrophe losses) that surpasses the trigger specified in the bond, the insured receive loss coverage. If the insurer becomes a bankrupt as a result of a catastrophe, the insured losses contain two parts: the uncovered losses and the reserve shortage for the insurance claims. The individual losses of insured may be defined as:

$$IL(t) = L(t) \cdot (1 - \varphi) - min(0, R(t)). \tag{3}$$

where $L(t)$ are losses caused by a catastrophe at time t, $IL(t)$ denotes individual loss at time t, φ is insurance coverage rate, and $R(t)$) is the insurer's risk reserve at time t.

2.2 Application to Typhoon Risk in China

The use of the performance indicators (1)–(2) in a combination with (fast) Monte Carlo simulations can be easily illustrated by a simple example. Assume there is an insurer who is willing to issue a three-year cat-bond with 50% principal protected and 50% coupon rate protected to transfer its catastrophe risk. The risk free rate in the market is 2%, and the government is willing to offer 15 million dollars if catastrophe occurs with fee rate 2.1%. Losses are random, follow Weibull distribution, and cat-bond trigger is a catastrophe with losses exceeding 25 million US dollars. For each feasible solution (combination of coupon rate and issue volume), we perform 10,000 Monte Carlo simulations. For each simulation the losses of insured and the risk reserve of the insurer are calculated. On this basis, for each combination of coupon rate and issue volume we estimate two probabilities defined by constraints (1)–(2).

For choosing optimal parameters of cat-bond, we select all those combinations for which the constraints are fulfilled. These combinations are the dots depicted in Fig. 4. The horizontal axis shows the issue volume while the vertical axis displays the coupon rate. Each point in the Fig. 4 denotes acceptable combination of issue volume and coupon rate of this cat-bond. For the designed bond, although the original input scale of coupon rate is from 2.22% to 30% and of issue volume is from 10 million to 10 billion, we see that the model suggests the highest coupon rate the issuer can offer is 2.85%, and it cannot issue more than 46 million US dollar cat-bond. We illustrate the application of the model to typhoon risk in China where we get similar results. The frequency and severity of each typhoon are assumed to be independent. Probability distributions of typhoon frequency and severity are estimated independently based on data from 1949 through 2005 (National Bureau of Statistics of China 2006). Estimation is done in MATLAB (2010)

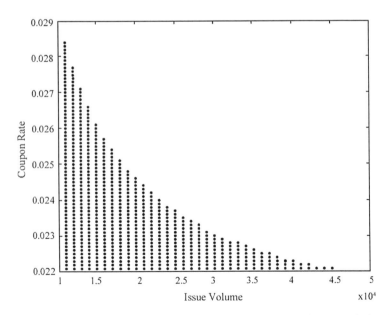

Fig. 4 Accepted coupon rate and issue volume combination of a simulated catastrophe bond

by distribution fitting toolbox, parameters are estimated by maximum likelihood method. Poisson, Binomial, Weibull, and Extreme value distributions are used to estimate the typhoon frequency. Kolmogorov–Smirnov (K–S) test identifies the distribution which explains the distribution of the typhoon occurrence time (per year) in China. As a results, typhoon frequency is best explained by Weibull distribution.

Exponential, Pareto, Gamma, Weibull and Log-normal distributions are tested to fit the typhoon severity distribution. Loss data were adjusted by GDP (base year is 2005). Pareto distribution turns to have much better goodness of fit (a p-value) than any other distribution, therefore, it is selected to explain the typhoon loss distribution (see more details in Liu et al. (2009)).

In this case study we assume that losses from typhoon are insured by a mandatory insurance; the insurer issues a half principal protection and half coupon protection catastrophe bond to transfer the typhoon risk; and there is no possibility for the insurer to get a contingent credit from the government. In this setting, the model derives combinations of issue volume (which may vary within the ranges from 100 billion to 10 trillion RMB) and coupon rates (in the ranges from 2% to 30%). The model analyzes possible combinations of the issue volume and the coupon rate starting from lowest values with step sizes 100 billion RMB and 0.5% for the volume and the rate, respectively, and selects about 26000 acceptable combinations that satisfy the safety constraints (1).

The results are displayed in Fig. 5. There are more than 20000 accepted combinations produced in this experiment. Each combination is represented as a

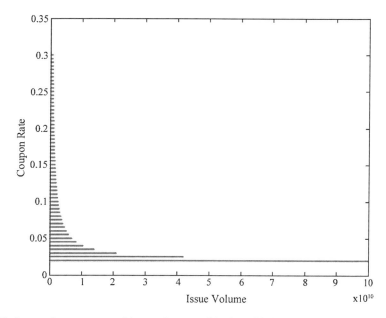

Fig. 5 Accepted coupon rate and issue volume combination of Chinese typhoon catastrophe bond

point in Fig. 5. The lines in Fig. 5 are not real lines but are composed of high-density discrete points. The horizontal axis shows the issue volume while the vertical axis - the coupon rate. Each point of the bars in Fig. 5 is an acceptable combination of issue volume and coupon rate for the cat-bond. There are clear negative relations between issue volume and coupon rate.

2.3 Moral Hazard

Let us consider 26000 acceptable combinations of issue volume and coupon rate derived from the model to analyze the dynamics of insurers risk reserve (Fig. 6). Each dark line in Fig. 6 matches a point in Fig. 5. These dark lines display the state of insurers risk reserve if he has a bond with corresponding coupon rate and issue volume from Fig. 5. According to typhoon losses data in China from 1990 to 2005 (National Bureau of Statistics of China 2006), typhoon catastrophes occurred in 1992, 1994, 1996, 1997, 2001 and 2003 where the losses in 1994 and 1996 had been extremely large exceeding 20 billion RMB. If an insurer issues a cat-bond with any of the 26000 combinations suggested by the model, it survives all the typhoon catastrophes within 16 years. Figure 6 shows that the insurer's risk reserve presents positive jumps increasing during the catastrophe years. The reason of these jumps is, for these years, the capital outflows stemming from the catastrophe insurance

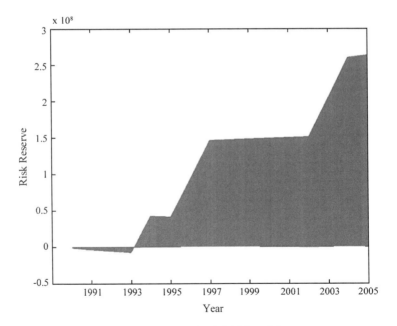

Fig. 6 Risk reserve dynamics with catastrophe bond (initial model)

Pricing Catastrophe Bonds under Safety Constraints

claim are much lower than the inflows due to compensation from cat-bond. This means that the insurer receives extra profits from the catastrophe event. These gains are due to a large issuance volume of cat-bond. Large profitability of catastrophes may decrease insurer's incentives to insure catastrophe losses and directly lead to insurer's moral hazard. To account for this effect, we include considerations of moral hazard in the initial model (1)–(2), which we discuss in the next section.

3 Model with Moral Hazard Safety Constraints

3.1 Moral Hazard Constraint

We revise the model (1)–(2) by including the considerations of moral hazard as an additional safety constraint. When the insurer applies for cat-bond issuance permission, the moral hazard issues are considered by government or regulation commission.

Formally, the moral hazard is defined as $B(\tau) > L(\tau) \cdot \varphi$, which means that once the trigger event occurs, the payment insurer gets from cat-bond is greater than the insurance claims.

The cat-bond payment $B(\tau)$ is estimated as the following:

$$B(t) = n \cdot B_0 \cdot [a \cdot (1 + r_f)^{t-T} + b \cdot r_c \cdot \sum_{i=0}^{T-t}(1 + r_f)^{-i}] \qquad (4)$$

where a is the portion of principal the bond holder will lose once a catastrophe occurs, b is the portion of coupon rate the bond holder will lose once a catastrophe occurs. A safety constraint related to moral hazard in addition to constraints (1) is formulated as follows:

$$P(B(\tau) > L(\tau) \cdot \varphi | \tau \in [0, T]) \leq \gamma \qquad (5)$$

meaning that the government or regulation commission will not provide the issuance permission unless the probability of moral hazard to insurer is less than a certain level γ. The constraint excludes bonds with large values $B(\tau)$ consistently with safety constraint (5).

3.2 Comparative Analysis of Results

To test the risk transfer efficiency and the moral hazard potential, we simulate the risk reserve behavior in the typhoon risk case in China (from 1980 to 2005) using the accepted combinations from the new model with parameter $\alpha = \beta = \gamma = 0.01$. The risk reserve trajectories are shown in Fig. 7.

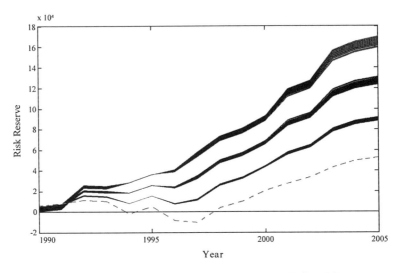

Fig. 7 Risk reserve movement of insurer with catastrophe bond (revised model)

The dash line in Fig. 7 shows the dynamics of insurer's risk reserve without cat-bond. In this situation, the insurer will become bankrupt in 1994. All the dark lines which have been concentrated to three bundles in Fig. 7 show the trajectories of risk reserves with cat-bond having one of the coupon rate and issue volume suggested by the revised model with moral hazard safety constraint. The risk reserve is positive through all years and does not increase much by the cat-bond issuance in catastrophe years. This means the moral hazard does not affect trajectories. These results illustrate that the revised model reduces or even eliminates the moral hazard risks of the original model. It is also clear that compared to the results of the original model, the moral hazard safety constraint narrows down the issue volume of cat-bond and reduces the number of acceptable "issue volume-coupon rate" combinations, what simplifies the decision making process for government, regulation commissions, and insurers.

4 Conclusions

The proposed model includes the issue volume as an output in cat bond pricing. The model overcomes the shortage of cat-bond trading data, and therefore it is applicable to countries with short cat-bond issuance history. Monte Carlo simulations allows to derive negative relationship between bond's issue volume and coupon rate. This relation can also be considered as the supply curve of cat-bond, which is useful for the issuer to make decisions on cat-bond issue scale.

The application of the model to typhoon risks in China indicates that the issue volume of the cat-bond's supply curve will be limited due to moral hazard risk safety constraint. Simulated risk reserve trajectories demonstrates that the model with moral hazard constraint better explains the relationship between bond's issue volume and coupon rate. However, there are still some challenges for future work. As catastrophe is a small probability event, Monte Carlo simulation requires fast versions. Otherwise, large sample sizes are time consuming. Besides, our future work needs to treat better inherent uncertainties of data: the required historical data may be missing. It is important also to develop a specific stochastic optimization method that would substantially reduce the time for selecting feasible robust cat-bond parameters.

Acknowledgements The original idea of this paper was formulated during the Young Scientists Summer Program (YSSP) 2008 at the International Institute for Applied Systems Analysis, IIASA, Laxenburg, Austria. The research was conducted in the Integrated Modeling Environment (IME) Project in cooperation with the Land Use & Agriculture (LUC) Program. We are grateful to Professor Yuri Ermoliev, Dr. Tatiana Ermolieva and Dr. Marek Makowski for their supervision and advice during the research work on this paper. We also give many thanks to the financial support from National Natural Science Foundation of China (Grant No. 70831001) and China Insurance Regulatory Commission Foundation (Grant No. QNA200813), which make this work possible.

References

Christensen, C. V. (1999). A New Model for Pricing Catastrophe Insurance Derivatives. CAF Working paper Series, No.28.

Christensen, C. V. (2000). Securitization of Insurance Risk. Working Paper.

Cox, S. H., & Pedersen, H. W. (2000). Catastrophe risk bonds. *North American Actuarial Journal, 4*(4), 56–82.

Cummins, J. D. (1993). An Asian Option Approach to the Valuation of Insurance Futures Contracts. Center for Financial Institutions Wording Papers, No.94-03.

Cummins, J. D., & Geman, H. (1995). Pricing catastrophe insurance futures and call spreads: an arbitrage approach. *Journal of fixed Income*, 46–57.

Embrechts, P. (1996). Actuarial vs. Financial Pricing of Insurance. Wharton Financial Institutions Center Working Paper, 96–112.

Embrechts, P., & Meister, S. (1995). Pricing Insurance Derivatives: The case of CAT Futures. Proceedings of the 1995 Bowles Symposium on Securization of Insurance Risk.

Ermolieva, T., Ermoliev, Y., MacDonald, G.J., Norkin, V.I., & Amendola, A. (2000). A system approach to management of catastrophic risks. *European Journal of Operational Research, 12*(2), 452–460.

Ermolieva, T., Ermoliev, Y., Fisher, G., & Galambos, I. (2003). The role of financial instruments in integrated catastrophic flood management. *Multinational Finance Journal, 7*(3&4), 207–230.

Geman, H., & Yor, M. (1993). Bessel processes, asian options, and perpetuities. *Mathematical Finance, 3*(4), 349–375.

Guy Carpenter (2008). The Catastrophe Bond Market at Year-End 2007: The Market Goes Mainstream. Guy Carpenter Annual Reports.

Han, L. Y., & Liu, S. (2007). Research on the innovation of catastrophe options in China. *Proceedings of the First International Conference on Risk Analysis and Crisis Response*, 473–478.

Jaffee, D. M., & Russell, T. (1997). Catastrophe insurance, capital markets, and uninsurable risks. *The Journal of Risk and Insurance, 64*(2), 205–230.

Lane, M. N. (1998). Price, risk and ratings for insurance-linked notes: evaluating their position in your portfolio. Derivatives Quarterly.

Lane, M. N. (2000). Pricing risk transfer transactions. *ASTIN bulletion, 30*(2), 259–293.

Lane, M. N. (2008). The Measurement of ILS Returns. Working paper, www.lanefinancialllc.com, 2008-08.

Lane, M., & Mahul, O. (2008). Catastrophe Risk Pricing. Policy Research Working Paper 4765. The World Bank.

Liu, S., Han, L. Y., Ermolieva, T., & Ermoliev, Y. (2009). Catastrophe Bond Pricing based on Behavior Model. Proceeding of Modelling, Simulation and Identification – 2009, Beijing, China.

MATLAB (2010). version 7.10.0. The MathWorks Inc.: Natick, Massachusetts.

Meyers, G., & Kollar, J. (1999). Catastrophe risk securitization insurer and investor perspectives. Casualty actuarial society "Securitization of Risk" discussion paper program Arlington, Virginia: Casualty Actuarial Society 223–272.

National Bureau of Statistics of China (2006). 1949–2005 China Statistical Yearbook. China Statistics Press: Beijing.

Romaniuk, M., & Ermolieva, T. (2005): Application of EDGE software and simulations for integrated catastrophe management. *International Journal of Knowledge and Systerms Sciences.*

Sheehan, K. P. (2003). Catastrophe securities and the market sharing of deposit insurance risk. *FDIC Banking Review, 15*(15).

Swiss, Re. (2007). Sigma. Swiss Reinsurance Report, www.swissre.com, 2007-03.

Wang, S. S. (2004). Cat Bond Pricing Using Probability Transforms. Insurance and the State of the Art in Cat Bond Pricing, 278, 19–29.

Printed by Publishers' Graphics LLC